SCUOLA NORMALE SUPERIORE

QUADERNI

Symmetry in Nature

A Volume in honour
of Luigi A. Radicati di Brozolo

TOMO II

PISA - 1989

Ristampa
Pisa 2005

ISBN: 978-88-7642-253-9

Contents of Tomo I

Contents of Tomo II

Some participants at the Symposium

Gilberto Bernardini, Gian Carlo Wick, Edoardo Amaldi

Rudolf Peierls, Feza Gürsey

Freeman Dyson

Louis Michel

Lecture-hall during the Symposium

Tsung Dao Lee, Ludwig Faddeev

John Archibald Wheeler

Luigi A. Radicati di Brozolo

At Villa Roncioni

Conformal Theories Associated with an Additive Charge

L.K. HADJIIVANOV - I.T. TODOROV (*)

To Luigi Radicati
on the occasion of
his 70th birthday

Abstract

Additive fusion rules yield an equation for the conformal weights Δ_ν of the type first discovered by Zamolodchikov and Fateev:

$$e^{i\pi(\Delta_\lambda+\Delta_\mu+\Delta_\nu+\Delta_{\lambda+\mu+\nu})} = e^{i\pi(\Delta_{\lambda+\mu}+\Delta_{\mu+\nu}+\Delta_{\lambda+\nu})}.$$

The general solution of this equation gives rise to a quadratic "mass formula" for Δ_ν that also covers the spectrum of conformal weights in minimal $(c_{p'p})-$ and $\hat{su}(2)_k$ current algebra models which involve non-linear fusion rules. We offer an interpretation of this result invoking the quantum symmetry of the models.

1. - Introduction

Two-dimensional conformal quantum field theory (CQFT) has reached a level of maturity permitting to formulate it in concise mathematical terms. We shall state our problem using the formulation of Frenkel, Lepowsky, Meurman [1] (see also [2]) which we proceed to summarize. (For a more traditional approach to the subject see the recent reprint collection [3] as well the review [4].)

(*) Institute for Nuclear Research and Nuclear Energy Bulgarian Academy of Sciences, Sofia

1784

1a. - *Graded local chiral algebras*

Let $\mathcal{V}_0$ be a pre-Hilbert space with a distinguished positive "conformal energy" operator L_0 in it of integer finitely degenerate spectrum

$$(1.1a) \qquad \mathcal{V}_0 = \bigoplus_{n=0}^{\infty} \mathcal{V}_0^{(n)}, \qquad (L_0 - n)\mathcal{V}_0^{(n)} = 0$$

$$(1.1b) \qquad \dim \mathcal{V}_0^{(0)} = 1, \qquad \dim \mathcal{V}_0^{(n)} < \infty.$$

(Such a starting point will lead us to a chiral algebra consisting of integer spin Bose fields. One can also consider Fermi fields (of spin $s \in \frac{1}{2} + \mathbb{Z}_+$) allowing for half-integer eigenvalues of L_0 in $\mathcal{V}_0$; we would write

$$(1.1c) \qquad \mathcal{V}_0 = \bigoplus_{\nu=0,\frac{1}{2},1\cdots} \mathcal{V}_0^{(\nu)} = \mathcal{V}_0^B \oplus \mathcal{V}_0^F$$

in that case. The splitting in two sectors follows the eigenvalues ± 1 of the superselection operator $e^{2\pi i L_0}$ which acts as the identity in the bosonic subspace $\mathcal{V}_0^B$ given by Eq. (1.1a).)

The unique (normalized) state $|0 > \in \mathcal{V}_0^{(0)}$ is the *vacuum*. We define a local system of chiral fields by associating to each finite energy vector $v \in \mathcal{V}_0$ an operator valued field $Y(v, z), z \in \mathbb{C}$ such that

$$(1.2) \qquad Y(v, z)|0 > = e^{L_{-1}z}|v >$$

and for any four finite energy states u, v, v_1, v_2 the matrix element

$$(1.3) \qquad f(z, \varsigma) = < v_1|Y(u, z)Y(v, \varsigma)|v_2 >$$

is a holomorphic function in $|z| > |\varsigma| > 0$ with rational continuation satisfying in the Bose-Fermi case (1.1c)

$$(1.4) \qquad f(\varsigma, z) = (-1)^{4\mu\nu} f(z, \varsigma).$$

This definition is justified by the following *uniqueness theorem* ([2] Sec. 2 Theorem 1): If two fields $Y_i(v, z), i = 1, 2$ of a given local system satisfy (1.2) when applied to the vacuum then they coincide.

Furthermore we assume the existence of a vector

$$(1.5) \qquad \omega \in \mathcal{V}_0^{(2)} \text{ such that } T(z) = Y(\omega, z) = \sum_{n \in \mathbb{Z}} L_n z^{-n-2}$$

is the stress energy tensor that generates the Virasoro algebra Vir:

$$(1.6) \qquad [L_n, L_m] = (n - m)L_{n+m} + \frac{c}{12}n(n^2 - 1)\delta_{n+m}$$

where the central charge c is a positive number belonging to the spectrum of unitary irreducible representations of Vir.

Eq. (1.2) and the uniqueness theorem imply the translation covariance law

$$(1.7) \qquad e^{\varsigma L_{-1}} Y(v, z) e^{-\varsigma L_{-1}} = Y(v, z + \varsigma)$$

and the operator product expansion (OPE), or duality relation

$$(1.8) \qquad Y(u, z) Y(v, \varsigma) = Y(Y(u, z - \varsigma) v, \varsigma)$$

(see [2] Sec. 2, Proposition 2 and Theorem 3; the right hand side of (1.8) can be defined by a convergent power series in $z - \varsigma$ for $|z| > |\varsigma| > |z - \varsigma| \geq 0$). The OPE algebra generated by the above local system will be called *the local chiral algebra* $\mathcal{A} = \mathcal{A}(\mathcal{V}_0)$. It has a unit, since according to the uniqueness theorem

$$(1.9) \qquad Y(|0>, z) = 1 .$$

Furthermore, we assume that there is an antiinvolution $*$ in $\mathcal{V}_0$ which preserves the grading and renders $\mathcal{A}$ a star algebra. If $v = v^{(s)} \in \mathcal{V}_0^{(s)}$ has a definite conformal weight (or spin) $s \in \mathbf{Z}_+$ then $Y(v, z)$ has a mode expansion which generalizes (1.5):

$$(1.10a) \qquad Y(v, z) = \sum_{n \in \mathbf{Z}} Y_n(v) z^{-n-s}, \quad [Y_n(v), L_0] = n Y_n(v)$$

and the star operation in $\mathcal{A}$ is defined to satisfy

$$(1.10b) \qquad Y_n(v)^* = Y_{-n}(v)^*.$$

The hermiticity of the stress anergy tensore means that $\omega^* = \omega$ so that $L_n^* = L_{-n}$. If $|m> \in \mathcal{V}_0^{(m)}$ then energy positivity implies

$$(1.10c) \qquad Y_n(v)|m> = 0 \text{ for } n > m.$$

1b. - *Local QFT representations of $\mathcal{A}$*

We consider positive energy graded $\mathcal{A}$-modules

$$(1.11) \qquad \mathcal{V}_i = \bigoplus_{n=0}^{\infty} \mathcal{V}_i^{(n)}, \quad (L_0 - \Delta_i - n) \mathcal{V}_i^{(n)} = 0, \quad \Delta_i > 0$$

(restricting our attention to a Bose chiral algebra corresponding to (1.1a)) such that the matrix elements of $Y(v, z)$ in $\mathcal{V}_i$ are still single valued meromorphic functions of z. A *rational* CQFT is characterized by the property that there are only a finite number of such $\mathcal{A}$-modules giving rise to local positive energy representations of $\mathcal{A}$.

A vector $v_i \in \mathcal{V}_i$ is called a *lowest weight* (or primary) *state* of $\mathcal{A}$ if

(1.12a) $(L_0 - \Delta_i)|v_i> = 0 = L_n|v_i>, \quad n > 0$

(1.12b) $Y(v^{(s)}, z)|v_i> = \sum_{n=0}^{\infty} Y_{-n} z^{n-s}|v_i> \quad \text{for } v^{(s)} \in \mathcal{V}_0^{(s)}.$

For any such v_i there is a *primary field* $\phi_i(z) = Y(v_i, z)$ which intertwines between the vacuum sector and $\mathcal{V}_i$ so that the relations (1.2) and (1.8) remain valid (for $v = v_i$). The correlation functions of primary fields $< 0|\phi_1(z_1)\phi_2(z_2)\cdots\phi_n(z_n)|0 >$ are again holomorphic functions for $|z_1| > |z_2| > \cdots > |z_n|$ but their analytic continuation outside this primitive domain is, in general, a multivalued function of the z's.

1c. - *Fusion rules involving an additive charge*

In general, a primary field $\phi_i(z)$ maps $\mathcal{V}_j$ (for $j \neq 0$) into a sum of $\mathcal{A}$-modules $\mathcal{V}_k$. If however ϕ_i carries an additive charge and if $\mathcal{V}_j$ is uniquely determined by the eigenvalue of this charge we have the simple *fusion rule*

(1.13) $\phi_i(z)\mathcal{V}_j \subset \mathcal{V}_{i+j}$

where we have indexed fields and spaces by the numbers i, j of charge units in the correspoding $\mathcal{A}$-module. There are two simple examples of rational CQFT's with such "linear" fusion rules. One is the $c = 1$ theory involving a $\hat{u}(1)$ current in which the chiral algebra $\mathcal{A} = \mathcal{A}(s)$ is generated by a pair of oppositely charged vertex operators

(1.14) $Y(\pm g, z) \quad s = \frac{1}{2}g^2(= 1, 2, \cdots), \qquad \mathcal{V}_0 = \underset{n \in \mathbb{Z}}{\oplus} \mathcal{V}_0(ng),$

each space $\mathcal{V}_0(ng)$ carrying an irreducible positive energy representation of $\hat{u}(1)$ see [5]. The $\mathcal{A}(s)$-primary fields are related bt a Klein transformation to the vertex operators

(1.15) $Y(\nu e, z), \quad e = g^{-1}, \quad 1 - s \leq \nu \leq s, \qquad (L_0 - \frac{\nu^2 e^2}{2})|\nu e> = 0.$

The fact that their number is finite is related to the observation that $Y(2se, z) = Y(g, z) \in \mathcal{A}(s)$. The second example is provided by the Zamolodchikov-Fateev $\mathbb{Z}_k$-parafermionic currents $\varphi_\nu(z)$ satisfying the fusion rule [6]

(1.16) $z_{12}^{\Delta_\mu + \Delta_\nu - \Delta_{\mu+\nu}} \varphi_\mu(z_1)\varphi_\nu(z_2) = C_{\mu\nu}\{\varphi_{\mu+\nu}(z_2) + \mathcal{O}(z_{12})\},$

$$C_{\mu\nu} = C_{\nu\mu} = \overline{C}_{\nu\mu}$$

(1.17)
$$0 \leq \mu, \nu, \mu + \nu \leq k, \qquad \Delta_\nu = \Delta_{k-\nu},$$
$$\Delta_0 = \Delta_k = 0, \qquad \varphi_0 = \varphi_k = 1, \qquad C_\nu k - \nu = 1.$$

The local chiral algebra is generated in this case by the integer spin conformal fields entering the OPE $\varphi_\nu(z_1)\varphi_{k-\nu}(z_2)$.

Eq. (1.16) is typical for theories with an additive charge. It says that each OPE gives rise to a silgle branch-cut and yields the following equation for the conformal weights [6,7,8]:

(1.18)
$$e^{i\pi(\Delta_\lambda + \Delta_\mu + \Delta_\nu + \Delta_{\lambda+\mu+\nu})} = e^{i\pi(\Delta_{\lambda+\mu} + \Delta_{\mu+\nu} + \Delta_{\lambda+\nu})}.$$

This is the first equation for the conformal weights derived (back in 1985) from the monodromy properties of correlation functions. Remarkably, the general solution of this equation incorporates practically all other cases treated recently by similar methods [9,10]: the minimal conformal models [11] and the $\hat{su}(2)_k$-current algebra models [12,13].

The objective of the present paper is twofold. First, in Sec. 2, we spell out the above fact (deriving anew Eq. (1.18) for Δ_μ and finding its general solution). Secondly, we address, in Sec. 3, the question: why does an equation for Δ_ν derived under the assumption that there is an additive charge also apply to apparently nonlinear fusion rules? Our answer to this question makes use of the recent proposal [14,15] to identify the internal symmetry of the $\hat{su}(2)_k$ and minimal models with the quantum group $U_q(sl(2))$.

2. - Derivation of Eq. (1.18) for the conformal weights Δ_ν from the fusion rule (1.16)

The fusion rule (1.16) implies (for $|z_1| > |z_2|$) the commutation (exchange) relation

(2.1)
$$\varphi_\nu(z_2)\varphi_\mu(z_1) = \vartheta(\mu, \nu)\varphi_\mu(z_1)\varphi_\nu(z_2)$$

with

(2.2)
$$\vartheta(\mu, \nu) = e^{i\pi(\Delta_{\mu+\nu} - \Delta_\mu - \Delta_\nu)}.$$

Let $|z_1| > |z_2| > |z_3|$; shifting $\varphi_\nu(z_3)$ through $\varphi_\lambda(z_1)\varphi_\mu(z_2)$ before and after the application of the OPE (1.16) we find the equation

(2.3)
$$\vartheta(\lambda, \nu)\vartheta(\mu, \nu) = \vartheta(\lambda + \mu, \nu).$$

Substituting (2.2) in (2.3) we recover Eq. (1.18). (Zamolodchikov and Fateev have originally obtained a slightly weaker condition [6] - with $e^{i\pi}$ substituted by $e^{2i\pi}$ - since they used monodromy rather than braid relations.)

Proceeding to the solution of (1.18) we find it advantageous to work with the phase factors ϑ which, according to (2.2), have the symmetry property

(2.4) $$\vartheta(\mu, \nu) = \vartheta(\nu, \mu).$$

Using (2.3) for $\mu = 1$ we derive the recurrence relation $\vartheta(\lambda, \mu)\vartheta(1, \nu) = \vartheta(\lambda + 1, \nu)$ which implies $\vartheta(\mu, \nu) = [\vartheta(1, \nu)]^\mu$. Using then the symmetry (2.4) and applying once more the recurrence relation we end up with the formula

(2.5) $$\vartheta(\mu, \nu) = e^{i\pi\delta\mu\nu}$$

where δ differs from $\Delta_2 - 2\Delta_1$ by an even integer. The general solution for Δ_ν is determined (modulo an even number) as a quadratic function of ν whose coefficients are expressed in terms of Δ_1 and δ:

(2.6) $$\Delta_\nu = \nu\Delta_1 + \frac{\nu(\nu - 1)}{2}\delta + 2M_\nu, \quad M_\nu \in \mathbb{Z}, M_0 = 0 = M_1.$$

We now consider the examples of parafermions characterized by Eq. (1.17). Since $\varphi_k = 1$ commutes with all fields we deduce that $\vartheta(k, 1) = 1$ and hence, according to (2.5)

(2.7) $$k\delta = -2p \quad (p \in \mathbb{Z}).$$

It then follows from $\Delta_\nu = \Delta_{k-\nu}$ that

(2.8) $$(k - 2\nu)\left(\Delta_1 - p\left(1 - \frac{1}{k}\right)\right) = 2(M_\nu - M_{k-\nu}).$$

Inserting in (2.6) $\Delta_1 = p\left(1 - \frac{1}{k}\right)$ we recover the general solution of [6]

(2.9) $$\Delta_\nu = p\nu\left(1 - \frac{\nu}{k}\right) + 2m_\nu, \quad m_\nu = m_{k-\nu} = M_\nu.$$

It is easy to show that by suitable redefinitions of m_ν and p (breaking the condition $m_1 = 0$) one can incorporate into (2.9) all other solutions of (2.8). The integers m_ν are restricted by the inequalities

(2.10) $$\Delta_\nu \leq \nu\frac{k - \nu}{k - 1}\Delta_1, \quad \nu = 1, 2, \cdots, k$$

proven in [7] (by studying the large z_0 behaviour of $(z_{01} \cdots z_{0k-\nu})^{\Delta_\nu + \Delta_1 - \Delta_{\mu+1}} < 0|\varphi_\nu(z_0)\,\varphi_1(z_1) \cdots \varphi_1(z_{k-\nu})|0 >)$. The first term in the right hand side of (2.9) saturates (2.10) so that we obtain

(2.11) $$m_\nu \leq \nu\frac{k - \nu}{k - 1}m_1$$

The standard parafermions correspond to $m_\nu = 0$, $p = 1$.

REMARK. The only known examples of generalized parafermions which require $m_\nu \neq 0$ are $k = 4$ Klein transformed $\hat{u}(1)$ vertex operators corresponding to $c = 1$. There is an infinite series of such $\mathbb{Z}_4$-parafermions with $\Delta_1 = \Delta_3 = \frac{2n+1}{4}, \Delta_2 = 1(p = 3 - 2n, m_1 = m_3 = n - 1 = m_2)$, $n = 0, 1, 2 \cdots$ (cf. [7]). The analysis of the large z_0 behaviour of the product $(z_{01}z_{02}\cdots z_{0\nu+1})^{2\Delta_1}z_0^{\Delta_\nu+\Delta_1-\Delta_{\nu+1}} < 0|\varphi_1^*(z_0)\varphi_1(z_1)\cdots\varphi_1(z_{\nu+1})\varphi_\nu^*(0)|0 >$ allows to establish [7] the inequality

$$(2.12) \qquad \Delta_\nu \leq \nu^2 \Delta_1.$$

This inequality is saturated by the $\hat{u}(1)$-models (whose spectrum is displayed by Eq. (1.15)). We can obtain the spectrum of lowest weights for the $A(s)$-chiral algebra $\Delta_\nu = \frac{\nu^2}{4s}$ from the general formula (2.6) by imposing charge conjugation invariance, $\Delta_\nu = \Delta_{-\nu}$ and the condition $\Delta_{2s} = s$, provided that all integers M_ν in (2.6) vanish.

As noted in the introduction Eq. (2.6) also covers the spectrum of primary conformal weights of the $\hat{su}(2)_k$-current algebra,

$$(2.13) \qquad \Delta_\ell = \frac{\ell(\ell+2)}{4(k+2)} \qquad (\ell = 2I = 0, 1, \cdots, k)$$

and of the first line of the Kac table of weights of a $c_{p'p}$-minimal conformal model:

$$(2.14) \qquad c_{p'p} = 1 - 6\frac{(p - p')^2}{pp'}, \Delta_\nu = \frac{\nu}{4}\{(\nu+2)\frac{p'}{p} - 2\}, \nu = 0, 1, \cdots, p - 2$$

($p'p$ coprimes) although the fusion rules in both cases appear to be "nonlinear".

3. - A quantum group interpretation of Eqs. (2.13,14)

The association of the models of primary weights (2.13,14) with the quantum group $U_q \equiv U_q$ (sl (2)) [16,17] came after the observation made by Pasquier (see [18] and references to earlier work cited there) that the expansion formula for the product of "quantum dimensions"

$$(3.1) \qquad [l_1 + 1]_q[l_2 + 1]_q = \sum_{\substack{l=|l_1-l_2| \\ l - |l_1 - l_2| \text{ even}}}^{l_m} [l + 1]_q$$

where

$$(3.2) \qquad [x]_q \frac{q^x - q^{-x}}{q - q^{-1}}$$

$$(3.3) \qquad l_m = \min(l_1 + l_2, 2k - l_1 - l_2)$$

valid for

(3.4) $$q^{k+2} = -1$$

parallels the fusion rule of either of the two models. (Speaking about the $c_{p'p}$-model of weights (2.14) one has to substitute ℓ by ν and $k+2$ by p.) Further work [19,20,21,22] diplayed deep going analogies between Clebsch-Gordan expansions of finite dimensional representations of U_q and fusion and duality properties of the corresponding conformal models. It led finally to the proposal [14,15] to introduce an internal quantum number space for chiral conformal fields and to interpret U_q as a quantum (first kind gauge) symmetry group of the models under consideration.

Here we shall use the observation that the quantum symmetry allows to "linearize" the conformal fusion rules and thus to understand the success of our simple minded (additive charge) approach in recovering the weight formulae (2.13,14).

Both models are characterized by the presence of a "step operator" $\varphi = \varphi_1$ of minimal (positive) conformal weight. Following [14,15] we shall regard φ as an U_q-doublet $\varphi_\alpha, \alpha = \pm\frac{1}{2}$ (of "quantum isospin" $\frac{1}{2}$). In order to fix the ideas we shall use the primary field φ of dimension $\Delta = \Delta_1$ (2.14) of a minimal conformal model (thus avoiding the discussion of charge degeneracy inherent to the $\hat{su}(2)_k$-case). The small distance vacuum OPE of the product of two φ's can be written in the form [15]

(3.5) $$z_{12}^{2\Delta}\varphi_\alpha(z_1)\varphi_\beta(z_2)|0> = \{C_{\alpha\beta} + A_{12}z_{12}^{\Delta_2}f_{\alpha\beta}^a\phi_a(z_2)\}(1 + O(z_{12}))|0>$$

where C in the U_q-invariant tensor

(3.6) $$(C_{\alpha\beta}) = \begin{pmatrix} 0 & -q^{-\frac{1}{2}} \\ q^{\frac{1}{2}} & 0 \end{pmatrix} \quad \text{(i.e., } C_{-\frac{1}{2}\frac{1}{2}} = q^{\frac{1}{2}}, \det C = 1),$$

$f_{\alpha\beta}^a$ are the "quantum Wigner coefficients" [18] for the addition of two isospins $\frac{1}{2}$:

(3.7) $$f_{\alpha\beta}^a = f_{\alpha\alpha-\alpha}^a\delta_{\alpha+\beta}^a, \quad f_{\frac{1}{2}\frac{1}{2}}^1 = Q^{\frac{1}{4}} = f_{-\frac{1}{2}-\frac{1}{2}}^{-1}, \quad (f_{\alpha\beta}^0) = \begin{pmatrix} 0 & q^{\frac{1}{2}} \\ q^{-\frac{1}{2}} & 0 \end{pmatrix}$$

Q being the (irrational extension of the) "number of states in the Potts model"

(3.8) $$Q^{\frac{1}{2}} = q + q^{-1} \quad (> 0),$$

ϕ_a is the "isospin 1" primary field of weight Δ_2, normalized Δ_2, normalized by

(3.9) $$z_{12}^{2\Delta_2} <0|\phi_a(z_1)\phi_6(z_2)|0> = G_{ab} = \begin{pmatrix} 0 & 0 & q^{-1} \\ 0 & -1 & 0 \\ q & 0 & 0 \end{pmatrix}_{ab},$$

A_{12} is a real structure constant.

The permutation of two φ's can be expressed in terms of the $\mathbb{C}^2 \otimes \mathbb{C}^2$ braid matrix

$$(3.10) \qquad b^{\alpha'\beta'}_{\alpha\beta} = -C_{\alpha\beta}C^{\alpha'\beta'} - q^{-1}\delta^{\alpha'}_\alpha \delta^{\beta'}_\beta \quad (C^{\alpha\sigma}C_{\sigma\beta} = \delta^\alpha_\beta)$$

with eigenvalues q and $-q^{-1}$:

$$(3.11) \qquad b^{\alpha'\beta'}_{\alpha\beta}C_{\alpha'\beta'} = qC_{\alpha\beta}, b^{\alpha'\beta'}_{\alpha\beta}f^a_{\alpha'\beta'} = -q^{-1}f^a_{\alpha\beta}.$$

The ratio $-q^2$ of its eigenvalues can be related to the conformal weight Δ_2 of ϕ; we find (cf. Eq. (4.6) of [20])

$$(3.12) \qquad q^2 = -\varepsilon e^{i\pi\Delta_2}(\varepsilon = 1 \text{ for } (2.14), \varepsilon = -1 \text{ for } (2.13)).$$

Eq. (3.5) effectively splits (due to (3.11)) the fusion rule for the product of two φ's (written, conventionally, in the form $\varphi \times \varphi = 1 + \phi$). For instance, the components $\varphi_{\frac{1}{2}}$ have an additive fusion rule (the third projection of the quantum isospin being preserved). Thus we are indeed entitled to apply the argument leading to the general weight formula (2.6).

Given Δ_2 we can use (3.12) to determine q with the result

$$(3.13) \qquad q(k) = e^{i\frac{\pi}{k+2}}, \quad q(p',p) = e^{i\pi\frac{p'-p}{p}}$$

($q(p',p)$ satisfies $q^p = -1$ for odd $p - p'$ only, which incorporates the unitary minimal series with $p = p' + 1$). Conversely, given q (e.g. from the modular transformation formula for conformal characters - cf. [21] Sec. 3) Eq. (3.12) can be used to deduce one relation for the parameters of the weight formula (2.6). To obtain another such relation we note that the field $\phi_A(z) = \phi_A(z, p-2)(1-\frac{p}{2} \le A \le \frac{p}{2}-1)$ transforming under an U_q representation of quantum dimension one (for odd $p - p'$)

$$(3.14) \qquad [p-1]_q = \frac{q^{p-1} - q^{1-p}}{q - q^{-1}} = \frac{-q^{-1} - (-q)}{q - q^{-1}} = 1 \text{ for } q^p = -1$$

satisfies the simple exchange relation

$$(3.15) \qquad \phi_A(z_2)\phi_B(z_1) = (-1)^{2s}\phi_A(z_1)\phi_B(z_2)$$

where s is the (integer or half-integer) spin of ϕ:

$$(3.16) \qquad s = \frac{(p-2)(p'-2)}{4} \qquad (p - p' \text{ odd}).$$

(The role of the field ϕ with maximal conformal weight s in a given unitary c_m-model / for $p' = m + 2, p = m + 3$ / which generates an extended chiral algebra was pointed out in [23]).

To summarize: Eqs. (3.12,13) and (3.15,16) imply

(3.17)
$$e^{i\frac{\pi}{2}(\Delta_2-1)} = e^{i\pi\frac{p'-p}{p}}$$

(3.18)
$$e^{-i\pi\Delta_{p-2}} = e^{-i\pi\delta}.$$

The solution of the type (2.6) of these equations allows, for $M_\nu = 0$, to determine the parameters Δ_1 and δ:

(3.19)
$$\Delta_2 = 2\Delta_1 + \delta = 2\frac{p'}{p} - 1, \qquad \delta = \frac{p'}{2p}$$

thus reproducing (2.14). Similarly, using (3.12) for $\varepsilon = -1$ and a condition of type (3.15) with $(-1)^{2\delta}$ substituted by $e^{-i\pi\frac{k}{2}}$ we obtain the parameters in (2.6) for the $\hat{su}(2)_k$-model:

(3.20)
$$\Delta_2(k) = 2\Delta_1 + \delta = \frac{2}{k+2}, \qquad \Delta_k(k) = k(\Delta_1 + \frac{k-1}{2}\delta) = \frac{k}{4}$$

recovering Eq. (2.13).

We remark that the additivity of the fusion rules in the $\hat{su}(2)_k$ case could be understood by using the field of maximal J_0^3-charge in each isospin multiplet. For the minimal models, however, one does need an additional structure: either the sceening charges of the Coulomb gas picture or the quantum (U_q-) symmetry exploited in this section.

Concluding, we note that all know to us rational CQFT's involve quadratic "mass formulae" signaling linearizable fusion rules.

The authors thank Roman Paunov for useful discussions.

REFERENCES

[1]   I.B. FRENKEL - J. LEPOWSKY - A. MEURMAN, Vertex operator calculus, in: *Mathematical Aspects of String Theory*, Proc. 1986 Conference, San Diego, ed. by S. - T. Yau (World Scientific, Singapore 1987) pp. 150-188; *Vertex Operator Algebras and the Monster* (Acad. Press. Boston, N.Y. 1988), see, in particular, Appendix: Complex realization of vertex operator algebras, pp. 461-482.

[2]   P. GODDARD, Meromorphic conformal field theory, Cambridge preprint DAMPT 89-1, Proceedings of CIRM Conference on *Infinite Dimensional Lie Algebras and Lie Groups* (Luminy, July 1988) ed. by V. Kac.

[3]   *Conformal Invariance and Applications to Statistical Mechanics*, Eds. C. ITZYKSON H. SALEUR - J.-B. ZUBER (World Scientific, Singapore 1988) 980 p.

[4]   P. FURLAN - G.M. SOTKOV - I.T. TODOROV, Two-dimensional conformal quantum field theory, SISSA/ISAS Internal Report 60/88/E.P., Riv. Nuovo Cim. (1989).

[5]   D. BUCHHOLZ - G. MACK - I. TODOROV, Nucl. Phys. B (Proc. Suppl.) **5B** (1988) 20.

[6]   A.B. ZAMOLODCHIKOV - V.A. FATEEV, ZhETF **89** (1985) 380 (transl.: Sov. Phys. JETP **62** (1985) 215; see also [3], pp. 203-213).

[7]   P. FURLAN - R.P. PAUNOV - I.T. TODOROV, Extended $U(1)$-conformal field theories in two dimensions, Trieste INFN preprint (1989).

[8]   I.T. TODOROV, Rational conformal theories involving a $U(1)$ current algebra, Bures-sur-Yvette preprint IHES/P/89/30, to appear in the Proceedings of the Vth Summer School "Particles and Fields", Saõ Paulo, Brazil (1989), ed. by M. Gomes.

[9]   K.H. REHREN, Commun. Math. Phys. **116** (1988) 675.

[10] K.H. REHREN - B. SCHROER, Nucl. Phys. **B312** (1989) 715.

[11] A.A. BELAVIN, A.M. POLYAKOV - A.B. ZAMOLODCHIKOV, Nucl. Phys. **B241** (1984) 333 (see also [3], pp. 5-52).

[12] V.G. KNIZHNIK - A.B. ZAMOLODCHIKOV, Nucl. Phys. **B247** (1984) 83.

[13] A. TSUCHIYA - Y. KANIE, Lett. Math. Phys. **13** (1987) 303; *Conformal Field Theory and Solvable Lattice Models*, Advanced Studies in Pure Mathematics **16** (1988) 297-372.

[14] G. MOORE - N. RESHETIKHIN, A comment on quantum group symmetry in conformal field theory, Princeton preprint IASSNS-HEP-89/18.

[15] D. BUCHHOLZ - I.B. FRENKEL - G. MACK - I.T. TODOROV, Chiral algebras with quantum group symmetry (in preparation).

[16] L.D. FADDEEV - N. YU. RESHETIKHIN - M.I. TAKHATJAN, LOMI preprint E 14-87; Algebra and Analysis **1** (1989) 178 (in Russian).

[17] V.G. DRINFEL'D, Quantum groups, *Proceed. International Congress of Mathematicians*, Berkeley, Cal. 1986 (Acad. Press 1987) vol. 1 pp. 798-820.

[18] V. PASQUIER, Commun. Math. Phys. **118** (1988) 355.

[19] G. MOORE - N. SEIBERG, Classical and quantum conformal field theory, Princeton preprint IASSNS-HEP-88/35 Commun. Math. Phys. (1989).

[20] L. ALVAREZ-GAUMÉ - C. GOMEZ - G. SIERRA, Hidden quantum symmetries in rational conformal field theories, Geneva preprint CERN-TH. 5129/88, UGVA-DPT 07/583/88.

[21] L. ALVAREZ-GAUMÉ - C. GOMEZ - G. SIERRA, Quantum group interpretation of some conformal field theories, CERN-TH. 5267/88.

[22] V. PASQUIER - H. SALEUR, Common structures between finite systems and conformal field theories through quantum groups, Saclay preprint SPhT/89-031 (submitted to Nucl. Phys. **B** FS).

[23] G.M. SOTKOV - I.T. TODOROV - M.S. STANISHKOV - V. TRIFONOV, Higher symmetries in conformal QFT models, in: *Topological and Geometric Methods in Field Theory*, Symposium in Espoo, Finland, Eds. J. Hietarinta, J. Westerholm (World Scientific, Singapore 1986) pp. 195-217.

Supersymmetric Algebra, Supersymmetric Space, and Invariant Theory

ROSA Q. HUANG (*) - GIAN CARLO ROTA (**) - JOEL A. STEIN (***)

1. - Introduction

Although Descartes and Galileo have taught us that the Great Book of Geometry is written in the language of Algebra, and although many of us have been making a living on their teachings, the more recent realization that the Algebra that is required for the writing of Geometry must be of either of *two* kinds has not yet been fully accepted. The story of the slowness on the part of mathematicians in acknowledging what may turn out to be a basic property of space has yet to be told. When told, it may not make flattering reading, from the time when Grassmann was laughed off as a crackpot in the past century, to the more recent days when physicists using supersymmetric variables were unjustly accused of notational sloppiness.

The dual symmetric/skew-symmetric nature of space was first noticed in the avant guarde of the intellect, in the study of the intimate constitution of matter, after the fact that elementary particles are of two kinds was experimentally verified. Gradually, it became clear that two kinds of algebraic systems, one commutative and one anticommutative, were required for the description of the roles of the two kinds of particles, and the striking analogies between the two algebras began to clamor for development. The appellates "super", "graded", and "Z_2-" were variously adopted to designate the kind of mathematics that purported to explain or unify such a striking duality. We thus witnessed the rise of supergravity, supermanifolds, graded Lie algebras, Z_2-graded algebras, supersymmetric algebras, etc. etc.

(*)   Massachusetts Institute of Technology
(**)   Massachusetts Institute of Technology, and University of Southern California
(***) California State University, Chico

Professor Rota's work was supported by National Science Foundation Grant MCS 8104855.

In comparison with these developments, which are at the forefront of mathematics in our day, our present objective is exceedingly modest. Our motivating question is the following: how is the classical concept of space, that is, the concept of projective space in n dimensions to be modified, when the algebraic apparatus that is used to describe it is made supersymmetric?

The answer we have found for this question is in some ways disappointing. No changes are necessary in the classical concept of space. There is no "superspace" that goes with supersymmetric algebra in the same way as ordinary projective space goes with exterior algebra.

Supersymmetric algebra turns out to be a more sophisticated, perhaps more powerful, way of describing the space we are already familiar with.

In arriving at this answer, we have attempted to follow a train of thought that reduces to the minimum the arbitrariness of choice of supersymmetric "analogs", one that, we hope, is least questionable on philosophical as well as on mathematical grounds.

Our starting point is the conclusion, which was arrived at near the end of the last century, that the properties of projective space are completely described by algebraic invariants. If we did not know what space "is", we could systematically reconstruct it starting from the algebra of invariants. This assertion is the gist of the two fundamental theorems of classical invariant theory. In fact, adopting the language of logic, upon which we shall heavily rely, the first and second fundamental theorems of invariant theory can be read as stating that the algebra of invariants is an adequate syntax of a logical system which has only one semantic model, namely, ordinary projective space.

Our program is to modify the algebra of invariants by introducing supersymmetric variables, while retaining all features of the classical theory, with at most changes of sign in the defining identities. Having done this, we proceed to investigate the possible models of such an algebra of "supersymmetric" invariants.

To our surprise, we find that the new algebra still has one and the same model, namely, exterior algebra in ordinary projective space. The role of positively signed variables is secondary; sooner or later, positively signed variables are polarized away, if one has to have numerically valued coordinates at the end. This is the main conclusion of the present work.

We did not seek out this conclusion. Our starting point was another. It was the study of the invariants of skew-symmetric tensors by a supersymmetric generalization of the symbolic method of nineteenth-century invariant theory, which we introduced some years ago. Thus, the body of the present work is the description of the symbolic method, which for reasons of exposition has been confined to skew-symmetric tensors, and of some of its possibilities, which, we would like to believe, throw some light on the classical theory of algebraic invariants.

2. - Synopsis

We shall motivate the algebra Bracket[L] of supersymmetric brackets over a signed alphabet L, starting from the notion of a determinant.

Let V be a vector space of dimension n over a field k, which for technical reasons we shall choose to be an infinite field. Having fixed a basis $e_1, e_2, \ldots, e_n$, let x_{ij} be the j-th coordinate of the vector x_i relative to the basis $e_1, e_2, \ldots, e_n$. We write

$$[x_1, x_2, \ldots, x_n] = \det(x_{ij})$$

to denote the determinant of an n by n matrix whose columns are the vectors $x_1, x_2, \ldots, x_n$. The determinant depends on the choice of basis $e_1, e_2, \ldots, e_n$ only to within a non-zero constant factor. Therefore, if we allow only unimodular changes of basis, the above determinant is independent of the choice of basis. In more sophisticated language, the determinant above is an invariant under the unimodular group. We shall call it the *bracket*, following Cayley. A vector space with a bracket is called a Peano space (as defined in [BBR]).

It is easy to characterize a bracket: it is the only non-degenerate multilinear skew-symmetric form with values in the field. Actually, there is another characterization of the bracket (that is, of determinants), which is purely *syntactic*, that is, which does not use the addition of vectors, but only identities satisfied by brackets filled with abstract symbols x_i's which are viewed as variables, in the spirit of universal algebra. Such a characterization is obtained by formulating the classical Laplace expansions for determinants in bracket-theoretic terms. This is done by systematically replacing x_{ij} by the bracket $[e_1, e_2, \ldots, e_{j-1}, x_i, e_{j+1}, \ldots, e_n]$ in any one of the Laplace identities. What one obtains after this replacement is identities like

$$[x_1, x_2, \ldots, x_n] \, [e_1, e_2, \ldots, e_n]$$
$$= \Sigma \, [e_j, x_2, \ldots, x_n] \, [e_1, e_2, \ldots, e_{j-1}, x_1, e_{j+1}, \ldots, e_n]$$

with j varying between 1 and n, and analogous ones when more than one of the e's are shuffled, which will be written down later. What matters now is that such an identity will remain valid for any choice of the x's and the e's.

If one now takes a polynomial algebra over k generated by abstract symbols $[x_1, x_2, \ldots, x_n]$, with the x's taken out of an abstract alphabet L^-, and if one imposes on these symbols the properties of skew-commutativity and the abstract Laplace identities such as the one above, one obtains the bracket ring Bracket[L^-]. The (classical) second fundamental theorem of invariant theory asserts that such an abstract ring can be "uniquely" represented as a subring of a polynomial ring, where each bracket is actually a determinant whose elements are variables. In other words, one obtains a syntactic characterization of the notion of a determinant.

This fundamental result (first proved under somewhat restrictive conditions by Ernesto Pascal) is supplemented by another result, known as the first

fundamental theorem of invariant theory, which states that every property of
sets of vectors in V which is independent of the choice of a coordinate system
can be stated by the vanishing (or non-vanishing) of polynomials in the brackets
formed out of such vectors.

These two fundamental results, taken together, can be viewed from the
point of view of logic, as stating that the abstract bracket ring is an adequate
syntax having as its only semantic model a vector space with a bracket, namely,
a Peano space. In this way, classical invariant theory can be regarded as a
analog of the predicate calculus, where the bracket is the basic predicate, where
the identities in the bracket ring play the role of rules of inference, and where
the theorems are the classical theorems of projective geometry. For example,
Desargues's theorem can be elegantly presented in this way.

This analogy suggests several further developments of classical invariant
theory: for example, an analog of Herbrand's theorem for the bracket ring,
and a classification of theories expressible by the bracket ring in terms of
decidability, finite axiomatizability, etc. It would be nice if some of these
notions of predicate logic, when transferred to the bracket ring, turned out to
answer, or to coincide with, questions of invariant theory not only for vectors,
but for various kinds of tensors. Remarkably, this turns out to be the case, but
only after the supersymmetric generalization is carried out. This generalization
we now proceed to describe.

Because the bracket is skew-symmetric, we can regard it as a linear map
from the exterior algebra Super$[L^-]$ generated by the alphabet L^-, to the scalar
field k. Writing it in the form $[w]$, for w in Super$[L^-]$, this map will have the
property that $[w] = 0$ unless w is an element of degree n in Super$[L^-]$. The
fundamental observation is that the abstract Laplace identities that define the
bracket algebra can be written as Hopf algebra identities in the Hopf algebra
Super$[L^-]$; in fact, the Hopf algebra notation (the so-called Sweedler notation)
makes the Laplace identities much less formidable notationally. They become

(*) $\Sigma\ [w\ w'_{(1)}][w'_{(2)} w''] = 0,$

where w, w' and w'' are homogeneous elements of Super$[L^-]$, and where *it is*
assumed that w' is of degree greater than n.

For example, the Laplace identity we have written out above is obtained
by taking $w = x_2 x_3 \ldots x_n$, $w' = x_1 e_1 e_2 \ldots e_n$, and $w'' = 1$ (notice that exterior
products are written by juxtaposition, rather than by the more common wedge
notation).

To summarize: all that is needed to develop the abstract algebra of brackets
is the Hopf algebra structure of the exterior algebra. Once one realizes this, one
also realizes how to extend the theory to the supersymmetric case. One replaces
the alphabet L^- by a signed alphabet $L = L^+ \cup L^-$, where L^+ and L^- are
disjoint. The supersymmetric algebra Super$[L]$ is now a tensor product of an
exterior algebra and an algebra of divided powers (that is, a commutative algebra;
more about this later); that is, in plain language, the supersymmetric algebra

Super$[L]$ is an algebra generated by monomials containing some letters that anticommute and some letters that commute. The algebra of (supersymmetric) brackets is now generated by elements $[w]$, as w ranges over all monomials of degree n in Super$[L]$. The commutation rules for the bracket $[w]$ are subsumed into the commutation laws of the letters that make up the monomial w, and the Laplace identities are exactly the same as in (*), except for signs, again with the proviso that the word w' be of degree (=Length) strictly greater than n.

Having imposed these identities, one obtains an algebra which is the natural supersymmetric analog of the algebra of brackets.

We have mentioned the fact that the algebra of divided powers must be used in preference to the algebra of ordinary polynomials. Divided powers are a device for avoiding the appearance of unwanted factorials in computations. The idea is to define the divided powers $a^{(i)}$ of a variable a "as if" they were $a^i/i!$, and to compute accordingly. A characteristic-free theory cannot dispense with divided powers.

The two fundamental properties of the algebra of supersymmetric brackets Bracket$[L]$ are, first, invariance under polarization, and second, the generalization of the Young straightening algorithm that can be proved for them. We discuss these properties in turn, before picking up the main thread of the discussion.

Polarizations, one of the great war horses of the Nineteenth Century, as developed and skillfully used by Alfredo Capelli, are a formal device for implementing substitutions of variables. The polarization $D(b, a)$ is a derivation which replaces the variable a by the variable b. Because of supersymmetry, such derivations can be even or odd; because of divided powers, one is forced to take divided powers of positive derivations, rather than ordinary powers. One easily adjusts to these quirks.

One shows that the algebra Bracket$[L]$ is invariant under polarization. This is actually a strong statement. It means (following an idea due to Andrea Brini) that the Laplace identities (*) can be replaced by the equivalent assumption that any monomial $[w][w'] \ldots [w'']$ which contains more than n occurrences of a positive letter a is identically 0. All other identities can be obtained by polarizing a positive letter into a given word. This device is especially powerful in applications to the classical bracket algebra (for identities with negatively signed letters only, that is), where it can be used to obtain identities satisfied by minors of a matrix.

The second fundamental property of the supersymmetric bracket algebra is the fact that superstandard Young tableaux form an integral basis. A monomial in the bracket algebra can be viewed as a tableau Young$[W, W', \ldots, W'']$, where the words $W, W', \ldots, W''$ are of length n and are taken from the free monoid generated by the alphabet L (without regard to sign). The i-th letters of each word form a column of the matrix associated with the tableau, obtained by placing the word one underneath the other in the plane. The tableau is said to be *superstandard* when the alphabet L is ordered, and in each row of the matrix two successive letters a and b are in relation $a \leq b$ in general, and $a < b$ if both letters are negatively signed; similarly, if a and b are successive letters in a column of the matrix, then $a \leq b$ holds, and $a < b$ if both letters are positively signed.

The superstandard basis theorem is a substantial generalization of the classical case, due to Young, Littlewood and Hodge, in which all letters are negatively signed.

Both of these properties are used in establishing the fundamental fact, and the main result of this paper, that a Peano space is the only model for a supersymmetric bracket algebra. The role of the positively signed letters is to represent "symbolically" skew-symmetric tensors in $\mathrm{Ext}(V)$. This symbolic representation, foreshadowed by Weitzenböck and developed in [GRS87] is an analog of the symbolic representation developed in the nineteenth century for symmetric tensors. It was a strange twist of fate that prevented nineteenth-century invariant theorists from realizing that the symbolic method they had so successfully used in the commutative case is actually more efficient in the skew-symmetric case, as we hope to have shown in the present as well as in previous papers.

The representation of the supersymmetric bracket algebra Bracket[L] in an ordinary Peano space is based upon a simple observation, which we wish to stress here, since in the more rigorous treatment given in the text it tends to be obscured by notational necessities.

What is to de done? A bracket, say $[a^{(3)}bc^{(5)}de]$, where a, b and c are positive letters, and where d, and e are negatively signed letters, is to be mapped into vectors and tensors in a Peano space. The divided power $a^{(3)}$ is to be mapped into a skew-symmetric tensor t of step 3, and the divided power $c^{(5)}$ is to be mapped into a skew-symmetric tensor t' of step 5, the remaining letters being mapped into vectors v, v' and v''. The only conceivable way of doing this is to map the bracket into the Peano space bracket $[t\ v\ t'\ v'\ v'']$. But this seems at first impossible. We have the commutation relations $a^{(3)}c^{(5)} = c^{(5)}a^{(3)}$, but $t\ t' = -t'\ t$, a seeming contradiction. Actually, one maps the bracket $[a^{(3)}bc^{(5)}de]$ to $[t\ v\ t'\ v'\ v'']$ with the letters in alphabetical order. If the letters inside a bracket are not in alphabetical order, then one restores the aphabetical order first, and then one maps. If letters in different brackets are in the wrong alphabetical order, one uses the straightening algorithm to restore alphabetical order, and then one maps the brackets into the corresponding Laplace convolution of brackets in the Peano space (actually, the device we use dispenses with the use of straightening, by performing an intermediate polarization of positive letters into negatively signed letters). The amazing fact is that this device actually works, and thus leads to a symbolic calculus for skew-symmetric tensors. We conclude with a generalization to the supersymmetric ambiance of the first fundamental theorem of invariant theory, under an important restriction: that of multilinearity. We conjecture that this restriction can be removed by suitable divided powers techniques applied to skew-symmetric tensors. If this conjecture is true, as we feel confident it is, then our guess that the invariant theory of skew-symmetric tensors is simpler than the classical invariant theory of forms would be vindicated.

In the present paper the exposition is limited to the barest sketches. We have omitted the introduction of full-fledged logical notation, in particular, of the quantifiers of the predicate calculus. In the statement of the generalization of the two fundamental theorems to supersymmetric bracket algebras, we have omitted a statement of the supersymmetric generalization of Gram's theorem, which would have required a lengthy detour into the theory of concomitants. We have also stopped short of a discussion of the symbolic method for supersymmetric tensors, which opens a new chapter in the theory (though one that becomes all the more inevitable in the present logic-oriented point of view).

We hope thereby to have lightened the exposition to the point where the reader can reconstruct there results alone. Similarly, we have only given bare sketches of proofs which would have been repetitions of arguments we have already expounded elsewhere.

Theorems 1 and 2, as well as the two equivalent definitions of the supersymmetric bracket algebra Bracket[L], are believed to be new. Other results are closely related to classical results in the literature, although we would like to believe that the present exposition casts a different light on facts that have been known since the nineteenth century.

3. - The supersymmetric algebra Super[L]

The supersymmetric algebra Super[L] (defined in [RS86] and in [GRS87]) is a generalization of the ordinary algebra of polynomials in a set L of variables. Our variables will be of two kinds: positively signed and negatively signed: $L = L^+ \cup L^-$ (neutral variables will not explicitly occur in this paper). Such a set L of variables will be called a *proper signed set*. The theory of supersymmetric algebras is developed in [RS86], [GRS87], [RS89]. We shall briefly recall the main idea.

Positively signed variables are the least familiar: they are the *divided powers*. To every positively signed variable a we assign a sequence $a^{(i)}$, $i = 0, 1, 2, \ldots$ of divided powers, that behave algebraically "as if" $a^{(i)}$ were equal to $a^i/i!$, with $a^{(0)} = 1$ and $a^{(1)} = a$. We have the "rules"

$$a^{(i)} a^{(j)} = \binom{i+j}{i} a^{(i+j)}$$

$$(a^{(i)})^{(j)} = \frac{(ij)!}{j!(i!)^j} a^{(ij)}$$

$$(a+b)^{(i)} = \sum_{j+k=i} a^{(j)} b^{(k)}$$

Positively signed variables, subject to these conventions, generate a commutative algebra, the *divided powers algebra* in the positively signed letters in L^+. Negatively signed variables generate an exterior algebra, that is, they anticommute: if a and b belong to L^-, then $ab = -ba$. Positively signed variables commute with negatively signed variables.

The algebra over the field k that is obtained by these "rules" is the *supersymmetric algebra* Super$[L]$ of the signed alphabet L. It is the tensor product of the divided powers algebra of L^+ and the exterior algebra of L^-. In other words, it is the algebra generated by all monomials in the letters in L, subject to the above commutation conventions (for full details, read [RS86] or [GRS], pp. 1-11). For instance, if b and c are negative letters, and all other letters are positive, then we have

$$a^{(2)} bc \; d^{(3)} a^{(4)} = -15 \; a^{(6)} d^{(3)} cb.$$

The *Length* of a monomial in Super$[L]$ is its total degree, where divided powers are weighed accordingly; for example, the Length of the above monomial equals 11. If w is a monomial in Super$[L]$, the parity of w will be 0 or 1 according as the Length of w is even or odd.

We shall be using polarization operators $D(b, a)$ as in [RS89]. Recall that there are signed derivations in the supersymmetric algebra Super$[L]$ which replace a letter a by a letter b, defined as follows. When the letters a and b are of the same sign, the polarization is a positive derivation, and when the letters are of different signs, the polarization is a negative derivation.

A polarization operator $D(b, a)$ of Super$[L]$ is a linear operator defined recursively by the rules

(a)   If $D(b, a)$ is a negative polarization, then for arbitrary monomials w and w' we set

$D(b, a)1 = 0$, $D(b, a)c = 0$ if $c \neq a$, $D(b, a) \; a^{(k)} = b \; a^{(k-1)}$ if a is positively signed, $D(b, a)a = b$ if a is negatively signed;

$$D(b, a)(ww') = (D(b, a)w)w' + (-1)^{|w|} \; w \; D(b, a)w'$$

(b)   If $D(b, a)$ is a positive polarization, then we set

$D(b, a)1 = 0$, $D(b, a)c = 0$ if $c \neq a$, $D(b, a)a^{(k)} = b \; a^{(k-1)}$ if a is positively signed, and $D(b, a)a = b$ if a is negatively signed. Divided powers satisfy

$$D^{(k)}(b, a)(ww') = \Sigma \; (D^{(i)}(b, a)w) \; D^{(j)}(b, a)w'$$

where the sum ranges over all non-negative integers i and j such that $i + j = k$. We further set

$$D^{(0)}(b, a) \; w = w$$

and

$$D^{(k)}(b, a) \; a^{(r)} = 0 \text{ if } r < k$$

$$D^{(k)}(b, a) \; a^{(r)} = b^{(k)} a^{(r-k)} \text{ if } r \geq k.$$

If $D(b, a)$ is a negative polarization, we set $D^{(k)}(b, a) = D(b, a)$ if $k = 1$ and $D^{(k)}(b, a) = 0$ for $k > 1$. Again, set $D^{(0)}(b, a)$ to be the identity operator.

Positive polarizations are abstractions from the classical differential operators

$$\sum_{i=1}^{n} b_i \frac{\partial}{\partial a_i}$$

whereas negative polarizations have the property that $D(b, a)^2 = 0$. They are *boundary operators* in the sense of homological algebra, and can be used to define complexes.

Let w be a word in Super$[L]$, and let a be any positive letter. Then there is a unique operator $T(w)$ (the "word polarization operator" defined in [RS89]) in Super$[L]$, a product of polarization operators, such that $T(w) \, a^{(n)} = w$. When the dependence of the word polarization operator $T(w)$ upon the letter a is needed, we write $T(w, a)$ in place of $T(w)$.

The word polarization operator $T(w)$ is uniquely defined by the following requirements: (1) $T(w \, w') = T(w)T(w')$; (2) $T(b) = D(b, a)$ if b is a letter of L, and (3) $T(b^{(r)}) = D^{(r)}(b, a)$ if b is a positive letter.

The supersymmetric algebra Super$[L]$ has te structure of a Hopf algebra. The coproduct of an element p of Super$[L]$ is expressed by the Sweedler notation

$$\Delta \, p = \Sigma_i \; p_{(1)} \otimes p_{(2)}.$$

The antipode s of the Hopf algebra Super$[L]$ is defined as follows. If w is a monomial of Length k, then $s(w) = (-1)^k w$; extend by linearity. Recall that the antipode and coproduct of a Hopf algebra are related by the identity

$$\Sigma \; s(w_{(1)}) \; w_{(2)} = \Sigma \; w_{(1)} \; s(w_{(2)})$$

equals 0 if Length$(w) > 0$, and equals w otherwise.

4. - The algebra of braces

Let $\{L\}$ be the alphabet whose members are all monomials w in Super$[L]$. A letter in $\{L\}$ will be denoted by $\{w\}$. The parity $|\{w\}|$ of the letter $\{w\}$ is defined to be the parity of the integer $|w| + $ Length(w). We denote by Brace$\{L\}$ the associative algebra over the field k generated by the alphabet $\{L\}$, subject to the following commutation relations:

$$\{w\}\{w'\} = (-1)^{|\{w\}| \, |\{w'\}|} \{w'\} \{w\},$$

and

$$\{w\}\{w\} = 0 \text{ if } |w| + \text{ Length}(w) \text{ is odd}.$$

If p, q and r are elements of Super$[L]$, and if α is a scalar (that is, an element of k), then the element $\{p\}$ of Brace$\{L\}$ is defined recursively by the rules

$$\{q + r\} = \{q\} + \{r\} \text{ if } p = q + r, \quad \{\alpha p\} = \alpha \{p\}.$$

If $D(b, a)$ is a polarization operator of Super$[L]$, the linear operator $\mathbf{D(b, a)}$ of Brace$\{L\}$ is defined recursively by the rules

$$\mathbf{D(b, a)} \{p\} = \{D(b, a)p\}$$

for p in Super$[L]$, and:

(a)   If $D(b, a)$ is a negative polarization, then for arbitrary monomials m and m' in Brace$\{L\}$ we set

$$\mathbf{D(b, a)}(m \; m') = (\mathbf{D(b, a)}m) \; m' + (-1)^{|m|} \; m \; \mathbf{D(b, a)} \; m',$$

where the parity of the monomial $m = \{w\}\{w'\} \ldots \{w''\}$ in Brace$[L]$ is the sum $|\{w\}| + |\{w'\}| + \ldots + |\{w''\}|$.

(b)   If $D(b, a)$ is a positive polarization, then

$$\mathbf{D}^{(k)}(b, a)(m \; m') = \Sigma \; (\mathbf{D}^{(i)}(b, a)m) \; D^{(j)}(b, a) \; m'$$

where $i + j = k$, and

$$\mathbf{D}^{(k)}(b, a)\{w\} = \{D^{(k)}(b, a) \; w\}.$$

When both a and b are negatively signed, then $\mathbf{D}^{(k)}(b, a)\{w\} = 0$ for $k > 1$. For example, if a and b are negative, we have

$$\mathbf{D}^{(2)}(b, a)(\{aw\}\{aw'\}) = \{bw\}\{bw'\},$$

and, if both a and b are positively signed (assuming the letter a does not appear in w), then

$$\mathbf{D}^{(k)}(b, a)\{a^{(r)}w\} = 0 \text{ if } r < k$$

$$\mathbf{D}^{(k)}(b, a)\{a^{(r)}w\} = \{b^{(k)}a^{(r-k)}w\} \text{ if } r \geq k.$$

An element of the algebra of braces will sometimes be called a brace polynomial. An element of the brace algebra which is a product of braces will be called a brace monomial.

A brace polynomial is said to be homogeneous if it is the sum of monomials in each of which every letter occurs with the same content, but different letters may occur with different contents.

To avoid unnecessary minus signs, it turns out to be convenient to introduce the *vertical notation* for monomials. The monomial vert$\{w, w', \ldots w''\}$ is defined as

$$\mathbf{T(w, a)T(w', b)} \ldots \mathbf{T(w'', c)}(\{a^{(n)}\} \; \{b^{(n)}\} \ldots \{c^{(n)}\}),$$

where $T(\mathbf{w}, \mathbf{a})$ is defined from $T(w, a)$ in the obvious way. For example, $\text{vert}\{w\} = \{w\}$ and $\text{vert}\{w, w'\} = \text{sign}(\text{Length}(w) \ |w'|) \ \{w\} \ \{w'\}$.

The above monomial differs at most by a sign from the monomial $\{w\}\{w'\} \ldots \{w''\}$.

Let p be a homogeneous brace polynomial. If y is a letter not occurring in p, then the polynomial q obtained by substituting y in place of the letter x is defined as follows. If k is the degree (that is, the number of occurrences) of the letter x in p, the substitution of y for x is performed by taking $q = \mathbf{D}^{(k)}(\mathbf{y}, \mathbf{x})p$. For example, in vert-notation we have

$$\mathbf{D}^{(k)}(\mathbf{b}, \mathbf{a}) \ \text{vert}\{w, w'\} = \Sigma \ (-1)^{j|w|} \ \text{vert}\{D^{(i)}(b, a)w, D^{(j)}w'\}$$

when $D(b, a)$ is a negative polarization, and

$$\mathbf{D}^{(k)}(\mathbf{b}, \mathbf{a}) \ \text{vert}\{w, w'\} = \Sigma \ \text{vert}\{D^{(i)}(b, a)w, D^{(j)}w'\}$$

when $D(b, a)$ is a positive polarization.

5. - The bracket algebra

With L a proper signed alphabet, we proceed to define an algebraic system that connects the notation of a supersymmetric algebra with the logic of projective geometry. This is the algebra of brackets over the Supersymmetric algebra $\text{Super}[L]$.

The algebra $\text{Bracket}[L]$ of *dimension* n over the field k (dependence on the field k and on the dimension n will not be indicated) is the quotient of the algebra $\text{Brace}\{L\}$ obtained by imposing the congruence relation obtained from the following identities, where $w, w', \ldots, w''$ are any monomials in $\text{Super}[L]$:

(1) $\{w\} = 0$ if $\text{Length}(w) \neq n$;

(2) $\{w\}\{w'\} \ldots \{w''\} = 0$ whenever any positive letter a of L occurs more than n times in the monomial $\{w\}\{w'\} \ldots \{w''\}$ of $\text{Brace}\{L\}$.

(3) Let $\{w\}\{w'\} \ldots \{w''\}$ be a monomial in $\text{Brace}\{L\}$ in which some positive letter a occurs more than n times, and let b and c be any two letters of L. Then

$$\mathbf{D}^{(k)}(\mathbf{c}, \mathbf{b})(\{w\}\{w'\} \ldots \{w''\}) = 0.$$

The canonical image of an element p of $\text{Brace}\{L\}$ in $\text{Bracket}[L]$ will be denoted by $\text{Image}(p)$. In particular, $\text{Image}(\{w\}) = [w]$, and $\text{Image}(\{w\}\{w'\} \ldots \{w''\}) = [w][w'] \ldots [w'']$. It is easily shown that every polarization operator $\mathbf{D}(\mathbf{b}, \mathbf{a})$ induces a polarization operator on $\text{Bracket }[L]$, which will again be denoted by $D(b, a)$, and which satisfies the same identities as $\mathbf{D}(\mathbf{b}, \mathbf{a})$. We write $\|w\| = |\{w\}|$ when w is a word of length n.

The following property of the bracket algebra is easily established:

PROPOSITION 1. *Let p be an element of* Brace$\{L\}$, *and suppose that* Image$(p) = 0$. *Then* Image$(\mathbf{D}^{(k)}(\mathbf{b}, \mathbf{a})p) = 0$ *for any polarization* $D(b, a)$ *and for any positive integer k. More generally, for any brace polynomial p, we have* Image$(\mathbf{D}^{(k)}(\mathbf{b}, \mathbf{a})p) = D^{(k)}(b, a)$ Image(p).

An element p of the bracket algebra will sometimes be called a bracket polynomial. A product of brackets will sometimes be called a bracket monomial. Homogeneous bracket polynomials are defined as the images of homogeneous brace polynomials.

To avoid unnecessary minus signs, it turns out to be convenient again to introduce the *vertical notation* for monomials. The monomial vert$[w, w', \ldots w'']$ is defined as the image in Bracket$[L]$ of vert$\{w, w', \ldots w''\}$, that is, as

$$T(w, a)T(w', b) \ldots T(w'', c)([a^{(n)}] \; [b^{(n)}] \ldots [c^{(n)}]),$$

where $T(w, a)$ is defined in the obvious way from $\mathbf{T}(\mathbf{w}, \mathbf{a})$.

6. - The straightening laws

We proceed to establish the main results holding for the algebra of brackets.

PROPOSITION 1. *Let w be a monomial of Length greater than n in* Super$[L]$, *and let w' be any monomial in* Super$[L]$. *Then*

(*) $\Sigma \pm [w_{(1)}][w_{(2)} w'] = \Sigma$ vert$[w_{(1)}, w_{(2)} w'] = 0$.

Note that the vert-notation for monomials gives an automatic rule for the sign on the far left side.

PROOF. Suppose Length$(w) = p > n$. Let $T(w, a)a^{(p)} = w$, and let $T(w', b)b^{(2n-p)} = w'$. Now the left side of (*) is the image in Bracket$[L]$ of the brace polynomial

$$\mathbf{T}(\mathbf{w}, \mathbf{a})\mathbf{T}(\mathbf{w'}, \mathbf{b}) \; \{a^{(n)}\}\{a^{(p-n)}b^{(2n-p)}\} = \Sigma \text{ vert}\{w_{(1)}, w_{(2)} w'\},$$

and the image of the above brace polynomial is evidently 0, since it is a polarization of a brace monomial containing more than n occurrences of the positive letter a.

PROPOSITION 2 (Exchange property). *Let u, v and w be monomials in* Super$[L]$. *Then the following identity holds*:

$$\Sigma \pm [u \; v_{(1)}][v_{(2)} \; w] = \Sigma \pm [v \; u_{(1)}][s(u_{(2)})w].$$

More precisely, the signs are determined as follows:

$$\Sigma \; \text{vert}[u \; v_{(1)}, v_{(2)} w] = (-1)^{|u||v|} \Sigma \; \text{vert}[v \; u_{(1)}, s(u_{(2)}) w].$$

PROOF. Starting with the identity of [RS89]

$$\Sigma u \; v_{(1)} \otimes v_{(2)} = (-1)^{|u||v|} \Sigma \; (v \; u_{(1)})_{(1)} \otimes (v \; u_{(1)})_{(2)} \; s(u_{(2)})$$

we find

$$\Sigma \; \text{vert}\{u \; v_{(1)}, v_{(2)} w\} = (-1)^{|u||v|} \Sigma \; \text{vert}\{(v \; u_{(1)})_{(1)}, \; (v \; u_{(1)})_{(2)} s(u_{(2)}) w\}.$$

By the preceding proposition, if $\text{Length}((v u_{(1)})_{(2)}) > 0$, the image in $\text{Bracket}[L]$ of the corresponding term on the right side equals 0, and therefore the right side is congruent to

$$(-1)^{|u||v|} \Sigma \; \text{vert}\{v \; u_{(1)}, s(u_{(2)}) w\},$$

as desired.

PROPOSITION 3. *The quotient of the algebra of braces* $\text{Brace}\{L\}$ *by the congruence relation generated by the following identities*:

(1) $\{w\} = 0$ if $\text{Length}(w) \neq n$;

(2)   *The exchange identity*

$$\Sigma \; \text{vert}\{u \; v_{(1)}, v_{(2)} w\} = (-1)^{|u||v|} \Sigma \; \text{vert}\{v \; u_{(1)}, s(u_{(2)}) w\}$$

is naturally isomorphic to the bracket algebra $\text{Bracket}[L]$.

PROOF. We first show that if m is a brace monomial containing more than n occurrences of the letter a, then m is congruent to 0. Suppose that $m = \text{vert}\{a^{(k)} w, a^{(r)} w'\}$, where $k + r > n$. The exchange identity gives

$$\text{vert}\{a^{(k)} w, a^{(r)} w'\} = \Sigma \; \text{vert}\{a^{(k+r)} w_{(1)}, s(w_{(2)}) w'\} = 0,$$

since $k + r > n$.

If the bracket monomial m contains more than two factors, one extends the preceding argument by induction.

Next we prove invariance under polarizations. If p is congruent to 0 in the quotient algebra defined above, then we show that

$$\mathbf{D}^{(k)}(\mathbf{b}, \mathbf{a}) p$$

is congruent to 0 in the quotient algebra. This follows from noting that polarization in $\text{Super}[L]$ preserves the degree of a monomial in $\text{Super}[L]$ and that polarization commutes with coproducts and the antipode map:

$$D^{(k)}(b, a) \Sigma \; w_{(1)} \otimes w_{(2)} = \Sigma \; (D^{(k)}(b, a) w)_{(1)} \otimes (D^{(k)}(b, a) w)_{(2)}$$

and

$$D^{(k)}(b,a)s(w) = s(D^{(k)}(b,a)w).$$

In view of the identity satisfied by polarizations in Brace$\{L\}$

$$\mathbf{D}^{(k)}(\mathbf{b},\mathbf{a})(p\ q) = \Sigma \pm (\mathbf{D}^{(i)}(\mathbf{b},\mathbf{a})p)\ \mathbf{D}^{(j)}(\mathbf{b},\mathbf{a})q$$

(where $i + j = k$), we need only show that the defining identities in the Proposition are preserved under polarization. This we proceed to verify.

First note that polarizations preserve the length of a brace. Therefore, it is enough to verify that under polarization, exchange identities are mapped into linear combinations with integer coefficients of exchange identities. We have two cases.

CASE 1. $D(b,a)$ is a positive polarization. Consider the expression

$$\mathbf{D}^{(k)}(\mathbf{b},\mathbf{a})\left(\Sigma\ \text{vert}\{u\ v_{(1)}, v_{(2)}w\} - (-1)^{|u||v|}\Sigma\ \text{vert}\{v\ u_{(1)}, s(u_{(2)})w\}\right).$$

Simplify this expression using the rules given previously for polarizing expressions of the form vert$\{w', w''\}$ in Brace$\{L\}$ and expressions of the form $m\ m'$ in Super$[L]$. The expansion which results is

$$\Sigma(\Sigma\ \text{vert}\{(D^{(i)}(b,a)u)D^{(j)}(b,a)v_{(1)}, (D^{(r)}(b,a)v_{(2)})D^{(t)}(b,a)w\}-$$

$$(-1)^{|u||v|}\Sigma\ \text{vert}\{(D^{(i)}(b,a)v)D^{(j)}(b,a)u_{(1)}, (D^{(r)}(b,a)s(u_{(2)}))D^{(t)}(b,a)w\})$$

where the outermost sum ranges over all non-negative i, j, r and t which sum to k. Since polarization commutes with the antipode map, we infer that the above expression equals

$$\Sigma(\Sigma\ \text{vert}\{(D^{(i)}(b,a)u)D^{(j)}(b,a)v_{(1)}, (D^{(r)}(b,a)v_{(2)})D^{(t)}(b,a)w\}-$$

$$(-1)^{|u||v|}\Sigma\ \text{vert}\{(D^{(i)}(b,a)v)D^{(j)}(b,a)u_{(1)}, s(D^{(r)}(b,a)u_{(2)})D^{(t)}(b,a)w\}).$$

Since polarization commutes with comultiplication, the latest expression is seen to equal

$$\Sigma(\Sigma\ \text{vert}\{(D^{(i)}(b,a)u)(D^{(j)}(b,a)v)_{(1)}, (D^{(j)}(b,a)v)_{(2)}D^{(t)}(b,a)w\}-$$

$$(-1)^{|u||v|}\Sigma\ \text{vert}\{(D^{(i)}(b,a)v)(D^{(j)}(b,a)u)_{(1)}, s((D^{(j)}(b,a)u)_{(2)})\ D^{(t)}(b,a)w\})$$

where the outermost sum now ranges over all i, j and t which sum to k. Since $D(b,a)$ is a positive polarization, we have that whenever $D^{(j)}(b,a)u$ and $D^{(i)}(b,a)v$ are non-zero, that $|u| = |D^{(j)}(b,a)u|$ and $|v| = |D^{(i)}(b,a)v|$.

The above sum therefore equals

$$\Sigma(\Sigma\ \text{vert}\{(D^{(i)}(b,a)u)(D^{(j)}(b,a)v)_{(1)}, (D^{(j)}(b,a)v)_{(2)}D^{(t)}(b,a)w\}$$

$$- \text{sign}(|D^{(j)}(b,a)u\|D^{(i)}(b,a)v|)\Sigma\ \text{vert}\{(D^{(i)}(b,a)v)(D^{(j)}(b,a)u)_{(1)},$$

$$s((D^{(j)}(b,a)u)_{(2)})D^{(t)}(b,a)w\}).$$

Interchanging i and j in the first innermost sum, the above expression is seen to equal

$$\Sigma(\Sigma \text{ vert}\{(D^{(j)}(b,a)u)(D^{(i)}(b,a)v)_{(1)}, (D^{(i)}(b,a)v)_{(2)}D^{(t)}(b,a)w\}$$

$$- \text{sign}(|D^{(j)}(b,a)u||D^{(i)}(b,a)v|)\Sigma \text{ vert}\{(D^{(i)}(b,a)v)(D^{(j)}(b,a)u)_{(1)},$$

$$s((D^{(j)}(b,a)u)_{(2)})D^{(t)}(b,a)w\}),$$

as desired.

CASE 2. $D(b,a)$ is a negative polarization. Reasoning as in Case 1 we have:

$$\mathbf{D(b,a)}(\Sigma \text{ vert}\{u\ v_{(1)}, v_{(2)}w\} - (-1)^{|u||v|}\Sigma \text{ vert}\{v\ u_{(1)}, s(u_{(2)})w\})$$

$$= \Sigma(\Sigma \text{ sign}(j|u| + r|u| + r|v_{(1)}| + t|u| + t|v|)$$

$$\text{vert}\{(D^{(i)}(b,a)u)D^{(j)}(b,a)v_{(1)}, (D^{(r)}(b,a)v_{(2)})D^{(t)}(b,a)w\}$$

$$- (-1)^{|u||v|}\Sigma \text{ sign}(j|v| + r|v| + r|u_{(1)}| + t|u| + t|v|)$$

$$\text{vert}\{(D^{(i)}(b,a)v)D^{(j)}(b,a)u_{(1)}, (D^{(r)}(b,a)s(u_{(2)}))D^{(t)}(b,a)w\})$$

where the outermost sum ranges over all non-negative integers i, j, r, t which sum to 1 (one). Since polarization commutes with the antipode map, this expression equals in turn

$$\Sigma(\Sigma \text{ sign}(j|u| + r|u| + r|v_{(1)}| + t|u| + t|v|)$$

$$\text{vert}\{(D^{(i)}(b,a)u)D^{(j)}(b,a)v_{(1)}, (D^{(r)}(b,a)v_{(2)})D^{(t)}(b,a)w\}$$

$$- (-1)^{|u||v|}\Sigma \text{ sign}(j|v| + r|v| + r|u_{(1)}| + t|u| + t|v|)$$

$$\text{vert}\{(D^{(i)}(b,a)v)D^{(j)}(b,a)u_{(1)}, s(D^{(r)}(b,a)u_{(2)})D^{(t)}(b,a)w\}),$$

and since polarization commutes with the coproduct, the last-given expression equals

$$\Sigma(\Sigma \text{ sign}(j|u| + t|u| + t|v|)$$

$$\text{vert}\{(D^{(i)}(b,a)u)(D^{(j)}(b,a)v)_{(1)}, (D^{(j)}(b,a)v)_{(2)}D^{(t)}(b,a)w\}$$

$$- (-1)^{|u||v|}\Sigma \text{ sign}(j|v| + t|u| + t|v|)$$

$$\text{vert}\{(D^{(i)}(b,a)v)(D^{(j)}(b,a)u)_{(1)}, s((D^{(j)}(b,a)u)_{(2)})D^{(t)}(b,a)w\}),$$

where the outermost sum now ranges over all non-negative integers i, j and t which sum to 1. Interchanging i and j in the first innermost sum, we see that the above expression equals

$$\Sigma(\Sigma \ \text{sign}(i|u| + t|u| + t|v|)$$

$$\text{vert}\{(D^{(j)}(b,a)u)(D^{(i)}(b,a)v)_{(1)}, (D^{(i)}(b,a)v)_{(2)}D^{(t)}(b,a)w\}$$

$$- (-1)^{|u||v|}\Sigma \ \text{sign}(j|v| + t|u| + t|v|)$$

$$\text{vert}\{(D^{(i)}(b,a)v)(D^{(j)}(b,a)u)_{(1)}, s((D^{(j)}(b,a)u)_{(2)})D^{(t)}(b,a)w\})$$

Noting that when $D^{(i)}(b,a)v$ and $D^{(j)}(b,a)u$ are non-zero, we have

$$\text{sign}(|D^{(i)}(b,a)v\|D^{(j)}(b,a)u|) = \text{sign}((i + |v|)(j + |u|))$$

and that $ij = 0$, we see that the above expression equals

$$\Sigma(\Sigma \ \text{sign}(i|u| + t|u| + t|v|)$$

$$\text{vert}\{(D^{(j)}(b,a)u)(D^{(i)}(b,a)v)_{(1)}, \ (D^{(i)}(b,a)v)_{(2)}D^{(t)}(b,a)w\}$$

$$- \text{sign}(|D^{(j)}(b,a)u\|D^{(i)}(b,a)v|)\Sigma \ \text{sign}(i|u| + t|u| + t|v|)$$

$$\text{vert}\{(D^{(i)}(b,a)v)(D^{(j)}(b,a)u)_{(1)}, \ s((D^{(j)}(b,a)u)_{(2)})D^{(t)}(b,a)w\})$$

$$= \Sigma \ \text{sign}(i|u| + t|u| + t|v|) \ (\Sigma \ \text{vert}\{(D^{(j)}(b,a)u)$$

$$(D^{(i)}(b,a)v)_{(1)}, (D^{(i)}(b,a)v)_{(2)}D^{(t)}(b,a)w\}$$

$$- \text{sign}(|D^{(j)}(b,a)u\|D^{(i)}(b,a)v|)\Sigma \ \text{vert}\{(D^{(i)}(b,a)v)$$

$$(D^{(j)}(b,a)u)_{(1)}, s((D^{(j)}(b,a)u)_{(2)})D^{(t)}(b,a)w\}),$$

as desired.

The proof is therefore complete.

7. - The logic of brackets

We develop a logical system in the algebra of braces, having axioms and rules of inference, which is equivalent, in an evident sense, to the bracket algebra. Recall that if p is an element of the algebra of braces, we write Image(p) to denote its canonical image in the bracket algebra. We say that a polynomial p is *true in the logic of brackets*, when Image(p) = 0.

The *rules of inference* of the *logic of brackets* are rules whereby given a set F of true polynomials in the algebra of braces, one obtains a new set F' of true polynomials in Brace$\{L\}$. These rules are a syntactical rendering of the identities that define the bracket algebra. They are the following:

(1)   Remove any polynomial from the set F.

(2)   Take any linear combination p' of elements p appearing in F, with arbitrary coefficients in Brace$\{L\}$, and add to F the polynomial p'.

(3)   Add to F any polynomial of the form $\{w\}$, where w is a monomial of Length unequal to n.

(4)   Add to F any monomial m in Brace$\{L\}$ in which at least one positive letter occurs with multiplicity greater than n.

(5)   If p is in F, add to F the polynomial $\mathbf{D}^{(k)}(\mathbf{b},\mathbf{a})p$, for any letter a and any letter b.
      Besides the above five rules of inference, we shall occasionally apply the following *derived* rule of inference:

(6)   Add to the set F a polynomial which has been simplified from a polynomial already in F by application of the exchange identity (as in Proposition 2 of the preceding section).

A sequence of subsets of such polynomials $F_1, F_2, \ldots F_e$ is said to be a *proof* of a polynomial p in the logic of brackets, when:

(1)   The set F_1 is empty.

(2)   The set F_e contains p.

(3)   The set F_{i+1} is obtained from the set F_i by application of one of rules of inference (1)-(5) above (and hence, also by application of rule (6)).

The following fundamental result is immediate from the preceding discussion:

PROPOSITION 1. *A polynomial p of* Brace$\{L\}$ *is true in the logic of brackets if and only if it has a proof.*

We next describe the *straightening algorithm*, which allows one to verify whether a polynomial p in Brace$\{L\}$ is true in the logic of brackets. The straightening algorithm is an effective decision procedure.

Let Mon(L) be the free monoid generated by the alphabet L. Elements of Mon(L) other than the identity will be called *words*, and will be denoted by W, W', W'', etc. There is a canonical map mon$(\ )$ which associates a monomial mon$(W) = w$ in Super$[L]$ to a word W in Mon(L). For example, mon$(aabcaaad) = a^{(2)}bca^{(3)}d$ if the letter a is positive.

A *Young tableau*, in symbols Young$\{W_1, W_2, \ldots, W_k\}$, defined by the sequence of words $W_1, W_2, \ldots, W_k$ in Mon(L), is the monomial

$$\text{Young}\{W_1, W_2, \ldots, W_k\} = \text{vert}\{\text{mon}(W_1), \text{mon}(W_2), \ldots, \text{mon}(W_k)\}$$

in Brace$\{L\}$. Similarly, a Young tableau Young$[W_1, W_2, \ldots, W_k]$ in Bracket$[L]$

is defined as Image(Young$\{W_1, W_2, \ldots, W_k\}$), or as

$$\text{Young}[W_1, W_2, \ldots, W_k] = \text{vert}[\text{mon}(W_1), \text{mon}(W_2), \ldots, \text{mon}(W_k)].$$

Such a Young tableau in Bracket$[L]$ is non-zero only if each of the words $W_1, W_2, \ldots, W_k$ is of length n. This condition will be tacitly assumed from now on.

Recall that the alphabet L is a linearly ordered set. A word W in Mon(L) is said to be *dressed* when $W = ab \ldots c$, with $a, b, \ldots, c$ in L, and $a \leq b \leq \ldots \leq c$, with equality between two successive letters only if both letters are positively signed. A Young tableau Young$\{W_1, W_2, \ldots, W_k\}$ is *dressed* when each of the words $W_1, W_2, \ldots, W_k$ is dressed. Note that if W is dressed, then mon$(W) \neq 0$. A polynomial p in Brace$\{L\}$ is *dressed*, when it is expressed as a linear combination with coefficients in k of dressed Young tableaux.

Two letters a and b of Young$\{W_1, W_2, \ldots, W_k\}$ are said to be *successive in the i-th column* when a is the i-th letter of the word W_j and b is the i-th letter of W_{j+1} for some j. Two successive letters a and b in the same column of Young$\{W_1, W_2, \ldots, W_k\}$ are said to be form a *violation* when $a > b$ or when $a = b$ and both a and b are positive letters.

A dressed Young tableau without violations of either kinf is said to be *superstandard*. In a superstandard Young tableau, any two successive letters a and b in the same column satisfy the condition $a < b$ if not both a and b are negative, and $a \leq b$ if both a and b are negative. Note that a superstandard Young tableau Young$\{W_1, W_2, \ldots, W_k\}$ is non-zero (see [GRS87]), and that any two consecutive letters a and b in each of the words $W_1, W_2, \ldots, W_k$ satisfy the conditions $a < b$ if both a and b are negative, and $a \leq b$ otherwise.

These conditions on rows and columns characterize superstandard Young tableaux.

A lexicographic order of Young tableaux is induced by the lexicographic order on Mon(L) by associating to the Young tableau Young$\{W_1, W_2, \ldots, W_k\}$ the word $W_1 W_2 \ldots W_k$.

A straightening algorithm for a dressed element p of brace$\{L\}$ is a sequence of dressed elements $p_1, p_2, \ldots, p_e$ of Brace$\{L\}$ with the following properties:

(1) $p_1 = p$;

(2) p_e is a linear combination of superstandard Young tableaux;

(3) p_{i+1} is obtained from p_i by application of the straightening rule (6) as follows:

(a)   Let Young$\{W_1, W_2, \ldots, W_k\}$ be the (dressed) Young tableau of smallest lexicographic order appearing in p_i containing a violation, if any.

(b)   Let j be the smallest integer for which the words W_j and W_{j+1} contain a violation.

(c)   Let $m + 1$ denote the first integer for which the $m + 1$-th letters of W_j and W_{j+1} are a violation.

If $W_j = a_1 a_2 \ldots a_n$ and $W_{j+1} = b_1 b_2 \ldots b_n$, we thus have $a_m < a_{m+1}$ and $b_m \leq b_{m+1} = \ldots = b_{m+t} < b_{m+t+1}$ and $b_{m+t} \leq a_{m+1}$. Note that if $a_{m+1} = b_{m+t}$, then the letter is positive.

Let the $u = \text{mon}(a_1 a_2 \ldots a_m)$, let $v = \text{mon}(b_1 b_2 \ldots b_{m+t} a_{m+1} a_{m+2} \ldots a_n)$ and let $w = \text{mon}(b_{m+t+1} b_{m+t+2} \ldots b_n)$ in $\text{Super}[L]$. The monomials u, v and w are non-zero, since they are images of dressed words.

Apply the exchange rule (6), thereby replacing the occurrence of the Young tableaux $\text{Young}\{W_1, W_2, \ldots, W_k\}$ in p_i by a linear combination of dressed Young tableaux, each of which is strictly smaller in lexicographic order. Repeat until no more violations occur.

PROPOSITION 2. *The straightening algorithm terminates in a finite number of steps; in other words, the integer e is well-defined.*

PROPOSITION 3. *In the algebra* Bracket$[L]$, *every dressed bracket monomial can be uniquely expressed as a linear combination with integer coefficients of superstandard Young tableaux.*

PROPOSITION 4. (Decision algorithm or the logic of brackets). *A polynomial p in* Brace$\{L\}$ *is true in the logic of brackets if and only if the straightening algorithm applied to p leads to $p_e = 0$.*

The details of the proofs of these results can be found in [GRS87].

From the preceding results we infer another derived rule of inference for the logic of brackets:

(7)   (*cancellation rule*) Let $\text{Young}\{W_1, W_2, \ldots, W_k\} + \text{Young}\{W_1, W'_2, \ldots, W'_k\} = 0$ in the logic of brackets, where the monomials $\text{Young}\{W_1, W_2, \ldots, W_k\}$ and $\text{Young}\{W_1, W'_2, \ldots, W'_k\}$ are superstandard. Then $\text{Young}\{W_2, \ldots, W_k\} + \text{Young}\{W'_2, \ldots, W'_k\} = 0$ in the logic of brackets.

The above rule is the key step in several of the arguments to follow. Notice that the cancellation rule allows one to cross out a factor, even though the bracket algebra over a supersymmetric alphabet is not an integral domain.

8. - Models: the classical theory

Let $L = L^-$ be a negative alphabet (in this section only).

Then the bracket algebra Bracket$[L]$ is commutative. Our objective is to define the notion of model for the logic of brackets.

Our models will be *Peano spaces* in the sense defined in [BBR], which we now recall.

We define a *bracket* (of step n) on the vector space V of dimension n

over the field k to be a non-degenerate n-linear k-valued form defined over the vector space V, in symbols, a function

$$x_1, x_2, \ldots, x_n \rightarrow [x_1, x_2, \ldots, x_n]$$

defined as the vectors $x_1, x_2, \ldots, x_n$ range over the vector space V, with the following properties:

(1) $[x_{1,2}, \ldots, x_n] = 0$ if at least two among the x's coincide. This implies that the bracket is a skew-symmetric multilinear form.

(2)   for every x, y in V and every choice of scalars α, β in k, we have

$$[x_1, x_2, \ldots, x_{i-1}, \ \alpha x + \beta y, \ x_{i+1}, \ldots, x_n]$$
$$= \alpha[x_1, \ldots, x_{i-1}, x, x_{i+1}, \ldots, x_n]$$
$$+ \beta[x_1, \ldots, x_{i-1}, y, x_{i+1}, \ldots, x_n];$$

(3)   There exists a sequence of vectors $e_1, e_2, \ldots, e_n$, necessarily a basis of V, such that $[e_1, e_2, \ldots, e_n] = 1$. Such a basis will be called a *unimodular* basis.

We define a *Peano space of step* n to be a pair $(V, [\ ])$, where V is a vector space of dimension n and $[\ ]$ is a bracket of step n over V. We shall denote a Peano space by the single letter V, leaving the bracket understood, whenever no confusion is possible.

A Peano space can be viewed geometrically as a vector space in which an oriented volume element is specified. The bracket $[x_1, x_2, \ldots, x_n]$ gives the signed volume of the parallelepiped spanned by the vectors x_i. The equality $[x_1, x_2, \ldots, x_n] = 0$ will hold if an only if the vectors $x_1, x_2, \ldots, x_n$ are linearly dependent.

Every bracket can be expressed as a determinant. Indeed, choose any unimodular basis $e_1, e_2, \ldots, e_n$. Let $x_{ij} = [e_1, e_2, \ldots, e_{j-1}, x_i, e_{j+1}, \ldots, e_n]$. Then it is easily verified that $[x_1, x_2, \ldots, x_n] = \det(x_{ij})$.

Such an expression for the bracket as a determinant is misleading. The bracket should be visualized as a object existing in its own right, independently of the choice of a coordinate system, much like an inner product defining a Hilbert space is to be visualized in its own right.

A *model* of the ring Bracket[L], with values in a Peano space V, is a function ϕ from the alphabet L to the Peano space V. The *induced map* Φ on brackets defined as

$$\Phi([a_1 a_2 \ldots a_n]) = [\phi(a_1), \phi(a_2), \ldots, \phi(a_n)]$$

extends to a homomorphism Φ of the bracket algebra Bracket[L] to the field k, which will called the *induced homomorphism* of the model ϕ. Note that the bracket on the right is the bracket of the Peano space V, whereas the bracket on the left is the bracket of the Bracket algebra Bracket$[L]$.

The fact that the mapping Φ is indeed a homomorphism is a consequence of the fact that the bracket in a Peano space satisfies the exchange identities. Since the bracket in a Peano space is a determinant, these exchange identities are nothing but the classical Laplace expansions of a determinant.

A model ϕ having the property that there exist letters $f_1, f_2, \ldots, f_n$ in L such that $\Phi([f_1, f_2, \ldots, f_n]) = 1$ is said to be a *unimodular model*. There is no loss of generality in assuming that a model is unimodular, and we shall do so from now on.

For any letter a in L other that $f_1, f_2, \ldots, f_n$, consider n variables, labelled a_j, where $1 \leq j \leq n$. Let Pol(L) be the polynomial algebra over the field k which is obtained by adjoining the variables a_j, where $1 \leq j \leq n$ and where a ranges over L.

We have

PROPOSITION 1. *The map*

$$[f_1, f_2, \ldots, f_{j-1}, a, f_{j+1}, \ldots, f_n] \text{ to } a_j$$

extends to an isomorphism of the quotient of Bracket$[L]$ by the equation $[f_1, f_2, \ldots, f_n] = 1$ *onto the polynomial algebra* Pol(L) *which maps* $[a_1 a_2 \ldots a_n]$ *to* $det(a_{ij})$.

PROPOSITION 2. *Let ϕ be a unimodular model of* Bracket$[L]$ *in the Peano space V. Then there is a unique map $\phi^\wedge$ of the polynomial algebra* Pol(L) *to the field k such that $\phi^\wedge(a_j)$ is the j-th coordinate of the vector $\phi(a)$ in the coordinate system $\phi(f_1), \phi(f_2), \ldots, \phi(f_n)$ in V.*

Indeed, we have

$$\Phi([f_1, f_2, \ldots, f_{j-1}, a, f_{j+1}, \ldots, f_n])$$
$$= [\phi(f_1), \phi(f_2), \ldots, \phi(f_{j-1}), \phi(a), \phi(f_{j+1}), \ldots, \phi(f_n)] = \phi^\wedge(a_j).$$

The following result states that every identity satisfied by a determinant is a consequence of the exchange identities:

PROPOSITION 3 (Second fundamental theorem of invariant theory, classical version). *Let p be a polynomial in* Bracket$[L]$ *such that $\Phi(p) = 0$ for every model ϕ. Then $p = 0$.*

Again by analogy with logic, we shall say that a polynomial p in Brace$\{L\}$ is *valid* for the logic of brackets if $\phi(\text{Image}(p)) = 0$ for every model ϕ.

The preceding proposition states that the polynomial p is valid if and only if it is true in the logic of brackets. It is therefore an analog for the logic of brackets of the Gödel completeness theorem of the predicate calculus.

Combined with the decision procedure provided by the straightening algorithm, the preceding result provides an effective algorithm for verifying any valid polynomial in a Peano space.

Our program will be complete, if we show that every valid polynomial in a Peano space which is independent of a choice of a coordinate system can be expressed in the logic of brackets. It will then follow that the logic of bracket is an *adequate language* for projective geometry.

Let $p(a, b, \ldots, c; f_1, f_2, \ldots, f_n)$ be a polynomial in the brackets

$$[f_1, f_2, \ldots, f_{j-1}, a, f_{j+1}, \ldots, f_n], \quad [f_1, f_2, \ldots, f_{j-1}, b, f_{j+1}, \ldots, f_n],$$

$$\ldots, [f_1, f_2, \ldots, f_{j-1}, c, f_{j+1}, \ldots, f_n],$$

which is homogeneous in the letters $a, b, \ldots, c$. We say that p is an *invariant* of the letters $a, b, \ldots, c$ when, if ϕ and ϕ' are unimodular models such that $\phi(a) = \phi'(a), \phi(b) = \phi'(b), \ldots, \phi(c) = \phi'(c)$ we have $\Phi(p) = \Phi'(p)$.

PROPOSITION 4 (First fundamental theorem of invariant theory, classical version). *Let p be an invariant of $a, b, \ldots, c$. Then there exists a bracket polynomial q, not containing $f_1, f_2, \ldots, f_n$, such that $p = [f_1, f_2, \ldots, f_n]^g q$ for some non-negative integer g.*

In other words: any invariantly defined expression $p = 0$ can be expressed by an equation $q = 0$, where q is a bracket polynomial not containing the coordinates.

The preceding result can be read as follows: let p be a polynomial in the coordinates of vectors $a, b, \ldots, c$ which takes the same value in every unimodular coordinate system. Then p can be expressed as a bracket polynomial. The first fundamental theorem of invariant theory is usually expressed in this form. Note that such a version of the theorem is, strictly speaking, incorrect. Any correct statement of the theorem requires the introduction of syntactic notions, as we have done by introducing the "abstract" algebra of brackets.

9. - Supersymmetric models

We shall now extend the concept of a model to the supersymmetric bracket algebra, where the alphabet L is now arbitrary. Despite the fact that we now have letters of two signatures, the models still turn out to be Peano spaces.

We briefly digress to recall the notion of a skew-symmetric tensor in a Peano space.

Classically, a skew-symmetric tensor of step k is a homogeneous element of degree k of the exterior algebra $\text{Ext}(V)$ over the vector space V. It is a linear combination of decomposable skew-symmetric tensors:

$$t = \Sigma \ t_i$$

where each t_i is of the form $v_1^{(i)} v_2^{(i)} \ldots v_k^{(i)}$ for some (anticommuting) vectors $v_1^{(i)}, v_2^{(i)}, \ldots, v_k^{(i)}$. It thus makes sense, in a Peano space, to write

$$[t, x_1, x_2, \ldots, x_{n-k}] = \Sigma \; [v_1^{(i)}, v_2^{(i)}, \ldots, v_k^{(i)}, x_1, x_2, \ldots, x_{n-k}]$$

for any vectors $x_1, x_2, \ldots, x_{n-k}$. In this way, the structure of a Peano space allows us to associate a skew-symmetric multilinear form of order $n - k$ to a skew-symmetric tensor of step k. Reasoning along these lines, one sees that expressions like

$$\Sigma [t_{(1)} t'][t_{(2)} t'']$$

make sense, where t' and t'' are skew-symmetric tensors of any step. In other words, one can use the bracket notation for skew-symmetric tensors, and this we shall do. A skew-symmetric tensor of step k will be a multilinear form of order $n - k$, written $[t, x_1, x_2, \ldots, x_{n-k}]$. Such skew-symmetric forms form a Hopf algebra.

We now give the definition of a model of a supersymmetric bracket algebra Bracket[L].

A model of the algebra Bracket[L] in a Peano space $(V, [\;])$ is a function ϕ assigning a vector $\phi(a)$ to every negatively signed letter a, and assigning a skew-symmetric tensor t of step k for some integer k such that $1 \leq k \leq n$ to every positive letter.

If a (positive) letter a has been assigned a tensor $t = \phi(a)$ of step k, we say that the letter a has *arity* k. Negatively signed letters are assigned arity 1.

Our objective is now to extend the map ϕ to an induced map Φ of Bracket[L] to k, much as in the classical case.

First, consider a single bracket $[w]$ in Bracket[L]. If the word w contains only negative letters define $\Phi([w])$ and extend to all of Bracket[L^-] as in the preceding Section.

If the word w contains any letter with a multiplicity unequal to its arity, set $\Phi([w]) = 0$.

If $[w] = [a^{(k)} b_1 b_2 \ldots b_{n-k}]$, where a is a letter of arity k and $b_1, b_2, \ldots, b_{n-k}$ are negative letters, set $\Phi([w]) = [t, \phi(b_1), \phi(b_2), \ldots, \phi(b_{n-k})]$.

Next, we define Φ on a monomial $m = [w][w'] \ldots [w'']$. Proceed as follows:

(1)   Set $\Phi(m) = 0$ if some letter in m occurs with multiplicity not equal to its arity.

(2)   For each positive letter a of arity k occuring in m such that $\phi(a) = t$ and

$$t = \Sigma \; t_i$$

where each t_i is decomposable, say of the form $v_1^{(i)} v_2^{(i)} \ldots v_k^{(i)}$ for some vectors $v_1^{(i)}, v_2^{(i)}, \ldots, v_k^{(i)}$, choose k negatively signed letters $c_1^{(i)}, c_2^{(i)}, \ldots, c_k^{(i)}$, not occurring elsewhere, such that

$$[\phi(c_1^{(i)}), \phi(c_2^{(i)}), \ldots, \phi(c_k^{(i)}), x_1, x_2, \ldots, x_{n-k}] = [t_i, x_1, x_2, \ldots, x_{n-k}]$$

identically in the vectors $x_1, x_2, \ldots, x_{n-k}$.

Let $a, b, \ldots, d$ be a list of all the positively signed letters occurring in m, in the order $a < b < \ldots < d$. The word polarization operator

$$T(\Sigma c_1^{(i)} c_2^{(i)} \ldots c_k^{(i)}, a)$$

sends the word $a^{(k)}$ to the element $\Sigma c_1^{(i)} c_2^{(i)} \ldots c_k^{(i)}$. Similarly we define operators $T(\Sigma d_1^{(i)} d_2^{(i)} \ldots d_j^{(i)}, b), \ldots, T(\Sigma e_1^{(i)} e_2^{(i)} \ldots e_r^{(i)}, c)$ for other positive letters occurring in m, where $\phi(b)$ is the sum of decomposable tensors of step $j, \ldots,$ and $\phi(c)$ is the sum of decomposable tensors of step r. All newly introduced negatively signed letters $c, d, \ldots, e$ are assumed distinct and not occuring elsewhere.

The element of Bracket$[L]$ defined as

$$T(\Sigma c_1^{(i)} c_2^{(i)} \ldots c_k^{(i)}, a) \; T(\Sigma d_1^{(i)} d_2^{(i)} \ldots d_j^{(i)}, b)$$

$$\ldots T(\Sigma e_1^{(i)} e_2^{(i)} \ldots e_r^{(i)}, c)([w][w'] \ldots [w''])$$

contains only negatively signed letters, and

$$\Phi(T(\Sigma c_1^{(i)} c_2^{(i)} \ldots c_k^{(i)}, a) \; T(\Sigma d_1^{(i)} d_2^{(i)} \ldots d_j^{(i)}, b)$$

$$\ldots T(\Sigma e_1^{(i)} e_2^{(i)} \ldots e_r^{(i)}, c) \; ([w][w'] \ldots [w'']))$$

is therefore already defined. Furthermore, it does not depend on the choice of the newly introduced negatively signed letters. We can therefore define this element to be the image

$$\Phi([w][w'] \ldots [w'']).$$

This completes the definition of the induced map Φ associated with a model ϕ.

The preceding definition is easier that it has been made to sound. It can be summarized as follows. To find the image of a letter a of arity k, which appears in different terms of a monomial, one polarizes the "tensor" $a^{(k)}$ to a sum of decomposable skew-symmetric tensors, and then one uses the definition of Φ for negatively signed letters.

It can be verified that

$$\Phi(m \; m') = (-1)^{|m||m'|} \; \Phi(m'm)$$

for any two monomials m and m'.

EXAMPLE 1. If $a, b, \ldots, c$ are positive letters of arity one, and if $a < b < \ldots < c$ then $\Phi([ba \ldots c]) = \Phi([ab \ldots c]) = \Phi([ca \ldots b]) = \ldots = [\phi(a), \phi(b), \ldots, \phi(c)]$.

EXAMPLE 2. Suppose all letters in $[w][w'] \ldots [w'']$ are positively signed and have arity 1. Suppose furthermore that $w = \text{Mon}(W)$, $w' = \text{Mon}(W'), \ldots, w'' =$

Mon(W'') and that the letters of the word $W\ W'\ldots W''$ occur in increasing order. Then $\Phi([w][w']\ldots[w'']) = \Phi([w])\Phi([w'])\ldots\Phi([w''])$ if n is even, and $\Phi([w][w']\ldots[w'']) = (-1)^{k(k-1)/2}\Phi([w])\Phi([w'])\ldots\Phi([w''])$, where k is the number of brackets, if n is odd.

EXAMPLE 3. $\Phi([a^{(n)}][b^{(n)}]\ldots[c^{(n)}]) = \Phi([a^{(n)}])\Phi([b^{(n)}])\ldots\Phi([c^{(n)}])$ if n is even, and $\Phi([a^{(n)}][b^{(n)}]\ldots[c^{(n)}]) = (-1)^{k(k-1)/2}\Phi([a^{(n)}])\Phi([b^{(n)}])\ldots\Phi([c^{(n)}])$, where k is the number of brackets, if n is odd, and where $a < b < \ldots < c$.

EXAMPLE 4. If $\phi(a) = t$, $\phi(b) = t'$ and $\phi(c) = t''$, each letter having arity 2, and if $n = 3$, then $\Phi([a^{(2)}b][bc^{(2)}]) = -\Sigma[tt'_{(1)}][t'_{(2)}t'']$, and $\Phi([a^{(2)}c][cb^{(2)}]) = -\Sigma[t\ t''_{(1)}][t''_{(2)}t']$.

EXAMPLE 5. With the proper arities, we have $\Phi([a^{(2)}b][dc^{(2)}]) = -[tx][yt'']$, where $\phi(b) = x$ and $\phi(d) = y$, and $\Phi([a^{(2)}d][bc^{(2)}]) = [ty][xt'']$, the order being alphabetical, and where all letters are positively signed.

We say that a polynomial p in Brace$\{L\}$ is *valid* in the logic of brackets if and only if $\Phi(\text{Image}(p)) = 0$ for all models ϕ. Our objective is to prove that every statement about skew-symmetric tensors which is independent[*] of a choice of a coordinate system, that is, every *geometric* statement, can be expressed in the logic of brackets. This program requires, first, an extension of the completeness theorem to the supersymmetric case, which amounts to an extension of the second fundamental theorem of invariant theory to the supersymmetric case, and second, an extension of the first fundamental theorem to the supersymmetric case, stating that all occurrences of coordinates can be "eliminated" in any invariant statement. More precisely, we have:

THEOREM 1 (Completeness theorem for supersymmetric bracket algebras). *A formula p in Brace$\{L\}$ is valid in the logic of supersymmetric brackets if and only if the element Image(p) equals 0 in Bracket$[L]$, that is, if and only if the polynomial p is true in the logic of brackets.*

To state the second result, we require the notion of an *invariant* of a set of letters $a, b, \ldots, c$ (positive or negative) of preassigned arities $i, j, \ldots, k$ (with the understanding that negative letters are assigned arity 1).

If a is a letter of arity i, we define the *Plücker coordinates* of the letter a relative the negatively signed letters $f_1, f_2, \ldots, f_n$ to be

$$[a^{(i)}f_1f_2\ldots f_{h_1-1}f_{h_1+1}\ldots f_{h_i-1}f_{h_i+1}\ldots f_n]$$

running through all i-tuples obtained by omitting any i of the letters $f_1, f_2, \ldots, f_n$.

Notice that the Plücker coordinates are elements of Bracket$[L]$, not elements of the field k.

An *invariant* is a bracket polynomial $p(a, b, \ldots, c; f_1, f_2, \ldots, f_n)$ in the Plücker coordinates of the letters $a, b, \ldots, c$, homogeneous in the letters $a, b, \ldots, c$, such that for all unimodular models ϕ and ϕ' such that $\phi(a) = \phi'(a), \phi(b) = \phi'(b), \ldots, \phi(c) = \phi'(c)$ we have $\Phi(p) = \Phi'(p)$.

We can now state

THEOREM 2 (First fundamental theorem for supersymmetric brackets). *Let p be an invariant of the letters $a, b, \ldots, c$ of arities $i, j, \ldots, k$. Then there exists a bracket polynomial $q(a, b, \ldots, c)$ such that*

$$p(a, b, \ldots, c; f_1, f_2, \ldots, f_n) = [f_1, f_2, \ldots, f_n]^g \, q(a, b, \ldots, c)$$

for some non-negative integer g.

The proof of both theorems is given in two steps. First, one polarizes all positively signed letters into negatively signed letters. The resulting expression will be in Bracket$[L^-]$. One then applies the classical version of the theorem.

BIBLIOGRAPHY

[W]   H. WEYL, *The Classical Groups*, Princeton University Press, 1946.

[T]   H.W. TURNBULL, *The theory of determinants, matrices and invariants*, 3rd edition, Dover, 1960.

[C]   A. CHURCH, *Introduction to Mathematical Logic*, Princeton University Press, 1958.

[BBR85]   M. BARNABEI - A. BRINI - G.C. ROTA, *On the exterior calculus of invariant theory*, Journal of algebra, vol. 96 (1985), 120-160.

[RS86]   G.C. ROTA - J.A. STEIN, *Symbolic method in invariant theory*, Proceedings of the National Academy of Sciences, vol. 83 (1986), 844-847.

[GRS]   F. GROSSHANS - G.C. ROTA - J.A. STEIN, *Invariant theory and superalgebras*, (1987), American Mathematical Society, Providence, Rhode Island, 1987.

[BPT88]   A. BRINI - A. PALARETI - A.G.B. TEOLIS, *Gordan-Capelli series in superalgebras*, Proceedings of the National Academy of Sciences, vol. 85 (1988), 1330-1333.

[RS89]   G.C. ROTA - J.A. STEIN, *Standard basis in supersymplectic algebras*, Proceedings of the National Academy of Sciences, vol. 86 (1989), 2521-2524.

The Role of Dynamic Symmetries in Molecular Structure

F. IACHELLO (*)

Abstract

The role of dynamic symmetries in molecular structure is briefly discussed. The algebraic theory of bonds and bond-bond interactions is reviewed and some selected examples presented.

1. - Introduction

Dynamic symmetry is a far reaching concept that has been used in a variety of problems in physics. It originated in the non-relativistic hydrogen atom and found its most publicized application in the spectroscopy of light hadrons. The fundamental properties of dynamic symmetries are:

(i)   the Hamiltonian, H, (or mass operator, M) describing the system has an algebraic structure $\mathcal{G}$, i.e. it lies in the enveloping algebra of $\mathcal{G}$; the algebra $\mathcal{G}$ is called the spectrum generating algebra of H;

(ii) the Hamiltonian (or mass operator) contains only invariant (Casimir) operators of $\mathcal{G}$ and its subalgebras, $\mathcal{G} \supset \mathcal{G}' \supset \mathcal{G}'' \supset \cdots$.

Obviously (ii) is a special case of (i). The importance of dynamic symmetries lies in the fact that whenever one such situation occurs, the eigenvalue problem for H (or M) can be solved in closed form. These closed solutions provide considerable insight into the problem, especially for complex situations, and can be compared with experiment in a straightforward way. For a Hamiltonian of the type

$$(1.1) \qquad H = \alpha C(\mathcal{G}) + \alpha' C(\mathcal{G}') + \alpha'' C(\mathcal{G}'') + \cdots,$$

(*)   Center for Theoretical Physics and A.W. Wright Nuclear Structure Laboratory Yale University, New Haven, Connecticut 06511

where $C(\mathcal{G})$ denotes generically a Casimir operator od $\mathcal{G}$, the eigenvalues are simply given by

(1.2) $$E = \alpha < C(\mathcal{G}) > + \alpha' < C(\mathcal{G}') + \alpha'' < C(\mathcal{G}'') > + \cdots,$$

where $< C(\mathcal{G}) >$ is the expectation value of $C(\mathcal{G})$ in the appropriate representation of $\mathcal{G}$. The evaluation of E thus involves only the knowledge of the expectation values of the Casimir operators. These are all known for the classical Lie algebras.

2. - Dynamic symmetries in hadronic physics

One of the most interesting applications of the concept of dynamic symmetry is to hadronic spectroscopy. To this subject Luigi Radicati contributed greatly and it is thus appropriate to review it here. Gürsey and Radicati [1] suggested a classification of hadrons in terms of the algebra $SU(6)$. This algebra is the enlargement of the flavor $SU_f(3)$ algebra of Gell-Mann [2] and Ne'eman [3] to include spin. $SU(6)$ and its subalgebras provide the classification scheme

(2.1)

$$
\begin{vmatrix}
SU(6) & \supset & SU_f(3) \otimes SU_s(2) \supset SU_I(2) \otimes U_Y(1) \otimes SU_s(2) \\
\downarrow & & \downarrow \qquad\quad \downarrow \qquad\quad \downarrow \qquad \downarrow \qquad \downarrow \\
[\lambda_1,\lambda_2,\lambda_3,\lambda_4,\lambda_5] & & (\lambda_f,\mu_f) \qquad s \qquad\quad I \qquad Y \\
& \supset SO_I(2) \otimes & U_Y(1) \otimes SO_s(2) \\
& \downarrow & \qquad\quad \downarrow \\
& I_3 & \qquad\quad S_3
\end{vmatrix}
$$

where I have denoted by (λ_f, μ_f) the $SU_f(3)$ quantum numbers (Elliott notation [4]). If one assumes further that there is a dynamic symmetry corresponding to the chain (2.1), i.e. that M can be written as

(2.2)
$$M = M_0 + a C_2(SU_f(3)) + b C_2(SU_s(2)) + c C_1(U_Y(1)) +$$
$$+ d \left[C_2(SU_I(2)) - \frac{1}{4}(C_1(U_Y(1)))^2 \right],$$

one obtains the Gürsey-Radicati mass formula

(2.3)
$$M(\lambda_f, \mu_f, S, S_3, I, I_3, Y) = M_0 + a \left(\lambda_f^2 + \mu_f^2 + \lambda_f \mu_f + 3\lambda_f + 3\mu_f \right) +$$
$$+ b S(S+1) + cY + d \left[I(I+1) - Y^2/4 \right],$$

where the dependence on the $SU_f(3)$ quantum numbers has been written down explicity. In Eq. (2.2), $C_2(\mathcal{G})$ and $C_1(\mathcal{G})$ denote the quadratic and linear invariants of $\mathcal{G}$. Eq. (2.3) applies to a given $SU(6)$ representation and can be used to describe the low-lying spectrum of baryons. It has five parameters, M_0, a, b, c, d. An appropriate choice of the parameters leads to the results shown in Fig. 1, which provides an excellent example of dynamic symmetry in hadronic physics. For the particular representation shown in Fig. 1, characterized by the $SU(6)$ representation $(3, 0, 0, 0, 0)$ with dimension 56 and by the orbital angular momentum and parity, $L^P = 0^+$, only four parameters are relevant since only one $SU_f(3)$ representation appears for a given spin. The results shown in Fig. 1 are obtained by choosing the four parameters as to fit the levels marked by a star in the figure. It is clear from this figure, and from similar ones that describe mesons, that the Gürsey-Radicati $SU(6)$ provides an excellent zeroth-order approximations to hadronic spectra, with an average deviation of the order of 30 MeV.

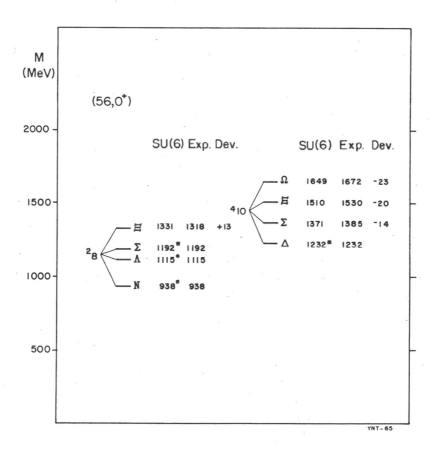

Fig. 1 - Classification of the spectrum of baryons with $L^P = 0^+$ according to the representation $(3, 0, 0, 0, 0)$ of the Gürsey-Radicati $SU(6)$. The parameters in the mass formula (2.3) are chosen as to reproduce the levels marked by a star.

3. - Dynamic symmetries in molecular physics

The same concept of dynamic symmetry used by Gürsey and Radicati in the description of hadrons has been extensively exploited in recent years in the study of molecular spectra. These applications are based on an algebraic model of molecular structure (vibron model) which will be now briefly reviewed.

3.1 - *The fundamental spectrum generating algebra*

It has been suggested [5] that the spectrum generating algebra of diatomic molecules in $U(4)$. This algebra can be constructed explicity by introducing four boson creation and annihilation operators, $b_\alpha^\dagger, b_\alpha (\alpha = 1, \cdots, 4)$, divided into a scalar $\sigma^\dagger$ and a vector $\pi_\mu^\dagger (\mu = 0, \pm 1)$, and taking all their bilinear products,

$$(3.1) \qquad \mathcal{G} : G_{\alpha\beta} = b_\alpha^\dagger b_\beta \qquad \alpha, \beta = 1, \cdots, 4.$$

The bosons are called vibrons, hence the name vibron model given to this approach. The Hamiltonian operator is constructed from number conserving products of creation and annihilation operators as

$$(3.2) \qquad H = E_0 + \sum_{\alpha\beta} \epsilon_{\alpha\beta} b_\alpha^\dagger b_\beta + \frac{1}{2} \sum_{\alpha\beta\gamma\delta} u_{\alpha\beta\gamma\delta} b_\alpha^\dagger b_\beta^\dagger b_\gamma b_\delta.$$

The basis states are

$$(3.3) \qquad \mathcal{B} : b_\alpha^\dagger b_\beta^\dagger \cdots |0>,$$

where $|0>$ denotes the boson vacuum. The algebra $U(4)$ admits two dynamic symmetries corresponding to the chains of algebras

$$U(3) \supset 0(3) \supset 0(2) \qquad \text{(I)},$$

$$(3.4) \qquad U(4) \Big\langle$$

$$0(4) \supset 0(3) \supset 0(2) \qquad \text{(II)}.$$

Introducing the appropriate representations, one can construct two possible classification schemes. The first is

$$(3.5) \qquad \left| \begin{array}{cccc} U(4) & \supset & U(3) & \supset & 0(3) & \supset & 0(2) \\ \downarrow & & \downarrow & & \downarrow & & \downarrow \\ N & & n_\pi & & L & & M_L \end{array} \right\rangle,$$

while the second is

(3.6)
$$\left. \begin{matrix} U(4) & \supset & O(4) & \supset & 0(3) & \supset & 0(2) \\ \downarrow & & \downarrow & & \downarrow & & \downarrow \\ N & & \omega & & L & & M_L \end{matrix} \right),$$

The quantum number N (a fixed number for each molecule) characterizes the representations of $U(4)$, which, for a system of bosons, are totally symmetric and thus labelled by only one number. The two chains in Eq. (3.4) describe soft and rigid molecules respectively [5]. The large majority of molecules are rigid and thus the second chain is the appropriate one. The quantum numbers ω, L, M_L are obtained from N using simple rules. The quantum number ω is given by $\omega = N, N-2, \cdots, 1$ or 0, (N odd or even), while $L = \omega, \omega-1, \cdots, 1, 0$ and $-L \leq M_L \leq +L$.

In the special case in which the Hamiltonian H can be written in terms of Casimir operators of one of the two chains, one has a dynamic symmetry. For chain II, this implies that

(3.7)
$$H^{(\mathrm{II})} = E_0 + AC_2(0(4)) + BC_2(0(3)).$$

The Casimir operator of 0(2) is not included, unless the molecule is placed in an external field. The Casimir operator of $U(4)$ can be absorbed into E_0. The eigenvalues of $H^{(\mathrm{II})}$ can be obtained by taking the expectation value of $H^{(\mathrm{II})}$ in the representation Eq. (3.6) and are given by

(3.8)
$$E^{(\mathrm{II})}(N, \omega, L, M_L) = E_0 + A\omega(\omega+2) + BL(L+1).$$

Spectra of several diatomic molecules can be described quite accurately by Eq. (3.8). An example is shown in Fig. 2. Here, the various 0(4) representations are labelled by the vibrational quantum number v instead of ω. These two quantum numbers are related by

(3.9)
$$v = \frac{N - \omega}{2}.$$

The dynamic symmetry II provides a simple way to describe rotation-vibration spectra of molecules.

3.2 - Spectrum generating algebras for complex molecules

The description of the spectrum of diatomic molecules in terms of the algebra $\mathcal{G} \equiv U(4)$ reviewed in the previous section can be generalized to complex polyatomic molecules by taking appropriate combinations of $\mathcal{G}$. Consider a molecule with ν atoms and thus $\nu - 1$ bond degrees of freedom. (One bond

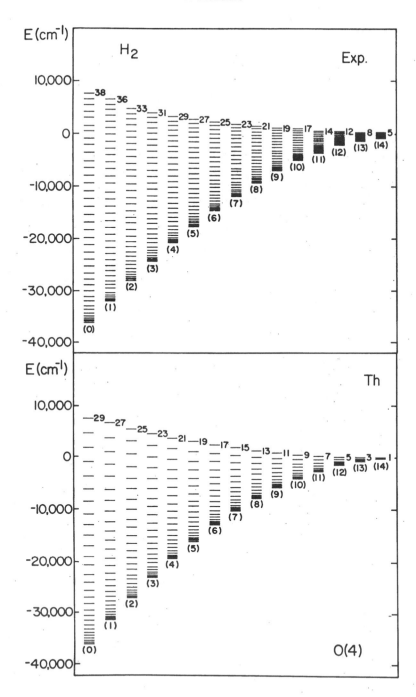

Fig. 2 - Comparison between the rotation-vibration spectrum of the H_2 molecule in its electronic ground state, $^1\Sigma_g^+$, and that calculated using Eq. (3.8) (dynamic O(4) symmetry).

degree of freedom disappears after removal of the center of mass motion of the molecule). Describe each bond, i, by an algebra $\mathcal{G}_i \equiv U_i(4)$, constructed from boson creation and annihilation operators as before

$$(3.10) \qquad \mathcal{G}_i : G^i_{\alpha\beta} = b^\dagger_{\alpha i} b_{\beta i}; \quad \alpha, \beta = 1, \cdots, 4; \quad i = 1, \cdots, \nu - 1.$$

The Hamiltonian operator for a complex molecule can be written as

$$(3.11) \qquad H = \sum_i H_i + \sum_{i<j} V_{ij},$$

where H_i is the Hamiltonian of each individual bond given by Eq. (3.2) with an index i attached to all creation and annihilation operators and V_{ij} is the bond-bond interaction. This interaction is constructed from all number conserving terms of the type

$$(3.12) \qquad V_{ij} = \sum_{\alpha\beta\gamma\delta} w^{ij}_{\alpha\beta\gamma\delta} b^\dagger_{\alpha i} b_{\beta i} b^\dagger_{\gamma j} b_{\delta j}.$$

The basis states are

$$(3.13) \qquad \mathcal{B} : b^\dagger_{\alpha i} b^\dagger_{\beta i} \cdots b^\dagger_{\gamma j} b^\dagger_{\delta j} \cdots |0>,$$

and span the space

$$(3.14) \qquad \prod_{i=1}^{\nu-1} \otimes U_i(4) = U_1(4) \otimes U_2(4) \otimes \cdots \otimes U_{\nu-1}(4).$$

Within this space one can construct several dynamic symmetries corresponding to different physical situations. A simple case is that of triatomic molecules [6], $\nu = 3$. The product space is here $U_1(4) \otimes U_2(4)$ and an important symmetry, often observed in practice, is that corresponding to the chain

$$(3.15) \qquad \left| \begin{array}{cccccccc} U_1(4) \otimes U_2(4) \supset 0_1(4) \otimes 0_2(4) \supset & 0(4) & \supset 0(3) \supset 0(2) \\ \downarrow & \downarrow & \downarrow & \downarrow & \downarrow & \downarrow & \downarrow & \downarrow \\ N_1 & N_2 & \omega_1 & \omega_2 & (\tau_1, \tau_2) & L & M_L \end{array} \right).$$

The boson numbers N_1 and N_2 characterize the representations of $U_1(4)$ and $U_2(4)$ and are fixed for each molecule. The other quantum numbers $\omega_1, \omega_2, (\tau_1, \tau_2), L, M_L$ are given by simple rules [6] in terms of N_1 and N_2.

By writing the Hamiltonian in terms of Casimir operators of the chain (3.15),

$$(3.16) \quad H = E_0 + A_1 C_2(0_1(4)) + A_2 C_2(0_2(4)) + A C_2(0(4)) + B C_2(0(3)),$$

and evaluating its expectation value in the states (3.15) one obtains the energy formula

$$(3.17) \quad \begin{aligned} E(N_1, N_2, \omega_1, \omega_2, \tau_1, \tau_2, L, M_L) &= E_0 + A_1 \omega_1 (\omega_1 + 2) \\ &+ A_2 \omega_2 (\omega_2 + 2) + + A \left[\tau(\tau_1 + 2) + \tau_2^2 \right] + BL(L+1). \end{aligned}$$

Eq. (3.17) provides an excellent description of several linear triatomic molecules. An example is shown in Fig. 3. Here, the various states are labelled by the molecular notation $v_1, v_2^{\ell_2}, v_3$ rather than by the algebraic quantum numbers $\omega_1, \omega_2, \tau_1, \tau_2$. These two sets of quantum numbers are related by

$$(3.18) \quad \begin{aligned} \omega_1 &= N_1 - 2v_1 \quad ; \quad \omega_2 = N_2 - 2v_2; \\ \tau_1 &= N_1 + N_2 - 2v_1 - v_2 - 2v_3 \quad ; \quad \tau_2 = \ell_2. \end{aligned}$$

3.3 - *Spectrum generating algebras for polymers*

Appropriate combinations of the fundamental algebras $U_i(4)$ have been used successfully to describe a variety of situations in small and medium size polyatomic molecules $(\nu \lesssim 12)$. These situations include linear and bent molecules with planar and a-planar structures. A further generalization of the method to include the description of polymers has also been suggested. Polymers can be considered as the continuum limit of long linear molecules. They can thus be described by considering the infinite dimensional product

$$(3.19) \quad \prod_{i=1}^{\infty} \otimes U_2(4) = U_1(4) \otimes U_2(4) \otimes \cdots \otimes U_\infty(4).$$

By devising methods to deal with the limit $\nu \to \infty$, one is able to derive results of particular importance in polymer physics.

4. - Dynamic symmetries and potential models

In view of their success, one may inquire why $U(4)$ and $0(4)$ are the appropriate spectrum generating algebra and dynamic symmetry of molecules. The reason seems to reside in the fact that interatomic interactions are, to a very good approximation, describable by Morse potentials [7]. These are potentials of the form

$$(4.1) \quad V(r) = V_0 \left(1 - e^{\alpha r} \right)^2.$$

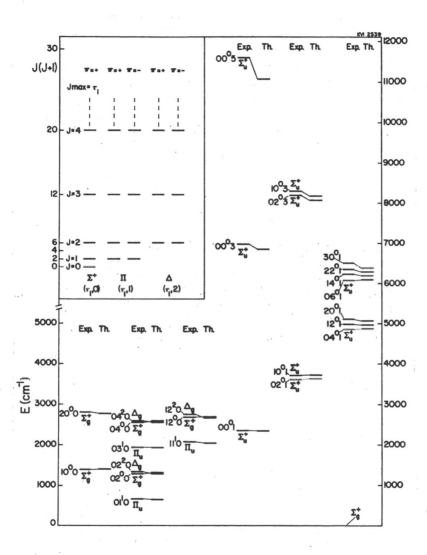

Fig. 3 - Comparison between the experimental vibrational spectrum of the CO_2 molecule in its ground electronic state Σ_g^+ and that calculated using Eq. (3.17). The insert shows the rotational spectrum built on top of each vibrational level. The vibrational states are labelled by the molecular notation $v_1, v_2^{\ell_2}, v_3$. The species type $\Sigma, \Pi, \cdots$ is also indicated.

Surprisingly enough, the Morse potential (4.1) has, in one space dimension, an *exact* dynamic symmetry [8] characterized by the chain $U(2) \supset 0(2)$. It can be shown [5] that, in three space dimensions, the same potential has an *approximate* dynamic symmetry characterized by the chain $U(4) \supset 0(4)$. In general, the solutions of the Schrödinger equation with Morse potentials in n space dimensions are associated with the chain of algebras $U(n+1) \supset 0(n+1)$. The accuracy of the approximations depends on the values of V_0 and α. A mapping of the Hamiltonian operator onto the algebraic space (3.1) keeping only two terms, as in Eq. (3.7), produces results that are typically accurate within $1^0/_{00}$. Better accuracies can be obtained either by breaking the dynamic symmetry or by introducing higher order terms constructed with three, four, ... generators $G_{\alpha\beta}$. The experimental observation of approximate $0(4)$ dynamic symmetries in molecules can thus be traced to the nature of the interatomic potential.

5. - Conclusions

The use of algebraic methods has led, in recent years, to major advances in molecular physics. By means of these methods it has been possible to attack and solve problems, such as those encountered in the structure of complex polyatomic molecules, that are very difficult, if not impossible, to treat otherwise.

A particularly important role in the study of these complex systems has been played by the exploitation of dynamic symmetries. These are special situations that occur whenever the Hamiltonian contains only invariant operators of the algebra $\mathcal{G}$. The concept of dynamic symmetry was used by Franzini and Radicati [9] in the description of ground state energies of nuclei within the context of Wigner supermultiplet theory, by Gürsey and Radicati [1] again in the description of hadronic spectra, and, more recently, by Arima and myself [10] in an extensive study of collective nuclear spectra.

The further successful application of this concept to molecular spectra, reviewed here, is a clear indication of its universal value. Although the origins of dynamic symmetries may be diverse, (Morse interatomic potentials in molecular physics, pairing-type interactions in nuclear physics, spin-flavor interactions in hadronic physics), its distinctive features appear in all branches of physics and provide important clues for a deep understanding of physical phenomena.

6. - Acknowledgements

It is a special pleasure to dedicate this article to Luigi Radicati in the occasion of his 70th birthday. His pioneering work [1,9] on the use of dynamic symmetries has been of much inspiration for the developments described here.

This work was performed in part under D.O.E. Contract No. DE-AC-02-75ER03074.

REFERENCES

[1] F. Gürsey - L.A. Radicati, Phys. Rev. Lett. **13**, 173 (1964).

[2] M. Gell-Mann, Phys. Rev. **125**, 1067 (1962).

[3] Y. Ne'eman, Nucl. Phys. **26**, 222 (1961).

[4] J.P. Elliott, Proc. Roy. Soc. **A245**, 128 (1958).

[5] F. Iachello - R.D. Levine - J. Chem. Phys. **77**, 3066 (1987).

[6] O. Van Roosmalen - F. Iachello - R.D. Levine - A.E.L. Dieperink, J. Chem. Phys. **79**, 2515 (1983).

[7] P.M. Morse, Phys. Rev. **34**, 57 (1929).

[8] R.D. Levine - C.E. Wulfman, Chem. Phys. Lett. **60**, 372 (1979).

[9] P. Franzini - L.A. Radicati, Phys. Lett. **6**, 322 (1963).

[10] F. Iachello - A. Arima, *"The Interacting Boson Model"*, Cambridge University Press, Cambridge, 1987.

The Casimir Effect in a Confocal Optical Resonator

E. IACOPINI (*)

Abstract

We discuss the possibility of observing the Casimir force at macroscopic distances (few centimetres), using a confocal optical resonator.

1. - Introduction

In the classical picture of electrodynamics, it is conceivable to have a system of macroscopic bodies in which the electric and magnetic fields are zero everywhere. The electromagnetic energy and, consequently, the e.m. forces acting in the system would then be identically zero.

In the quantum mechanical description of such a system, one cannot avoid to take into account the zero-point energy $1/2\hbar\omega$ of each normal mode of the quantized field, which is a consequence of the Heisenberg uncertainty principle.

The modes and, therefore, the zero-point e.m. energy of the system depend, in general, on the boundary conditions, i.e. on the geometric configuration of the system. Therefore, in this picture, non zero forces of electromagnetic nature always act among the costituent bodies.

In the case, for example, of an uncharged parallel plate capacitor, the attractive force per unit surface (Casimir pressure) between the two plates at distance d is given by [1,2]

$$(1) \qquad P_c = 1/240\pi^2 \hbar c d^{-4}.$$

Taking into account the power law $P_c \propto d^{-4}$, it was originally suggested by Casimir [3] that the repulsive Poincare's stress

$$(2) \qquad P_p = e^2/(8\pi r^4)$$

(*)   Scuola Normale Superiore

for an electron, considered as an ideal perfectly conducting sphere, would be balanced by the Casimir attractive force only for a well defined value of the electric charge. Under this assumption, the value of the fine structure constant α could be predicted.

However it was shown by Boyer and Schwinger that, for a spherical geometry, the Casimir force is repulsive [4] and, therefore, cannot balance the Poincare' stress.

The experimental confirmation of the existence of the Casimir force is quite poor. Because of the d^{-4} behaviour, only the sub-micron distance region has been investigated so far.

There are in fact experimental observations obtained by M.J. Sparnay [5] in the 0.5 - $2\mu m$ region, and by D. Tabor [6] in the 50 - $130nm$ region.

The difficulty is that the force is very small and, moreover, it is not easy to distinguish the Casimir contribution from the usual Van der Waals force.

In this paper, we first review the calculations of the Casimir force in the case of conducting parallel plates, which, for convenience, are considered as part of a rectangular cavity with perfectly conducting walls.

The case of a confocal resonator is then analysed and a method is suggested for using such a resonator to measure the Casimir force in the range of a few centimetres.

2. - The Casimir force in a rectangular cavity

The zero-point e.m. energy in a perfectly conducting rectangular cavity with perfectly conducting walls reads

(3)
$$W(L_1, L_2, L_3) = 1/2\hbar\Sigma_n\omega_n =$$
$$= 1/2\pi\hbar c\Sigma\left[(n_1/L_1)^2 + (n_2/L_2)^2 + (n_3/L_3)^2\right]^{1/2},$$

where L_1, L_2 and L_3 are the three cavity dimensions and the sum is performed on all the cavity modes.

There are two independent modes with the same frequency if n_1, n_2 and n_3 are non zero, and only one if one of the n's is zero (no modes if more than one n is zero).

The zero-point energy W, given by eq. (3) is infinite.

However, the relevant quantity is not W but the difference between W and the energy W_0 that would be in the cavity volume if we remove the cavity walls themselves, i.e. if we let L_1, L_2 and L_3 go to infinity,

(4) $$W_0(L_1, L_2, L_3) = \lim_{l_i \to \infty} (L_1 L_2 L_3 / 1_1 1_2 1_3) W(1_1, 1_2, 1_3).$$

The quantity

(5) $$U(L_1, L_2, L_3) = W(L_1, L_2, L_3) - W_0(L_1, L_2, L_3)$$

measures in fact the e.m. energy excess present in the cavity *only* because of the boundary conditions, and its first derivative with respect to any of the three coordinates is nothing but the Casimir force acting upon the two surfaces disposed perpendicularly to such a coordinate.

2a. - *The energy U*

We assume $L_1 = L_2 \equiv L >> L_3 \equiv d$. One has

(6) $$W(L, L, d) = 1/2\pi\hbar c\Sigma\left[(n_1^2 + n_2^2)/L^2 + n^2/d^2\right]^{1/2}.$$

Since $L >> d$ we will consider $n_1/L \equiv \xi$ and $n_2/L \equiv \eta$ as continous variables [1,2], i.e.

(7)
$$W(L, L, d) = 1/2\pi\hbar c \int [(n_1^2 + n_2^2)/L^2]^{1/2} dn_1 dn_2 +$$
$$+ \pi\hbar c \sum_{n>0} \int [(n_1^2 + n_2^2)/L^2 + n^2/d^2]^{1/2} dn_1 dn_2 =$$
$$= \pi\hbar c L^2 \Sigma_n'' \int_{\xi,\eta>0} d\xi d\eta [\xi^2 + \eta^2 + n^2/d^2]^{1/2},$$

where Σ_n'' is defined as follows

$$\Sigma_n'' a_n \equiv 1/2 a_0 + \sum_{n>0} a_n.$$

Using the polar adimensional coordinate

(8) $$\rho^2 \equiv d^2(\xi^2 + \eta^2),$$

after angular integration in the first quadrant $(\xi, \eta > 0)$, one obtains

(9)
$$W(L, L, d) = \pi\hbar c(L^2/d^3)\Sigma_n''(2\pi/4) \int_{\rho>0} \rho d\rho[\rho^2 + n^2]^{1/2} =$$
$$= 1/4\pi^2\hbar c(L^2/d^3)\Sigma_n'' \int_{\tau>0} d\tau[\tau + n^2]^{1/2},$$

where $\tau \equiv \rho^2$.

As far as W_0 is concerned, from its definition (4), one has

(10) $$W_0(L, L, d) = \lim_{L \to \infty} (d/L)W(L, L, L) = (d/L)W_0(L, L, L).$$

Using eq. (7) and definition (8), we obtain

$$W_0(L, L, d) = \pi \hbar c L^2 (d/L) \Sigma''_n \int\limits_{\xi, \eta > 0} d\xi d\eta [\xi^2 + \eta^2 + n^2/L^2]^{1/2} =$$

(11)

$$= \pi \hbar c (L/d) \Sigma''_n (\pi/2) \int\limits_{\rho > 0} \rho d\rho [\rho^2/d^2 + n^2/L^2]^{1/2}.$$

But in the same way as we have done for $n_1/L \equiv \xi$ and $n_2/L \equiv \eta$, we should consider also $n/L \equiv \varsigma$ as a continuous variable, and therefore one has

(12) $$W_0(L, L, d) = 1/2\pi^2 \hbar c (L^2/d) \int\limits_{\varsigma > 0} d\varsigma \int\limits_{\rho > 0} \rho d\rho [\rho^2/d^2 + \varsigma^2]^{1/2},$$

i.e., defining again $\tau \equiv \rho^2$ and $y \equiv \varsigma d$, we have

(13) $$W_0(L, L, d) = 1/4\pi^2 \hbar c (L^2/d^3) \int\limits_{y > 0} dy \int\limits_{\tau > 0} d\tau [\tau + y^2]^{1/2}.$$

From eq.s (5), (9) and (13), the energy excess $U(L, L, d)$ present in the rectangular cavity only because of the boundary conditions reads

(14) $$U(L, L, d) = 1/4\pi^2 \hbar c (L^2/d^3) C,$$

where the dimensionless constant C is defined as

(15) $$C = \Sigma''_n \int\limits_{\tau > 0} d\tau [\tau + n^2]^{1/2} - \int\limits_{y > 0} dy \int\limits_{\tau > 0} d\tau [\tau + y^2]^{1/2}.$$

The Casimir force per unit surface on the two $L \times L$ cavity walls in then given by

(16) $$P_c = 1/L^2 \partial U(L, L, d)/\partial d = -3/4 \; \pi^2 \hbar c d^{-4} C$$

and it is directed inward (outward) if P_c is $> 0 (< 0)$.

2b. - *The quantity C*

The dimensionless quantity C, defined by eq. (15), is the difference between two infinite quantities. Therefore, to evaluate C, we must devise some suitable criterion.

The leading physical idea of Casimir [1] was that, after all, the perfect ideal conductor does not exist. Any real conductor will, in fact, become trasparent to the e.m. radiation above some cut-off frequency Ω.

As a consequence, if we split both W and W_0 in the two contributions coming from the modes below and above the cut-off frequency Ω,

(17)
$$W = W(\omega < \Omega) + W(\omega > \Omega),$$
$$W_0 = W_0(\omega < \Omega) + W_0(\omega > \Omega),$$

since, for $\omega > \Omega$ the conductor is transparent, (i.e. $W(\omega > \Omega) = W_0(\omega > \Omega)$)we obtain

(18)
$$U = U(\omega < \Omega) = W(\omega < \Omega) - W_0(\omega < \Omega),$$

which is the difference between two finite quantities, since the number of normal modes below the cut-off frequency Ω is finite.

However, the trasparency of the conductor will not occur sharply at $\omega = \Omega$. In order to describe a real conductor, it is therefore more appropriate to introduce a suitable cut-off function $F(\omega/\Omega)$ such that

(19)
$$0 \le F(x) \le 1; \quad F(0) = 1; \quad F(1) = 1/2; \quad F(\infty) = 0.$$

The constant C becomes

(20) $\quad C = \Sigma_n'' \displaystyle\int_{\tau>0} d\tau [\tau + n^2]^{1/2} F(x(\tau, n)) - \int_{y>0} dy \int_{\tau>0} d\tau [\tau + y^2]^{1/2} F(x(\tau, y)),$

where

(21)
$$x(\tau, n) \equiv (\tau + n^2)^{1/2} (\Lambda/2d),$$

and $\Lambda \equiv 2\pi c/\Omega$ is the cut-off wavelength.

By introducing now the function

(22)
$$G(n) \equiv \int_{n^2}^{\infty} dw \sqrt{w} F(\Lambda\sqrt{w}/2d),$$

from eq. (20), we obtain

(23)
$$C = \Sigma_n'' G(n) - \int_0^{\infty} dy G(y).$$

To evaluate C from eq. (23), we can use the Euler-MacLaurin summation formula [7]. For this purpose, we recall that if a function f has its first $2n$ derivatives continous on an intercal (a, b), and we divide such an interval in m

equal parts of amplitude $\delta = (b - a)/m$, then it exists a $0 < \theta < 1$ depending on $f^{(2n)}$ in (a, b) for which

$$\sum_{0}^{m}{}_k f(a + k\delta) = 1/\delta \int_a^b f(t)dt + 1/2\{f(b) + f(a)\}+$$

$$+ \sum_{1}^{n-1}{}_k \delta^{2k-1}[(2k)!]^{-1}B_{2k}\{f^{(2k-1)}(b) - f^{(2k-1)}(a)\}+$$

$$+ \delta^{2n}[(2n)!]^{-1}B_{2n}\sum_{0}^{m-1}{}_k f^{(2n)}(a + k\delta + \delta\theta),$$

where B_n are the Bernoulli numbers ($B_0 = 1$, $B_2 = 1/6$, $B_4 = -1/30$, $B_6 = 1/42, \cdots$).

From this summation formula, in the special case where $a = 0, b = \infty, \delta = 1$ and for a function f which vanishes with all its derivatives in the limit $x \to \infty$, one obtains

$$\Sigma_n'' f(k) - \int_0^\infty dt f(t) = -\sum_{1}^{n-1}{}_k [(2k)!]^{-1}B_{2k} f^{(2k-1)}(0)+$$

$$+ [(2n)!]^{-1}B_{2n}\sum_{k\geq 0} f^{(2n)}(k + \theta),$$

for some θ between 0 and 1.

We will assume that the cut-off function $F(x)$ is a differentiable function which, together with all its derivatives, approaches zero at infinity, faster than x^2. In this hypothesis, the Euler-MacLaurin expansion of C reads

(24) $$C = \Sigma_n'' G(n) - \int_0^\infty dy G(y) =$$

$$= -\{[B_2/2!]G^{(1)}(0) + [B_4/4!]G^{(3)}(0) + \cdots [B_{2n-2}/(2n-2)!]G^{(2n-3)}(0) + \cdots\}+$$

$$+ [B_{2n}/(2n)!]\sum_{k\geq 0} G^{(2n)}(k + \theta),$$

and this quantity can be easily evaluated in the limit when $\Lambda << d$ (perfect conductor). From the definition (22), one has

(25) $$G^{(1)}(x) = -2x^2 F(\Lambda x/2d) = -2(2d/\Lambda)^2 y^2 F(y),$$

where we have defined $y \equiv \Lambda x/2d$.

From successive differentiations of eq. (25), we obtains

$$G^{(n+1)}(x) = -2(\Lambda/2d)^{n-2}[D^n y^2 F(y)]|_{y=\Lambda x/2d} =$$
$$= -2(\Lambda/2d)^{n-2}\{n(n-1)F^{(n-2)} + 2ny F^{(n-1)} + y^2 F^{(n)}\},$$

i.e.

$$G^{(1)}(0) = 0,$$

(26) $$G^{(3)}(0) = -4,$$

$$G^{(2k+3)}(0) = -2(\Lambda/2d)^{2k}(2k+2)(2k+1)F^{(2k)}(0), \quad (k>0).$$

Moreover, in the limit when $\Lambda << d$, one has

$$\sum_{k\geq0} G^{(n)}(k+\theta) = -2(\Lambda/2d)^{n-3}\sum_{k\geq0} D^{n-1}y^2 F(y)|_{y=\Lambda(k+\theta)/2d}$$

(27)
$$\approx -2(\Lambda/2d)^{n-3}\int_{\Lambda\theta/2d}^{\infty} D^{n-1}y^2 F(y)(2d/\Lambda)dy$$

$$= 2(\Lambda/2d)^{n-2}D^{n-1}y^2 F(y)|_{y=\Lambda\theta/2d}.$$

By substituting in eq. (24) the expressions (26) and (27), in the limit when $\Lambda/d \to 0$ (ideal conductor), we obtain for C the following expression

(28) $$\lim_{\Lambda/d\to0} C = -[B_4/4!](-4) = -1/180,$$

and using this value of C in eq. (16), we obtain for the Casimir pressure the well known result

(29) $$P_c = 1/240\pi^2\hbar c d^{-4}.$$

3. - The Casimir force in an optical resonator

Let us consider an ideal lossless optical resonator, made of two spherical mirrors, having the same radius of curvature R, placed at a distance d. By solving the self-consistent field equations in the cylindrical coordinate system (r, ϕ, z) which has the z-axis aligned with the resonator optical axis and the origin at the resonator centre, we obtain the cavity eigenmodes TEM_{pmq} [8,9].

The resonance condition reads

(30) $$2d/\lambda = q + 1/\pi(2p+m+1)\cos^{-1}(1-d/R),$$

and the field distribution on the mirrors is given by the real and the imaginary part of the complex quantity

$$(31) \qquad E(r, \phi, \pm d/2) = E_0 (r\sqrt{2}/r_0)^m L_p^m (2r^2/r_0^2) \exp(-r^2/r_0^2) \exp(im\phi),$$

where

$$(32) \qquad r_0 = (\lambda R/\pi)^{1/2} [d/(2R - d)]^{1/4}$$

is the spot size of the fundamental mode TEM_{00q} and $L_p^m(x)$ are the associated Laguerre polynomials.

For a confocal resonator, i.e. when $d = R$, the resonance condition (30) becomes

$$(33) \qquad 2d/\lambda = q + p + (m + 1)/2,$$

which implies that all the modes $TEM_{k-j,2j,q-k}, k \geq 0, 0 \leq j \leq k$ resonate at the *same* frequency

$$(34) \qquad \omega_{k-j,2j,q-k} = \frac{\pi c}{d} (q + 1/2) = \omega_{00q}.$$

Let us assume that the resonator is made of two interferential mirrors, perfectly reflecting between $\Omega - \Delta\Omega/2$ and $\Omega + \Delta\Omega/2$ and perfectly transparent outside that frequency interval.

The only modes which can resonate are those having

$$(35) \qquad \Omega - \Delta\Omega/2 \leq \omega_{pmq} \leq \Omega + \Delta\Omega/2,$$

therefore, the electromagnetic zero-point energy $W(d)$ associated to the resonator will be

$$(36) \qquad W(d) = 1/2\hbar\Sigma_{pmq}\omega_{pmq}\mathcal{F}(\omega_{pmq}),$$

where the weigth function $\mathcal{F}$ takes into account the condition (35). But since the modes $TEM_{k-j,2j,q-k}$ are degenerate with the mode TEM_{00q} one has

$$(37) \qquad \begin{aligned} W(d) &= 1/2\hbar \sum_{q>0} \sum_{0\leq k\leq q} \sum_{0\leq j\leq k} \omega_{k-j,2j,q-k}\mathcal{F}(\omega_{k-j,2j,q-k}) = \\ &= 1/2\hbar \sum_q \omega_{00q}\mathcal{F}(\omega_{00q})\mathcal{N}(\omega_{00q}), \end{aligned}$$

where $\mathcal{N}(\omega_{00q})$ is the number of transverse modes which resonate at $\omega = \omega_{00q} = (\pi c/d)(q + 1/2)$.

Since the mirrors have finite size, the diffraction losses will increase with the transverse mode numbers m, p.

According to the analysis done in ref. [8,9], we will take for $\mathcal{N}(\omega)$ the square of mirror's Fresnel number

(38)
$$\mathcal{N}(\omega) = (a^2\omega/8\pi cd)^2 \equiv (a^2/4d\lambda)^2,$$

where a is the mirror diameter and $\lambda \equiv 2\pi c/\omega$.

As far as the energy W_0 is concerned, from its definition (4), we have

(39)
$$W_0(d) = (1/\sigma^3) \lim_{\sigma \to \infty} W_\sigma(\sigma d),$$

where $W_\sigma(\sigma d)$ is the zero-point energy associated to a resonator obtained by scaling of the factor σ all the cavity dimensions.

In the limit when $\sigma \to \infty$, the number of pure longitudinal modes oscillating into the cavity will increase as the ratio between the mirror bandwidth $\Delta\Omega$ and the mode spacing $\pi c/\sigma d$, whereas the number $\mathcal{N}(\Omega)$ of transverse modes oscillating at $\omega \approx \Omega$ will increase as σ^2. One has

(40)
$$W_\sigma(\sigma d) = 1/2\hbar\Omega[(\Delta\Omega\sigma d/\pi c)(\sigma^2 a^2\Omega/8\pi c\sigma d)^2],$$

i.e.

(41)
$$W_0(d) = 1/4\hbar\Omega(\Delta\Omega d/\pi c)(a^2\Omega/8\pi cd)^2.$$

Let us assume now that the mirror bandwidth $\Delta\Omega$ is smaller than the mode spacing into the cavity, i.e.

(42)
$$\Delta\Omega < \pi c/d.$$

This means that there will be only one cavity eigenfrequency that will, possibly, resonate. This will happen if there exists an integer q such that

(43)
$$\Omega - \Delta\Omega/2 \leq (\pi c/d)(q + 1/2) \leq \Omega + \Delta\Omega/2.$$

The zero-point e.m. energy $W(d)$ associated to the cavity will, therefore, be zero if no cavity eigenfrequencies are fulfilling eq. (35), or be equal to

(44)
$$W(d) = 1/2\hbar\Omega\mathcal{N}(\Omega) = 1/2\hbar(\pi c/d)(q + 1/2)\mathcal{N}(\Omega) \equiv \mathcal{W}$$

if there is an integer q for which eq. (43) is satisfied.

But it is obvious from the same equation that one can switch from $W(d) = 0$ to $W(d) = \mathcal{W}$ (and viceversa) by only changing the mirror distance d of the quantity

(45)
$$\delta = \pi c/2\Omega \equiv \lambda/4,$$

where λ is the mirror central wavelength $\lambda \equiv 2\pi c/\Omega$.

The Casimir energy

(46)
$$U(d) = W(d) - W_0(d)$$

will, therefore, change by $\pm W$ for a mirror displacement of δ.

This energy change implies the existence of a Casimir force F_c between the mirrors, which is a periodical function of the distance, with period 2δ. The magnitude of F_c is given by

(47)
$$F_c \approx W/\delta = 4\pi(\hbar c/\lambda^2)\mathcal{N}(\Omega) = 4\pi(\hbar c/\lambda^2)(a^2/4\lambda d)^2.$$

3a. - Numerical example

To measure the Casimir force F_c which is very small, the parameters λ and $\mathcal{N}(\Omega)$ must be choosen so as to make it as large as possible.

The main limitation on λ and d comes from the condition (42), from which one has

(48)
$$\Delta\Omega/\Omega < \pi c/d\Omega = \lambda/2d.$$

With very good quality interferential mirrors, one can reach a $\Delta\Omega/\Omega \approx 10^{-5}$ at $\lambda \approx$ few microns, and this implies that the mirror distance d cannot exceed few centimetres.

Concerning the number $\mathcal{N}$ of degenerate modes, let us consider again eq. (30)

(49)
$$2d/\lambda = q + 1/\pi(2p + m + 1)\cos^{-1}(1 - d/R).$$

If the mirror distance d is not exactly equal to the mirror radius R, the mode degeneracy is partially removed: for a given q, only the modes TEM_{pmq} for which $2p + m = \text{cost}$ remain degenerate.

Let us assume

(50)
$$d = R(1 - \varepsilon).$$

To the lowest order in ε, one has

(51)
$$2d/\lambda = q + p + (m + 1)/2 - \varepsilon(2p + m + 1)/\pi.$$

The transverse modes will remain completely degenerate with the longitudinal ones only if

(52)
$$\varepsilon(2p + m + 1)/\pi << 1,$$

i.e. for a number of modes $\mathcal{N}$ bounded by

(53)
$$\mathcal{N} < (\pi/\varepsilon)^2.$$

Assuming $\varepsilon \approx 10^{-2}$ we can have a number of degenerate modes simultaneously resonating in the cavity up to $N \approx 10^5$.

We are now in the position to define a feasible experimental set-up to measure the Casimir force in a confocal cavity. We will assume

i)   a mirror distance $d \approx 5cm$;

ii)   a useful mirror diameter $a \approx 1cm$;

iii) a central mirror wavelength $\lambda \approx 2\mu m$.

In these conditions, from eq. (48) the requirements on the mirror bandwidth is $\Delta\Omega/\Omega < \lambda/2d = 2 \cdot 10^{-5}$, corresponding to a mirror wavelength resolution of $\Delta\lambda = 0.4\text{Å}$, which is feasible.

The number N of degenerate modes which can resonate simultaneously is given by eq. (38). One obtains $N \approx 6 \cdot 10^4$, well within the limitation imposed by the confocality condition (53).

The Casimir force F_c on the mirrors, as given by eq. (47), is

(54) $$F_c \approx 4\pi\hbar c/\lambda^2 N \approx 0.6 \text{ mdyne},$$

which is in the range of sensitivity of laboratory instrumentation.

We point out that the force F_c is modulated with a periodicity of $\lambda/2$ for a change of the lenght d of the cavity. This feature can be used to separate this effect from background perturbations.

As far as the thermal corrections are concerned, they are negligible provided $\hbar\Omega >> kT$. In our case, at room temperature, we have $\hbar\Omega = 0.62eV >> kT = 0.026eV$.

4. - Conclusions

The method suggested should allow to abserve the Casimir force in an optical resonator, for distances in the range of few centimetres.

The method can be implemented with existing good quality optical components and it should be fearly insensitive to the ordinary perturbative effect like, for instance, the Van der Waals forces.

Acknowledgments

I whish to thank Prof. L.A. Radicati for having raised my interest in the problem of the Casimir effect, Prof. E. Zavattini for many useful discussions on the subject and Prof. E. Polacco for criticism.

REFERENCES

[1]   H.B.G. CASIMIR, Proc. Kon. Ned. Akad. Wetenshap **51**, 793 (1948).

      H.B.G. CASIMIR - D. POLDER, Phys. Rev. **73**, 360 (1948).

[2]   T.H. BOYER, Ann. of Phys. **56**, 474 (1970).

      M. FIERZ, Helv. Phys. Acta **33**, 855 (1960).

[3]   H.B.G. CASIMIR, Physica **19**, 846 (1953).

[4]   T.H. BOYER, Phys. Rev. **174**, 1764 (1968).

      T.H. BOYER, Phys. Rev. **174**, 1631 (1968).

      K.A. MILTON - L.L. DeRAAD jr. - J. SHWINGER, Ann. of Phys. **115**, 338 (1978).

[5]   M.J. SPARNAY, Physica **24**, 751 (1958).

[6]   I.N. ISRAELACHVILI - D. TABOR, Proc. Roy. Soc. London **A331**, 19 (1972).

[7]   M. ABRAMOWITZ - I.A. STEGUN, Handbook of Mathematical functions Pag. 806.

[8]   G.D. BOYD - J.P. GORDON, Bell System Technical Journal **40**, 489 (1961).

[9]   G.D. BOYD - H. KOGELNIK, Bell System Technical Journal **41**, 1347 (1962).

Nuclear Volume Effects in the Spectrum of Atomic Oxygen

MASSIMO INGUSCIO - ANTONIO SASSO
GUGLIELMO MARIA TINO (*)

Abstract

We investigate isotope shift on the $^{16,17,18}O$ optical transitions by means of laser polarization spectroscopy. High resolution measurements provide the first experimental evidence for a nuclear field effect on such a light element. Observations are explained referring to the complete spherical symmetry of the ^{16}O nucleus.

1. - Introduction

It is well known in atomic physics that different isotopes of a particular element show relative displacements of spectral lines on account of distribution of nuclear charge over a finite region - a different distribution for different isotopes.

For the isotopes of light elements ($Z < 30$) the effect is very much smaller than the mass-dependent effect and until recently it has been possible to ignore it except as a factor in the interpretation of hyperfine structure intervals [1]. For instance, in the case of hydrogen, the correction in energy for $2s$ levels due to the finite size of the nucleus is 0.12 MHz in H and 0.73 MHz in D. However the precision of contemporary laser spectroscopy has reached the point where it is becoming necessary to be aware of the volume effect in the optical spectra [2]. The lighter the element, the more challenging is the measurement, also because of the larger Doppler broadening of spectral lines. Enough reasons for undertaking such measurements come from the fact that they stimulate the development of more powerful experimental methods and more refined theoretical treatments of the atomic-nuclear structure.

This report in honour of prof. L.A. Radicati comes from the University

(*) Dipartimento di Scienze Fisiche dell'Università di Napoli, I 80125 Napoli, Italy

where he started his career as a professor and deals with the experimental evidence we are obtaining for a nuclear volume contribution to the isotope shift in the spectrum of oxygen ($Z = 8$).

The small effect, otherwise negligible in such a light element, originates from the completely spherical structure of ^{16}O nucleus (doubly magic). This symmetry is altered by the addition of nucleons outside the closed shell. In ^{17}O we have a nuclear spin $I = 5/2$ and both a magnetic dipole ($-1.894\mu_N$) and an electric quadrupole ($-2.578 efm^2$) arise. The occurrence of three stable isotopes - ^{16}O, $^{17}O(0.04\%)$, $^{18}O(0.2\%)$ - is favourable since, as we shall see later, at least three isotopes are necessary to distinguish the volume effect from the predominant mass one. However the experiment requires the combination of high resolution with high sensitivity.

2. - Experimental

The difficulties for the investigation of oxygen start with the production of the atom from the molecule, in environmental conditions suitable for high resolution spectroscopy. Furthermore, excited atoms are necessary since no allowed optical transitions are accessible from the ground state. Only recently we made atomic oxygen accessible to cw laser sub-Doppler investigations by means of intermodulated optogalvanic spectroscopy [3]. Oxygen atoms could be produced from trace amounts of O_2 in a neon or argon substained radiofrequency disharge. Rather complicated collisional processes make the production very critical on the experimental parameters [4], however allowing fine structure and preliminary ^{16}O - ^{18}O isotope shift measurements [3,5]. Sub-Doppler resolution was obtained by having the sample interact with two counterpropagating beams. In this scheme only atoms with zero longitudinal velocity component can interact with both the laser beams. As a consequence the intermodulated detection scheme, based on the recording of the change of atomic *populations* caused by the interaction with both the beams, produces sub-Doppler signals. In the actual experimental situation, however, velocity changing collisions make it possible also for longitudinally moving atoms to interact with both the laser beams. This causes a broad pedestal which reduces both the sensitivity and the resolution of the Doppler-free signal. On the contrary, collisions destroy angular momentum orientation of the atoms, and the pedestal can be eliminated using a technique which monitors the atomic *orientation* and not only the level populations.

Circularly polarized light causes an optical orientation of the absorbing gas, hence inducing birefringence, as it was early discussed by Gozzini [6]. The laser version of the technique [7] is illustrated in Fig. 1. Radiation from a tunable narrow band ($\Delta\nu/\nu \sim 10^{-9}$) ring Dye-laser is splitted into two beams of different intensities, counterpropagating through the sample. The stronger beam is circularly polarized in order to create the orientation of a velocity selected group of atoms. The induced birefringence is probed by the weaker linearly polarized beam. High sensitivity is achieved detecting the transmitted intensity

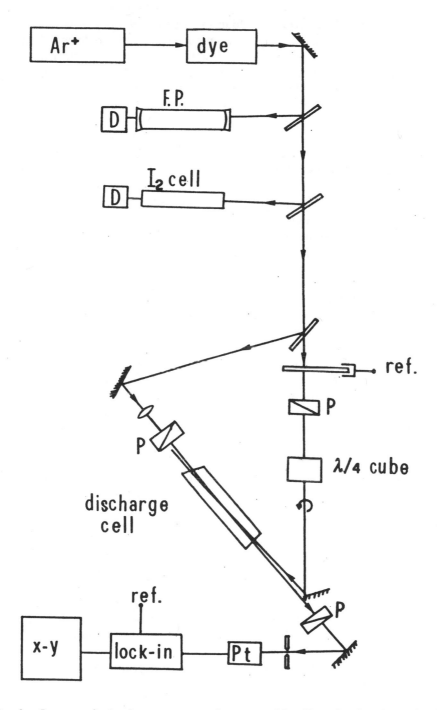

Fig. 1 - *Laser polarization apparatus for a sensitive Doppler-free investigation of the atomic oxygen spectrum.*

after a crossed linear polarizer with high extinction coefficient. Again, the signal will be given by only those oriented atoms interacting with both the strong and weak beams, i.e. by atoms with zero velocity component, and will occur when the laser frequency is tuned exactly at the center of the atomic transition. A small portion of the laser light is sent to a stable Fabry-Perot interferometer for calibration of the frequency scan, while the laser wavelength is determined by recording the absorption from a I_2 sample.

3. - Results and Discussion

A scheme of oxygen energy levels and investigated transitions is shown in Fig. 2. It is worth noting that we have chosen transitions involving s states, which are more likely to exhibit a nuclear volume contribution to the isotope shift, and for reference transitions not involving s states. A typical recording, using enriched samples, is shown in Fig. 3 b) for the transition at 615.8 nm; isotope shift values can be obtained with an accuracy better than 1%. The hyperfine structure of ^{17}O transition is not resolved, yelding a hyperfine separations comparable with the homogeneous width ($\sim$ 150 MHz). In order to show the powerfulness of the sub-Doppler technique, a conventional absorption profile is shown in a) for a natural abundance sample where it is evident that Doppler broadening should completely mask the isotope shifts. Similar recordings were obtained for other transitions and in Table I we summarize the isotope shift values measured for the transitions investigated searching for a nuclear volume effect. For demonstrating the better accuracy provided by laser polarization spectroscopy, we report also the ^{16}O - ^{18}O values available from intermodulated spectroscopy [5].

TABLE 1

transition	wavelenght (nm)	$^{16}O-^{18}O IS$ (MHz)	$^{16}O-^{17}O IS$ (MHz)	$^{16}O-^{18}O IS$, Ref. [5] (MHz)
$3p^5P_3-4d^5D_4$	615.8	1320(5)	686(10)	1310(40)
$3p^5P_3-5s^5S_2$	645.7	1163(9)	646(10)	1160(20)
$3p^3P_2-6s^3S_1$	604.6	1278(10)	723(10)	1300(40)

Table 1 - Measured isotope shifts (IS) values on oxygen atom.

In order to really evidence the contribution of the size of the nucleus to the observed shift, it is necessary to subtract the predominant mass effect. In multielectron atoms, the mass effect is customarily divided into a normal mass shift (NMS) and a, so called, specific mass shift (SMS). The NMS in due to the reduced mass correction, which can be easily computed. On the contrary the SMS, which originates from the influence of correlations in the motion of the electrons on the recoil energy, depends on the knowledge of the electronic wavefunctions and cannot be calculated with sufficient accuracy.

Following a generally used semiempirical expression, both the mass and field terms are taken as the product of a purely nuclear factor and a factor depending on the wavefunctions of the electronic states involved in the observed optical transitions. By means of this factorization, the isotope shift on a transition "a" for two isotopes α and β with mass number A_α and A_β is

$$(1) \qquad \Delta\sigma^a_{\alpha\beta} = K^a A_{\alpha\beta} + E^a C_{\alpha\beta}$$

In the first (mass) term, K^a is the factor depending on the electronic states of the transition and $A_{\alpha\beta} = (A_\beta - A_\alpha)/A_\beta A_\alpha$; in the second (field) term, $C_{\alpha\beta}$ is proportional to the difference of the mean square radii of the two isotopes and E^a is proportional to the difference of the electronic densities at the nucleus for the two levels of the transition.

It is useful to plot (King plot) the "modified isotope shift" (MIS)

$$(2) \qquad \mu^a_{\alpha\beta} = \Delta\sigma^a_{\alpha\beta}/A_{\alpha\beta}$$

for all the possible pairs of values A_α, A_β against the corresponding expression μ^b for another transition "b", by means of the following relation easily obtained by (1) and (2)

$$(3) \qquad \mu^a_{\alpha\beta} = (E^a/E^b)\mu^b_{\alpha\beta} + [K^a - K^b(E^a/E^b)]$$

In absence of a volume effect fot both the transitions $(E^a = E^b)$ all the points of such a plot overlap. On the contrary, in presence of a volume contribution, the plotted points would lie on a straight line, the slope of which yelding the ratio E^a/E^b of the field shifts of the two different transitions. This is evident in Fig. 4a, where we make the plot for a transition involving an s electron state versus another not involving s levels. The first transition, as reasonable, seems to be more nuclear size affected than the other. Two transitions both involving an s state are plotted in Fig. 4b. In this case the nuclear size contribution is shown to be comparable. A more precise determination of the relative effect will require a further refinement in the measurements, especially because of the existence of only three stable isotopes which will limit to only two the available points for an exact slope evaluation.

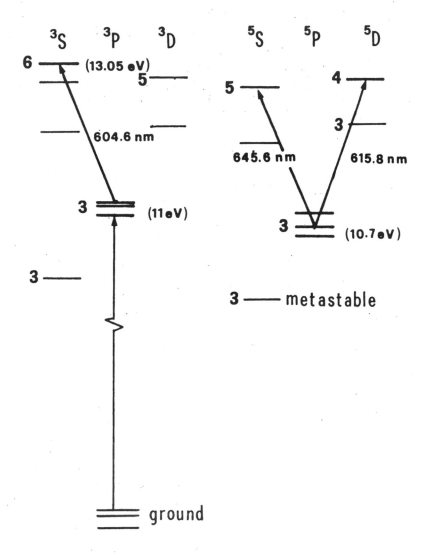

Fig. 2 - Partial level scheme of the oxygen atom. Transitions investigated in the present work, searching for a nuclear size contribution to the isotope shift, are indicated by arrows.

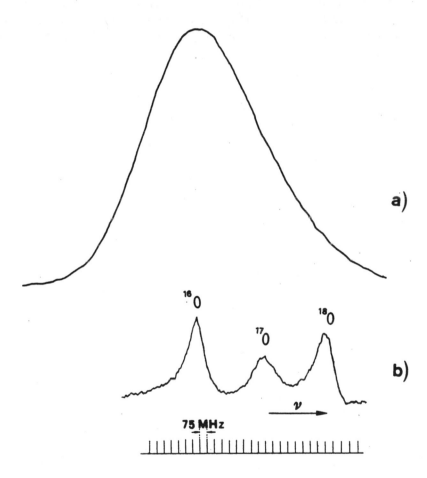

Fig. 3 - Laser polarization spectroscopy recording of the isotope shift of the $^5P_3$ *-* $^5D_4$ *transition at 615.8 nm, using an enriched sample b). In conventional absorption spectroscopy the effect should be completely masked by the large Doppler broadening recorded in a) for a natural abundance sample.*

4. - Conclusion

In conclusion, we have obtained first experimental evidence of a nuclear field effect on the optical spectrum of atomic oxygen. These results were possible thanks to the development of a high sensitivity laser polarization technique. Extension to other transitions is in progress for the confirmation of the present experimental evidence.

It is worth noting that oxygen is by far the lightest element for which this is now observed, apart for the particular case of hydrogen for which there is no specific mass effect and the normal one can be computed exactly. A precise investigation of the change of the completely spherical $(Z = 8)$ nuclear structure should now be possible.

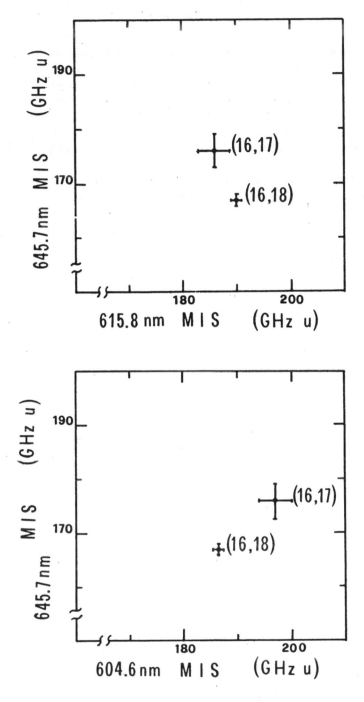

Fig. 4 - Analysis of the isotope shifts using "King plots". A nuclear size contribution is evidenced as explained in the text.

REFERENCES

[1]   E. ARIMONDO - M. INGUSCIO - P. VIOLINO, Rev. Mod. Phys. **49**, 81 (1977).

[2]   For the recent progresses in atomic spectroscopy see for instance: "*The Hydrogen Atom*", F. BASSANI - M. INGUSCIO - T.W. HÄNSCH, Eds, Spinger Verlag, 1989.

[3]   M. INGUSCIO - P. MINUTOLO - A. SASSO - G. TINO, Phys. Rev. **A37**, 4056 (1988).

[4]   A. SASSO - P. MINUTOLO - M.I. SCHISANO - G.M. TINO - M. INGUSCIO, J. Opt. Soc. Am. **B5**, 2417 (1988).

[5]   K. ERNST - P. MINUTOLO - A. SASSO - G.M. TINO - M. INGUSCIO, Opt. Lett. **14**, 554 (1989).

[6]   A. GOZZINI, Compt. Rend. Acad. Sci. **255**, 1905 (1962).

[7]   C. WIEMAN - T.W. HÄNSCH, Phys. Rev. Lett. **36**, 1170 (1976).

Beam Radiation in the Deep Quantum Regime

M. JACOB (*)

It is an honour for me to contribute to this Festschrift Volume. It is the occasion to express how much the friendship which Luigi has extended to me is dear to my heart. It has always been a great pleasure for me to meet him in various places in the world and, in particular, in Pisa and at CERN. His contributions to the study of the symmetry have touched in a prominent way that part of our field of research which is probably the deepest and the most adorned with aesthetic beauty.

This modest contribution appeals to Luigi's wide range of interest in physics. It touches a question which has some specific novelty, namely the application of Feynman graph calculations to a machine physics problem.

Electrodynamics is the best understood domain in our field. Yet its many subtleties are still hard for the human mind to grasp, and while many answers turn out to be simple and often of order α/π, they are reached by cumbersome paths which probably reflect more the complication of the interface between the physics and our understanding of the phenomena than the intrinsic complexity of Nature.

This essay illustrates that point. It describes the work which I have done recently in collaboration with Tai Tsun Wu on the question of radiation - one uses there the word "beamstrahlung" despite its etymological weakness, and pair production, when high density bunches of particles cross each other at very high energies.

Linear electron-positron colliders working in the TeV energy range appear to be very promising future machines. The very high fields present inside the colliding bunches induce a very strong radiation. It has some similarities with synchrotron radiation but dealt with in a quantum way, and also some similarities with the Weizsächer-Williams quantum fragmentation of electrons and photons.

The new results are very different from those met in the classical regime. They have yet a great specific simplicity which appears once all the work is done. The following is an introduction to our recent work on that question of quantum machine physics.

(*)   CERN - Geneva

1. - Introduction

As previously said this reports on work done at CERN in collaboration with T.T. Wu. The linear colliders under consideration are in the TeV range. Reaching such energies is difficult; it is, however, pointless if one does not also reach a large enough luminosity. The luminosity should increase as E^2, where E is the beam energy, if one wishes to collect a decent enough rate for the bench mark cross-section $e^+e^- \rightarrow \mu^+\mu^-$. A luminosity of 10^{33} is a must for $E \sim 1$ TeV.

The luminosity is practically determined by four parameters:

$$(1) \qquad\qquad L \sim \frac{N^2 f H}{R^2}$$

Increasing the number of particles per bunch N and the bunch crossing frequency f costs power. The possible gain through H, the pinch parameter, is limited. One has to play on R and one is talking about R at the level of $10^{-8} m$!

With large N and small R, the field inside the bunch $(\sim N/R)$ is very high. Incident particles radiate as they cross such a high field region. One is driven into the deep quantum regime where

$$(2) \qquad\qquad Y = \frac{\gamma^2}{m\rho_c} >> 1$$

where ρ_c is the radius of curvature.

Two questions are considered in this note, namely radiation by an electron traversing a positron bunch and pair formation by a photon traversing a positron (electron) bunch. The kinematics is defined in Figs. 1a and 1b, respectively. These pictures correspond to a classical approximation and the space location of the process is defined as the point of the stationary phase. The process is actually "spread" over a distance ℓ_c around that point, where ℓ_c will be defined later.

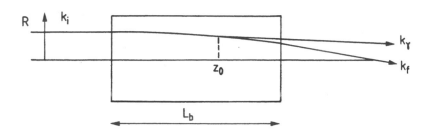

Figure 1a

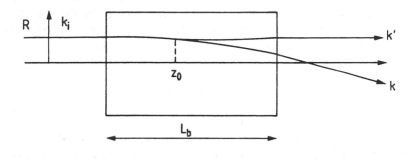

Figure 1b

The quantum mechanical calculation corresponds to the Feynman graphs of Figs. 2a and 2b respectively. The calculation is done to lowest order in α and thus corresponds to a distorted wave Bron approximation. The amplitude reads in both cases:

(3)
$$\overline{\psi}_f \gamma_\mu \psi_i \varepsilon^\mu$$

where ψ_i and ψ_f are the wave functions of the charged particles in the field of the bunch and ε is the polarization vector of the photon. The crosses on the graph correspond to interaction with the field in the bunch.

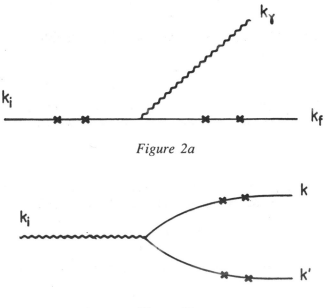

Figure 2a

Figure 2b

From (3) one calculates the radiation intensity $I(x)$, normalized to unit incident flux, where x is the fraction of the electron energy taken by the photon (beamstrahlung) or the fraction of the photon energy taken by the electron (pair formation). The spectrum found are rather hard, with a rather flat distribution in x. the fractional energy loss:

$$(4) \qquad \delta = \int_0^1 x I(x)\,dx$$

is considerable. Values at the level of 20% can be considered as typical.

The classical regime is well known. We concentrate here on the deep quantum regime, emphasizing its specific simplicity which this regime also possesses. It is clear that there is also an interesting intermediate regime which deserves attention.

2. - Main Results

Our interest in this problem started with the remark that the respective dominance of two important lengths is inverted as one goes from the classical regime to the deep quantum regime. In the classical regime, the coherent radiation length L_c:

$$(5) \qquad L_c \sim \left(\frac{N}{L_b}\right)^{-1} \left(\frac{\alpha}{R_m}\right)^{-1}$$

is much larger than the quantum mechanical radiative length L_e

$$(6) \qquad L_e \sim \frac{\gamma}{m}$$

In the deep quantum regime, into which one is naturally driven with a high energy, high luminosity machine, it is the converse which is true

$$(7) \qquad L_e \gg L_c$$

The radiation process changes in nature. We found that simplicity still prevails when one introduces a new coherent radiation length

$$(8) \qquad \ell_c = (L_c^2 L_e)^{\frac{1}{3}}$$

In the deep quantum regime $\ell_c \gg L_c$, and one indeed finds that

$$(9) \qquad Y = \left(\frac{\ell_c}{L_c}\right)^3$$

In practice the quantity

$$(10) \qquad \Lambda = \frac{L_b}{\ell_c}$$

remains large (10 to 100) $\Lambda \sim N\alpha D$, where D is the disruption parameter. To a first approximation, one can consider that different sections of the bunch, each one of them ℓ_c long, radiate incoherently. One may remark that ℓ_c is that particular combination of L_c and L_e which eliminates any explicit reference to the electron mass. This is natural since we are in a regime where the transverse momentum collected by the electron as it crosses the bunch is much larger than the electron mass

$$(11) \qquad \Delta_T >> m$$

The canonical parameter of the classical approach α/m has to drop out entirely! More specifically, one has

$$(12) \qquad \ell_c = \left(\frac{R^2 L_b^2 E}{(N\alpha)^2} \right)^{\frac{1}{3}}$$

This new regime leads naturally to a calculation based on Feynman graphs. With it we found two things:

(i) simplicity prevails once one introduces ℓ_c;

(ii) there are typical quantum efects, which one may refer to as radiation "before and after" bunch crossing, since they do not involve the bunch length L_b. Their leading contribution is proportional to $\alpha ln(L_e/\ell_c)$. This is connected with a Weizsächer-Williams fragmentation.

Our approach was first presented and exploited in four papers (M.J. and T.T.W.):

(i)   Quantum approach to beamstrahlung, Phys. Lett. B197 (1987) 253;

(ii) Quantum calculation of beamstrahlung: the spinless case, Nucl. Phys. B303 (1988) 373;

(iii) Quantum calculation of beamstrahlung: the Dirac case, Nucl. Phys. B303 (1988) 389;

(iv) Beamstrahlung in the multi-TeV (a general review), Proceedings of the Kazimierz Conference (88), WSPC.

Our attitude has been to go as far as possible analytically. For that reason we worked in the low D approximation. D is the disruption parameter. It is the ratio between the length of the bunch and the focal length corresponding to the focusing effect of a bunch on impiging particles. Our results are accurate provided that a linear expansion in D is acceptable. Radiation effects at large D can a priori only be larger than those found for low D.

To be more specific we start with a high energy approximation of the type

(13) $$A = A_0 + \frac{1}{E} A_1$$

for both the modulus and the non-trivial part of the phase of the wave functions. This sounds a priori very accurate as E is very large, but it turns out to be only a low D (linear in D) approximation.

The leading contributions to δ are then particularly simple. For beamstrahlung they read:

(14) $$\delta = K_e \frac{\alpha}{\pi} \ell_n \frac{L_e}{\ell_c} + K_i \frac{\alpha}{\pi} \frac{L_b}{\ell_c}$$

where K_e and K_i are both numerical coefficients of order one. Relation (14) refers to a uniform bunch for which ℓ_c is defined by (12). The radiation spectrum is rather hard. It has also a very simple expression. For the most important term, proportional to L_b, one finds

(15) $$I(x) = \left(\frac{1-x}{x} \right)^{\frac{2}{3}} \frac{1 + (1-x)^2}{1-x}$$

The first factor results from the intrinsic properties of radiation in the deep quantum regime. It is the one found in the spinless (Klein-Gordon) case. The second one is mere spinology and corresponds to the complications met when dealing with a Dirac particle.

In the regime considered ($\Delta_T \gg m$), only one helicity configuration matters, once parity conservation has been imposed. The neglected one is proportional to L_b/L_e, which is small in practice. For that reason, one can work in the (simpler) Klein-Gordon case and merely include a Dirac spin factor in the spectrum, at the end. The same applies for pair formation. In that case the shape of the spectrum is (Dirac)

(16) $$I(x) \sim \frac{(1-x)^2 + x^2}{(x(1-x))^{\frac{1}{3}}}$$

when the production intensity reads

(17) $$I(x) = K_p \frac{\alpha}{\pi} \frac{L_b}{\ell_c} \frac{(1-x)^2 + x^2}{(x(1-x))^{\frac{1}{3}}}$$

thus picking up only the leading terms proportional to L_b. Here K_p is also a numerical factor of order one and ℓ_c is the same as in the case of beamstrahlung (12). The calculation of pair production, prompted by a remark by W. Schnell, is reported in TH.5274/89, to be published in Phys. Lett. B.

We see that, in all cases, a small factor α/π is to a large extent compensated by a large factor Λ (10). Radiation losses and pair production are therefore both considerable effects. The deep quantum regime is characterized by an intense radiation with a rather hard spectrum.

This has some drawbacks:

(i)   a loss in effective power;

(ii) important background and a hazard to the machine and the detector;

(iii) an erosion of the resonance peaks which one may wish to find.

However:

(i)   for heavy Higgs search $(WW \rightarrow H)$ an important radiation is tolerable;

(ii) an electron-positron linear collider is also an intense photon-photon collider. This is also very interesting in heavy Higgs search $(\gamma\gamma \rightarrow H)$. Yet, the important hard photon flux turns partly into hazardous electron-positron pairs in the intense field of the bunch!

One may remark at this stage that ℓ_c includes a factor R^2/N^2, inversely proportional to the luminosity and a factor s, corresponding to the centre-of-mass energy square. Since the two factors should compensate each other, ℓ_c varies as $L_b^{-1/3}$.

The deep quantum regime calls for compact bunches. As emphasized by Blankenbecler and Drell, ribbon bunches should help in keeping the luminosity up while decreasing the field inside the bunch and hence the beamstrahlung losses. This however raises further complications for the machine.

Our latter work on that topic, in 1988, mainly concentrated on the effect of varying density in order to go beyond the poorly realistic constant density bunch considered in our former work. This has been reported in three papers:

(i)   Beamstrahlung with fluctuating charge density, to be published in Nucl. Phys. B; TH.5133/88;

(ii) Beamstrahlung for longitudinally non-uniform bunch, to be published in Nucl. Phys. B; TH.5192/88;

(iii) Beamstrahlung in high energy electron-positron linear colliders with non-uniform bunches, Phys. Lett. B216 (1989) 442.

The approach now consists of an expansion in Λ (10) which is a large parameter (10 to 100 in practice). One takes the Mellin transform of the radiation intensity

$$I(x) = \frac{\alpha}{\pi} K(x, \Lambda)$$

(18)
$$\overline{K}(\xi) = \int\limits_0^\infty K(x, \Lambda) \Lambda^{-1-\xi} d\Lambda$$

and calculates the residues at the poles met at $\Lambda = 1$ (leading term proportional to L_b), $\Lambda = 0$ (which includes the $\ell n L_e$ edge effects) and $\Lambda = -1$ [which is sensitive to rapid density variation within the bunch, proportional to $(\rho')^2$ and $\rho\rho''$]. This can be done though a series of successive analytic continuations as the different pole terms can be isolated one by one. The result is a rigorous treatment of edge effects and varying density for arbitrary bunch shapes.

The edge effects are fully calculated and include the $\ell n L_e$ term previously focused upon.

The terms corresponding to an integral over the bunch length (roughly speaking now proportional to L_b) are of two kinds:

(i) a contribution proportional to Λ (large) which merely multiplies K_i in (14) by the extra factor

(19)
$$\frac{\int \tilde{\rho}^{2/3}(z)\,dz}{\int \tilde{\rho}(z)\,dz}$$

where $\tilde{\rho}$ is the normalized varying density

(20)
$$\int \tilde{\rho}(z)\,dz = L_b$$

(ii) A contribution of relative order Λ^{-2} (hence small) which involves an integral over z of

(21)
$$R(z) = \tilde{\rho}^{-2/3}(z)\left\{\frac{3\tilde{\rho}(z)\tilde{\rho}''(z) - 4\tilde{\rho}'^2(z)}{\tilde{\rho}^2(z)}\right\}$$

A proper analytic continuation has to be made to deal with edge effects, when $\tilde{\rho}(z) \to 0$. An actual numerical value (a relatively small contribution) can thus easily be obtained for realistic bunch shapes.

As the density varies, so does $\ell_c(z)$ which has now to be defined locally. From (12) one sees that $\ell_c(z) \sim \tilde{\rho}^{-2/3}(z)$, hence extra technical complication at the edges when this quantity becomes large.

3. - Calculation

The reader is referred to the different papers previously itemized for a detailed presentation. We here merely stress a few general points.

The calculation proceeds as follows:

(i) calculate the wave functions. This is done in the high E (but actually low D) approximation, as previously said. One solves the Klein-Gordon (Dirac) equation in the field of the bunch.

(ii)   Study the conditions for the stationary phase for the production amplitude. In practice, most of the radiation originates from a limited zone corresponding to a limited stationary variation of the phase. The location of this zone is determined by the kinematics and the bunch properties. Two points are worth emphasizing:

(a)   The phase stationarity conditions are very different in the beamstrahlung case and in the pair production case. This is due to the fact that the main additive effect, associated with the phase of the electron wave function before and after radiation, hence throughout the whole bunch in beamstrahlung, cancels in the production case, where an electron and a positron propagate with almost collinear paths, after formation only. To be more specific the transverse co-ordinate for the stationary phase corresponds to

(22)
$$\frac{\vec{r}}{R} = - \left(\frac{2N\alpha}{R}\right)^{-1} (\vec{h}_{1\perp} + \vec{h}_{\gamma\perp})$$

for beamstrahlung (Fig. 1a), and to

(23)
$$\frac{\vec{r}}{R} = - \left(\frac{2N\alpha}{R}\right)^{-1} \left(\frac{3}{1-\tilde{z}}\right) \left(\frac{4x(1-x)}{(1-\tilde{z})^2} \frac{\vec{K}_2}{\varepsilon} + \vec{h}_\perp\right)$$

for pair production (Fig. 1b).

Here $2N\alpha/R$ is the maximum bending momentum which can be collected through bunch-crossing. In the second case, we have introduced

(24)
$$\Sigma = \frac{N\alpha}{R^2} \frac{Lb}{E} \sim D$$
$$\tilde{z} = \frac{2}{L_b} z$$
$$\vec{K}_\perp = \vec{h}_\perp \rightarrow \vec{h}\prime_\perp \sim \Sigma \, \vec{h}_\perp$$

Such a difference is understandable. We are however very far from a mere crossing relation between the two processes.

(b)   Once we have fixed $\vec{r}$ by (22) or (23), we find that there is no real point of stationary phase in z. One actually finds that

(25)
$$\frac{\partial \phi}{\partial z} = \frac{x}{8(1-x)E}(|m|^2 + 4m^2)$$

where $|m|^2$ is the modulus square of the radiation (production) amplitude average or summed over polarizations. This relation turns out to be very useful!

There are nevertheless two nearly (separation of order Λ^{-1}) complex points of stationary phase and, in bertween, a real point where $\partial\phi^2/\partial z^2 = 0$. It is therefore natural to expand around that point z_0, defined as the point of the stationary phase. The phase then varies as an odd cubic in $(z - z_0)$ around that point. The characteristic length for its variation defines the proper coherent length. It is ℓ_c (12), for both beamstrahlung and pair production. The deep quantum regime corresponds to values of ℓ_c significantly larger than L_c (9).

(iii)   The next step is to compute the relevant matrix elements in the neighbourhood of the stationary point, m, the values of which were anticipated in writing (25). Because of the different conditions on $\vec{r}$ (22) and (23), the radiation matrix elements and the pair production matrix elements are very different. One is again far from a simple crossing relation. The same relation (25) however holds!

(iv)   At this stage one can proceed in two different directions.

(a)   One calculates the full amplitude integrating the production amplitude calculated through steps (i), (ii) and (iii) over all space. The integral over the radial co-ordinate is done through the stationary phase method. The integral over z, with its odd cubic dependence of the phase, naturally leads to Airy functions. One then integrates the modulus square of the amplitude over phase space to reach the production rate $I(x)$.

(b)   One integrates over the radial co-ordinates and over phase space, reaching $I(x)$ as a double integral over z and z'. This may a priori seem to be more difficult but one can make great use of (25) to extract the actual rate from a much simpler integral, through the calculation of partial derivatives. The proportionality of the derivative of the phase in (25) to the matrix element square which appears in the rate (m is in most cases negligible in front of the matrix element which is proportional to bending momenta) clearly brings much simplification. One therefore shortcuts entirely Airy function and picks up with increasing simplicity terms of higher order in Λ (the leading ones!)

While we followed path (a) in our earlier approach, we switched to path (b) (it looked like a must) when dealing with varying densities. Insofar as the complicated z dependence of the production amplitude in pair formation can be formally compared to a varying density effect, the new method is particularly efficient in that case. For the leading term, proportional to Λ, the most complicated functions encountered are merely Γ functions. We thus shortcut the integration over Bessel functions of fractional orders met along path (a).

In the case of beamstrahlung studied along path (b), one finds an integral over $z - z'$ with a rapidly changing phase which limits its range in practice to ℓ_c. One is left with a trivial integral over $z + z'$ which simply gives an overall factor Λ.

In the case of pair production, the integrand has, as previously stressed, a rather complicated z, z' dependence. However, the phase space limits are also z, z' dependent, since bending occurs only after the pair is formed (beyond z_0). It turns out that the z, z' dependence brought by the phase space conditions magnificently cancels the z, z' dependence met with the computation of the matrix element, once the integral over $z - z'$ has imposed the strong conditions associated with the variation of the phase $(z - z' < \tilde{\ell}_c)$.

While crossing simplicity is totally lost in the intermediate steps, both when dealing with the matrix element and the phase space limits, it beautifully reappears at the end with the results for beamstrahlung (14) and for pair production (17) which show a very strong similarity. Only once the rate is obtained do we find for both terms of order $(\alpha/\pi)\Lambda$ and, in both cases, rather hard spectra.

The $\ell n L_e$ term has a pure quantum origin. It can be considered as resulting from the quantum dissociation of the electron into a photon and an electron, or of the photon into an electron-positron pair before entering the bunch (and after leaving the bunch in the former case), mass-shell conditions being imposed over a length ℓ_c at the edge of the bunch. It becomes quite sizeable as one increases the energy, but is still far from leading in the TeV range, where the term linear in Λ dominates.

This is where we are. We have still clearly to try to push our method to larger D and extend it to the intermediate regime $(\ell_c \sim L_c)$ where a good fraction of its simplicity should of course go away. The question of multiphoton production and of pair production with incident electrons clearly calls for further calculations.

Acknowledgements

In this note we have presented our own work. It is merely an introduction to our published work on that subject. It is clear however that the study of beamstrahlung has already a long history and that previous work could extract a lot from semi-classical approaches. The subject has received much attention recently and we feel that a new approach which starts with Feynman graphs has much appeal in the deep quantum regime. We have of course interacted with many of our colleagues engaged in similar calculations, R. Blankenbecler and S. Drell and P. Chen and K. Yokoga, at SLAC; J.S. Bell and M. Bell at CERN and V. Baier, V.M. Katkov and V.M. Strakharenko at Novosibirsk. Everyone follows a somewhat different approach and there is now a good consensus on most results. T.T. Wu acknowledges the support of the United States Department of Energy under grant DE-FG02-84ER40158.

The Role of Time-Reversal and Parity in the Magneto-Optics of Semiconductors

G.C. LA ROCCA - S. RODRIGUEZ (*)

1. - Introduction

As in other areas of physics, also in solid state theory, symmetry considerations are of fundamental importance. The symmetry of a perfect crystal is fully characterized by its space-group. An electron in a crystal is subject to a potential whose symmetry is lower than that of free space. For instance, while all measurable properties of a free particle are invariant under arbitrary translations, for a particle within a crystal only the translations of the appropriate Bravais lattice leave the system invariant. As a consequence, the momentum of a particle moving in a solid is not strictly conserved, instead, its wave vector defined modulo a vector of the reciprocal lattice is a constant of the motion. Similar considerations apply also to the rotation symmetry group: the subgroup of the orthogonal group leaving a crystal invariant is discrete. We are here interested in the role played by the operations of time-reversal and parity which are discrete symmetries for a free particle as well as for a particle in a solid. The Hamiltonian of an electron in a solid is time-reversal invariant, but may or may not have inversion symmetry. In the following, we study a peculiar effect accompanying some magneto-optical transitions in zinc blende semiconductors, such as GaAs and InSb; these crystals lack inversion symmetry.

This effect was observed by Dobrowolska et al. [1] and by Chen et. al. [2] who measured the far infrared magneto-absorption of n-type InSb in the ordinary Voigt configuration (i.e., the wave vector $\vec{q}$ of the incident radiation is perpendicular to the static magnetic field $\vec{B}_0$, the light being linearly polarized parallel to $\vec{B}_0$). The sample is kept at liquid helium temperature and the spectra are recorder at a fixed wavelength while the magnetic field is swept up to 6 T. The absorption line of interest to us corresponds to the spin-flip resonance of conduction electrons. It occurs at $B_0 = 4.15T$ for a wavelength of $118.8\ \mu m$. The most striking feature of the experimental results is that the intensity of this line changes by almost a factor of 2 upon reversal of either $\vec{q}$ or $\vec{B}_0$. It is also

(*)   Department of Physics, Purdue University, Wet Lafayette, Indiana 47907 USA

noteworthy that the intensity is strongly anisotropic, depending on the orientation of $\vec{B}_0$ and $\vec{q}$ with respect to the crystallographic axis, and is up to two orders of magnitude stronger than expected according to the simplest theory, i.e., taking the electron-photon interaction as due to the coupling of the intrinsic magnetic moment of the electron with the oscillatory magnetic field of the incoming radiation. These data have been explained in all details by a microscopic analysis that takes into consideration the complexity of the solid state environment [3,4]. Moreover, an analogous microscopic study predicts similar and even more pronounced effects for specific magneto-optical transitions of holes in the same crystals [5]. Before discussing the main points of this theoretical analysis, it is useful to give a macroscopic view of these phenomena, applicable to both $n-$ and $p-$type materials.

We consider the four most common experimental geometries: ordinary Voigt (OV), extraordinary Voigt (EV) and Faraday ($F_\pm$) configurations. In the Voigt configurations, the incident light propagates perpendicularly to $\vec{B}_0$ and is linearly polarized along $\vec{B}_0$ (OV) or at right angles to $\vec{B}_0$ (EV). In the Faraday configurations, the incident light propagates along $\vec{B}_0$ and is circularly polarized with positive (F_+) or negative (F_-) helicity; for electrons, the F_+ and F_- configurations are called the cyclotron resonance active (CRA) and inactive (CRI) geometries, respectively. From the macroscopic point of view, some of our conclusions can be deduced from the point group symmetry T_d of the zinc blende structure and from time-reversal invariance. All physical properties of the system remain unchanged when comparison is made between configurations related by a symmetry operation of the point group. Besides, by virtue of time-reversal symmetry, the simultaneous transformations $\vec{B}_0 \rightarrow -\vec{B}_0$ and $\vec{q} \rightarrow -\vec{q}$ lead to no change in the observable properties of the system; therefore, reversal of either $\vec{q}$ or $\vec{B}_0$ are equivalent. Consider, for instance, the EV geometry with $\vec{B}_0 \| [001]$; the polarization and propagation vectors lie in the (110) plane. Reversal of $\vec{q}$ is accomplished by a rotation of 180° about [001], a symmetry operation of T_d. Since the sense of the polarization vector is immaterial, there can be no observable change upon $\vec{q}$-reversal (or $\vec{B}_0$-reversal). A similar conclusion holds for the OV configuration with $\vec{B}_0 \| [001]$. These results are no longer valid when $\vec{B}_0 \| [110]$ or $\vec{B}_0 \| [111]$; in fact, no operation of T_d (including time-reversal) allows the identification of the $\vec{q}$-reversed (or $\vec{B}_0$-reversed) configurations. Finally, the $F_\pm$ geometries with $\vec{B}_0$ along [001], [110] or [111] are all invariant under $\vec{B}_0$-reversal (and, thus, also under $\vec{q}$-reversal) as follows from the application of the reflection in the ($1\bar{1}0$) dihedral plane, an element of the group T_d.

2. - Electric dipole and magnetic dipole transition amplitudes

In this section, we are concerned with the physical principles on which the microscopic theory explaining the effects described in Sec. 1 is based. We limit ourselves to the most significant aspects, omitting technical details. We consider the electron situation first and then discuss the main similarities and differences for the case of holes. The latter is more complicated owing to the degeneracy of the valence band.

In the lowest order approximation the Hamiltonian H describing a conduction electron in a zinc blende crystal in the presence of a magnetic field $(\vec{B}_0 = \nabla \times \vec{A}_0)$ is

$$(1) \qquad H_0 = \frac{\hbar^2 k^2}{2m^*} + \frac{1}{2} g_0 \mu_B \vec{\sigma} \cdot \vec{B}_0,$$

where $\vec{k} = -i\vec{\nabla} + \frac{e}{c\hbar}\vec{A}_0$, μ_B is the Bohr magneton, and m^* and g_0 are the effective mass and g-factor of the material, respectively. This Hamiltonian is the same as that for a free electron except for the renormalized mass and g factor; in particular, it is isotropic and even under parity. However, this simple form of H cannot account for the experimental data. In fact, the eigenvalues of H_0 are the familiar Landau levels spin-split by the Zeeman interaction. The corresponding eigenstates are the product of harmonic-oscillator-like orbitals and spin functions. In this approximation, the eigenstates have a definite parity and the electric dipole operator (ED) is odd, while the magnetic dipole operator (MD) is even. In fact, in the velocity gauge

$$(2) \qquad \begin{aligned} ED &= \frac{e}{c} A\hat{\epsilon} \cdot \vec{v} = \frac{e}{c} A \left(-\frac{i}{\hbar}\right) [\hat{\epsilon} \cdot \vec{r}, H] \\ &\simeq \frac{e}{c} A \left(\frac{-i}{\hbar}\right) [\hat{\epsilon} \cdot \vec{r}, H_0], \end{aligned}$$

and

$$(3) \qquad \begin{aligned} MD &= A(i\vec{q} \times \hat{\epsilon}) \cdot \frac{\partial H}{\partial \vec{B}_0} \\ &\simeq \frac{1}{2} g_0 \mu_B \vec{\sigma} \cdot (i\vec{q} \times \hat{\epsilon}) A, \end{aligned}$$

where A and $\hat{\epsilon}$ are the amplitude and polarization of the vector potential of the incident light. Thus, the parity conserving spin-flip transition, in which only the spin state changes, while the orbital state remains the same, can only be introduced by the magnetic dipole interaction. Its intensity, proportional to the magnitude squared of the matrix element of MD, turns out to be much smaller than observed, isotropic instead of anisotropic and, furthermore, invariant under reversal of $\vec{B}_0$ and $\vec{q}$, contrary to the experimental findings. We must, therefore, include in the electron Hamiltonian other modifications due to the solid state environment, besides the changes in the mass and g factor.

As a consequence of the physico-chemical differences between the two distinct species in the unit cell (say, In and Sb), the zinc blende structure lacks inversion symmetry. It has long been known [6] that for crystals having point group symmetry T_d, the first correction to H_0 is of the form

(4) $$H_A = \delta_0[\sigma_x(k_y k_x k_y - k_z k_x k_z) + c.p.],$$

where δ_0 is a material constant. Three features of H_A are noteworthy: it is odd under parity, it is a spin-orbit coupling interaction and it is anisotropic (the expression above is referred to the cubic axes). H_A is small compared to H_0 because it is cubic in the components of $\vec{k}$, because of its relativistic origin and because it is due to the difference between the anion and cation potentials, which is regarded as a correction to their average value. Still, because it has different symmetry properties than H_0, its inclusion in the Hamiltonian is of great importance leading as it does to a relaxation of the selection rules. In particular, H_A mizes states with different parities and spins so that the electric dipole amplitude can add to the spin resonance transition matrix element. This contribution (referred to as electric dipole spin resonance, or EDSR) is enough to explain the enhanced intensity and the complex anisotropy which are observed.

The change in intensity upon reversal of $\vec{B}_0$ or $\vec{q}$ can be understood as follows. The amplitude due to the magnetic dipole interaction also contributes to the spin resonance transition matrix element. It turns out that the electric dipole spin-flip amplitude proportional to H_A is nearly ten times larger than the magnetic dipole contribution; however, the latter cannot be neglected because of its interference with the former. The spin-flip transition probability ($|\nu\rangle \rightarrow |v'\rangle$) is, thus, proportional to

(5) $$I = |\langle \nu'|ED + MD|\nu\rangle|^2.$$

The magnetic dipole amplitude is linear in $\vec{q}$ and upon $\vec{q}$-reversal acquires a minus sign with respect to the electric dipole amplitude, which leads to a considerable change in the intensity [7]. Upon $\vec{B}_0$-reversal, instead, it is the electric dipole amplitude that changes sign, causing the same results. In fact, we can use time-reversal symmetry to compare the $\vec{B}_0$-reversed configurations, as follows. We denote by T the time reversal operator; any operator obtained from another by reversing $\vec{B}_0$ is designated by an upper bar. Since $\overline{H}_0 = TH_0T^{-1}$ and $\overline{H}_A = TH_AT^{-1}$, if $H|\nu\rangle = (H_0 + H_A)|\nu\rangle = E_\nu|\nu\rangle$, then the eigenvector of $\overline{H}$ with eigenvalue E_ν is $|T\nu\rangle$. The transition $|T\nu\rangle \rightarrow |T\nu'\rangle$ with $\vec{B}_0$ reversed occurs at the same energy as $|\nu\rangle \rightarrow |\nu'\rangle$ in the original configuration. Thus, upon $\vec{B}_0$-reversal, we must consider the electric dipole and magnetic dipole transition matrix elements betwenn $|T\nu\rangle$ and $|T\nu'\rangle$. Because $T(ED)T^{-1} = -(\overline{ED})$ and $T(MD)T^{-1} = (\overline{MD})$, we finally have

(6) $$|\langle T\nu'|\overline{ED} + \overline{MD}|\nu T\nu\rangle| = |\langle \nu'| - ED + MD|\nu\rangle|.$$

The situation for hole transitions is analogous, even though the four-fold degeneracy of the valence band brings about a few qualitative changes. The hole Hamiltonian (H^h) is conveniently expressed in terms of the $J = \frac{3}{2}$ angular momentum matrices. In lowest order approximation the hole Hamiltonian is [8]

$$(7) \qquad H_0^h = -\frac{\hbar^2}{2m}\left[\left(\gamma_1 + \frac{5}{2}\overline{\gamma}\right)k^2 - \frac{1}{2}\overline{\gamma}\sum_{i,j}\{k_i, k_j\}\{J_i, J_j\}\right] - \frac{e\hbar}{mc}K\vec{B}_0 \cdot \vec{J},$$

where $\{A, B\} = (AB + BA)$ and $\gamma_1, \overline{\gamma}$ and K are material parameters (just as m^* and g_0 for the electron case). We notice that H_0^h is even under parity. The first correction to H_0^h has the form [9]

$$(8) \qquad H_A^h = \frac{C}{\sqrt{3}}[\{J_x, J_y^2 - J_z^2\}k_x + \text{c.p.}],$$

C being a material constant. H_A^h is therefore odd. For holes, H_0^h and H_A^h play much the same role as H_0 and H_A did for electrons. The most significant difference is that while H_A is cubic in k, H_A^h is linear. As a consequence, while for electrons the ratio of the electric dipole and magnetic dipole spin-flip amplitudes is independent of the magnitude of $\vec{B}_0$, for holes this ratio can be changed by tuning the magnitude of $\vec{B}_0$ and, of course, the frequency of the light. In fact, for both electrons and holes $MD \propto q \propto B_0$, at resonance; but, while for electrons $ED \propto k^2 \propto R_0^{-2} \propto B_0$ (R_0 is the usual Landau lenght), for holes ED is independent of k and, thus, of B_0. Therefore, by varying B_0, it is possible in this case, to achieve completely destructive (or constructive) interference between electric dipole and magnetic dipole amplitudes.

Acknowledgments

This paper is dedicated to Professor Luigi A. Radicati di Brozolo on the occasion of his seventieth anniversary. It is a pleasure to thank him for his kind hospitality at the Scuola Normale Superiore in Pisa and for his stimulating comments on the subject of this work.

REFERENCES

[1] M. Dobrowolska - Y.-F. Chen - J.K. Furdyna - S. Rodriguez, Phys. Rev. Lett. **51**, 134 (1983).

[2] Y.-F. Chen - M. Dobrowolska - J.K. Furdyna - S. Rodriguez, Phys. Rev. **B32**, 890 (1985).

[3] S. Gopalan - J.K. Furdyna - S. Rodriguez, Phys. Rev. **B32**, 903 (1985).

[4] Namme Kim - G.C. La Rocca - S. Rodriguez, Phys. Rev. **B40**, 15 August 1989.

[5] F. Bassani - G.C. La Rocca - S. Rodriguez, Phys. Rev. **B37**, 6857 (1988); G.C. La Rocca - S. Rodriguez - F. Bassani, Phys. Rev. **B38**, 9819 (1988).

[6]   E.I. RASHBA - V.I. SHEKA, Sov. Phys. Solid State **3**, 1257 (1961); E.I. RASHBA -
      V.I. SHEKA, Sov. Phys. Solid. State **3**, 1357 (1961).

[7]   For example, we note that $(1 + 0.1)^2 = 1.21$ and $(1 - 0.1)^2 = 0.81$. Thus a
      difference of 10% in the interfering amplitudes gives rise to a 50% change in
      intensity.

[8]   J.M. LUTTINGER, Phys. Rev. **102**, 1030 (1956); A. BALDERESCHI - N.O. LIPARI, Phys.
      Rev. **B8**, 2697 (1973).

[9]   G. DRESSELHAUS, Phys. Rev. **100**, 580 (1955); E.O. KANE, J. Phys. Chem. Solids
      1, 249 (1957).

Deterministic Turbulence

PETER D. LAX

1. - Preface

Modern physics and mathematics are the twin offspring of Isaac Newton. Not identical and often beset by sibling rivarly, they maintained fraternal relations and thrived. The great mathematicians and physicist of the 18th and 19th centuries often made distinctive contributions to both subjects. It is only in our century that a significant group of mathematicians thought that mathematics is now completely free of its origins, and henceforth is governed by its own criteria of depth, and an internal compass to guide its future growth.

Such a view of mathematics was formulated by Bourbaki, and for a while, in the thirties, forties and fifties, dominated mathematical thought in France and the United States. Happily, this view of mathematics is rapidly disappearing, thanks mainly to brilliant discoveries in other fields that have invigorated mathematics with new ideas, new problems and new tools. Not surprisingly, many of these ideas came from physics, such as the field theory of Yang-Mills, which was used by Simon Donaldson to show the existence of nonstandard differential structures in four dimensional Euclidean space, or the Kadomtsev-Petviashvili equation in the theory of water waves that turned out to be the key for solving a hundred year old problem of the purest mathematics about Riemann surfaces. Faddeev and Pavlov have shown how to use the notions of scattering theory to study automorphic functions; and string theory, in the hands of E. Witten, has led to an infinite dimensional version of the Lefschetz fized point formula.

Mathematicians were not the only ones responsible for the temporary estrangement between mathematics and physics; physicists played a role also; of course they never felt that they could get along without mathematics, only that they could get along without mathematicians. Richard Feynman was a vocal advocate of this jaundiced view, more often in public than in private, and he was not alone. Friedrichs has described a chance meeting with Heisenberg during which he conveyed the gratitude of mathematicians to the creators of quantum mechanics for breathing life into the theory of operators in Hilbert space; Heisenberg allowed that this was so. Friedrichs then expressed his hope that mathematicians have, at least partly, repaid this debt. When Heisenberg looked puzzled, Friedrichs added that after all it was a mathematician who clarified the difference between a self adjoint operator and one that was merely symmetric. "What's the difference?", said Heisenberg.

This, too, is fading; physicists are much more receptive to mathematicians' way of looking at problems; Martin Kruskal, the creator of soliton theory, is after all a mathematician.

One of the institutions where the close relation of mathematics and physics was never broken is the Scuola Normale in Pisa; this is due in large measure to the wise guidance of Luigi Radicati, to whom this study is respectfully offered.

2. - Introduction

I have chosen the phrase *deterministic turbulence* to describe one possible route to turbulence. The examples presented below go that route, but are fueled by different mechanisms; so they are more analogous to turbulence.

There are - at least - three approaches to turbulence. One is to regard turbulent flow as an ensemble of flows created by a random field of force acting on an unstable or barely stable flow-field governed by some differential equation. The task for the theorist is:

a)   To prove that the variables of physical interest - velocity, pressure, density - have expected values.

b)   To derive laws governing the evolution of these expected values. These laws will involve expected values of functions of velocity, pressure, density and of their translates.

c)   To find approximations to the exact equations of evolution, accurate enough for engineering purposes, under conditions that can be ascertained, which involve only a finite number of variables.

Another approach to turbulence rests on the surmise - supported by experimental observations and computer experiments, and some mathematicial analysis - that turbulence commences when a hitherto smooth solution of the equations of fluid dynamics develops singularities. It is possible to generalize the notion of solution in way that makes sense as mathematics and physics, so that smooth solutions can be extended as generalized solutions beyond the time when smoothness breaks down. However, it could very well happen that extension is very far from being unique. In this case it is the task of the theorist to borrow or invent a principle of statistical mechanics to form a weighted average over all possible extensions, obtaining in this fashion expected values for velocity, pressure and density. The two further tasks b) and c) remain as before.

A third approach to turbulence is based on the physically significant parameters, such as Reynolds number, coefficient of heat conduction, etc. that appear in the equation. According to this scenario, as one of these parameters approaches a critical value, solutions become more and more oscillatory, with shorter and shorter wave lengths and with amplitudes that do not tend to zero. Such as a sequence does not converge in the ordinary i.e. strong sense. In this case the task of the theorist is to show that these solutions converge in a weak sense, such as the sense of distributions. These weak limits play the role of expected values of variables of physical interest. The two further tasks b) and c) remain as before.

In the next section we present three examples of equations whose solutions follow the third scenario. They are similar to each other and different from fluid dynamics, inasmuch as the physical parameter that tends to the critical value zero is not viscosity but dispersion.

3. - Examples of deterministic turbulence

A. The first example is the zero dispersion limit of the KdV equation, studied first by Lax and Levermore, and its more subtle aspects by Venakides. The equation is

$$u_t + u u_x + \epsilon^2 u_{xxx} = 0$$

subject to initial conditions

$$u(x, 0) = u_0(x).$$

The initial function is assumed smooth and to tend rapidly to zero as x tends to $\pm\infty$. The cases when $u_0(x)$ tends to different values as x tends to $+\infty$ and $-\infty$, and when $u_0(x)$ is periodic have similar resolution.

Denote the solution of the above problem by $u^{(\epsilon)}(x, t)$; what happens when ϵ tends to zero? As long as the limiting equation

$$u_t + u u_x = 0$$

has a smooth solution with initial value $u_0(x)$, $u^{(\epsilon)}(x, t)$ converges uniformly to that smooth solution. There is however a critical time t_0 whose value is determined by u_0 when the smooth solution develops singularities. Beyond that time $u^{(\epsilon)}(x, t)$ is oscillatory; more precisely, the x, t space is, for $t > t_0$, divided into two regions. In one of them, call it R_0, $u^{(\epsilon)}$ is uniformly smooth and converges weakly as ϵ tends to zero uniformly to a limit u that satisfies the limiting equation. In the other region $u^{(\epsilon)}$ is oscillatory, with wave length $O(\epsilon)$, and amplitude bounded away from 0. In this region $u^{(\epsilon)}$ converges as ϵ tends to zero to a limit $\bar{u}$.

Rewrite the KdV equation in conservation from:

$$u_t + \frac{1}{2}(u^2)_x + \epsilon^2 u_{xxx} = 0.$$

Taking the limit of this equations in the sense of distributions we obtain

$$\overline{u}_t + \frac{1}{2}(\overline{u^2})_x = 0$$

where $\overline{u^2}$ is the weak limit of u^2. Since in the oscillatory region $\overline{u^2} > \overline{u}^2$, we see that $\overline{u}$ does *not* satisfy the limit equation.

What equation does $\overline{u}$ satisfy? The oscillatory region can be divided into subregions $R_k, k = 1, 2, \cdots$, such that in each R_k the weak limit $\overline{u}$ can be expressed in terms of $2k + 1$ functions that together satisfy a system of hyperbolic equations of which they are the Riemann invariants. Any bounded time domain $t \leq T$ intersects only a finite number of regions R_k.

We know all this because the KdV equation is completely integrable, see [6], [7] and [8] for details. Thus in this case all the tasks a), b) and c) can be carried out successfully.

B. For small ϵ the KdV equation is an approximation of $u_t + u u_x = 0$. The second example also is an approximation of that equation, a semidiscrete one on a spatial lattice with mesh width Δ. Denoting the approximation to $u(k\Delta, t)$ by $u_k(t)$ we impose on u_k the differential equation

$$\frac{d}{dt}u_k + u_k \frac{u_{k+1} - u_{k-1}}{2\Delta} = 0,$$

and the initial condition

$$u_k(0) = u_0(k\Delta).$$

Note that by Taylor's theorem

$$\frac{u_{k+1} - u_{k-1}}{2\Delta} \simeq u_x + \frac{\Delta}{6}u_{xxx},$$

so that the differential difference equation looks like

$$u_t + u u_x + \frac{\Delta^2}{6}u\, u_{xxx} \simeq 0,$$

very much KdV-ish.

Denote by $u^{(\Delta)}(x, t)$ the function whose value in the mesh interval centered at $k\Delta$ equals $u_k(t)$. Goodman and Lax in [5] have investigated the behavior of $u^{(\Delta)}$ as Δ tends to zero, and found behavior very similar to $u^{(\epsilon)}$ when ϵ tends to zero in example A. The evidence presented there is mostly numerical, but as we point out, the system of ordinary differential equations satisfied by the u_k is *completely integrable*, and so a rigorous analysis is possible.

C. The last example is a full fledged difference approximation to the Lagrange form of the equations of compressible flow:

$$u_t + p_x = 0,$$

$$V_t - u_x = 0,$$

$$e_t + p u_x = 0.$$

Here u is velocity, p pressure, V specific volume and e specific internal energy. An equation of state connects e, p and V. The first two equations are the laws of conservation of momentum and mass, the third the work equation.

In 1944 v. Neumann set up difference equations as approximations to the differential equations in the following fashion. The termhodynamic quantities p, V and e were associated with the center of each mesh cell:

$$p_k^n, V_k^n \text{ and } e_k^n$$

denote approximations to p, V and e at $x = k\Delta x, t = n\Delta t$. Velocity is associated with cell boundaries:

$$u_{k+1/2}^{n+1/2}$$

denotes approximation to u at $x = (k + 1/2)\Delta x, t = (n + 1/2)\Delta t$. This choice permits the replacement of the differential equations by difference equations that are *centered*, i.e. where the truncation error is of third order in $\Delta x, \Delta t$. Using these difference equations and punched card equipment v. Neumann solved a simple initial value problem whose explicitly known solution contained a shock. The solutions of the difference equations were oscillatory behind the shock; v. Neumann suggested that these oscillations in velocity correspond to the irreversible conversion of kinetic energy into internal energy. He further conjectured that as Δt and Δx tend to zero so that $\frac{\Delta t}{\Delta x} = \lambda$ is kept constant, these oscillatory solutions tend in the weak sense to exact discontinuous solutions of the equations of compressible flow.

In [4] I counterconjectured that the weak limit does *not* satisfy the equations of compressible flow, very much in analogy with the weak limits of solutions of the KdV equations. Although the counterconjecture is very plausible, it seems difficult to prove. Instead, Tom Hou and the author have carried out a series of carefully controlled calculations which reveal that solutions of the difference equation with prescribed initial values tend to the solution of the differential equations of compressible flow with the same initial data as long as such a solution exists and is smooth. However for times larger than that, the approximate solutions are oscillatory in certain regions of the x, t plane. In these regions the solutions converge weakly as $\Delta x \to 0, \Delta t/\Delta x = \lambda$. The weak limits do not satisfy the equations of compressible flow.

BIBLIOGRAPHY

[1]   H. FLASCHKA - M.G. FOREST - D.W. MCLAUGLIN, *Multiphase averaging and the inverse spectral solution of the Koreweg-de Vries equation*, Comm. Pure Appl. Math. 33, 739-784 (1980).

[2]   H. FLASCHKA, *The Toda Lattice I*, Phys, Rev. **B9**, (1974) 1924-1925; II, Progr. Theoret. Phys. **51**, 703-716 (1974).

[3]   T. HOU - P.D. LAX, *Dispersive Approximations in Fluid Dynamics*, to appear, (1989).

[4]   P.D. LAX, *On Dispersive Difference Schemes*, Physica 180, North-Holland, Amsterdam, 250-252, (1986).

[5]   P.D. LAX - J. GOODMAN, *On Dispersive Difference Schemes I*, Comm. Pure Appl. Math. **41** 591-613, (1988).

[6]   P.D. LAX - C.D. LEVERMORE, *The Small Dispersion Limit of the KdV Equation*, Comm. Pure Appl. Math., **36**, I 253-290, II 571-593, III 809-830, (1980).

[7]   S. VENAKIDES, *The Zero dispersion limit of the Periodic KdV Equation* AMS Transaction, Vol. 301, 189-226, (1987).

[8]   S. VENAKIDES, *High Order Lax-Livermore Theory I*, to appear, Comm. Pur and Appl. Math. (1989).

The s-Channel Theory of Superconductivity [†]

T. D. LEE (*)

Foreward

Luigi Radicati is a theoretical physicist of exceptional power, and a gentle person of great integrity. Our friendship has spanned more than three decades. It is with warmth and appreciation that I dedicate this paper to him.

1. - Introduction

In the usual "cold" superconductors, one has, as in the BCS theory,

(a) the coherence length ξ varying from about 10^4Å for type 1 to about 300Å for type 2;

(b) the critical temperature T_c approximately proportional to the inverse of the square root of the isotope mass;

(c) the gap energy $\Delta \sim 1.76\kappa T_C \sim 10^{-3} eV$.

The more recently discovered "warm" superconductors [1,2] have different characteristics:

(a') a much shorter [3,4] coherence length ξ, only ~ 10Å, the same order as the size of the lattice cell,

(b') no isotope effect (but T_c increases with the carrier density), and

(c') $\Delta \sim \kappa T_c \sim 10^{-2} eV$.

In this paper, we discuss the basic features of a new theory [5] of superconductivity, based essentially *only* on the experimental input (a').

The observation of such a small "coherence length" ξ indicates that the pairing between electrons, or holes, in warm superconductors is reasonably localized in the coordinate space. Hence, the pair-state can be well approximated

[†]   This research was supported in part by the U.S. Department of Energy.

(*)   Columbia University, New York, N.Y. 10027

by a phenomenological local boson field $\phi(\vec{r})$, whose mass is $\approx 2m_e$ and whose elementary charge unit is $2e$, where m_e and e are the mass and charge of an electron. It follows then that the transition

(1.1) $2e \rightarrow \phi \rightarrow 2e$

must occur, in which e denotes either an electron or a hole; furthermore, the localization of ϕ implies that phenomena at distances larger than the physical extension of ϕ (which is $O(\xi)$) are insensitive to the interior of ϕ. Since ξ is of the same order as the scale of a lattice unit cell, it becomes possible to develop a phenomenological theory of superconductivity based *only* on the local character of ϕ.

Of course, physics at large does depend on several overall properties: the spin of ϕ, the stability of an individual ϕ-quantum, the isotropicity and homogeneity (or their absence) of the space containing ϕ and so on. The situation is analogous to that in particle physics: the smallness of the radii of pions, ρ-mesons, kaons, $\cdots$ makes it possible for us to handle much of the dynamics without any reference to their internal structure, such as quark-antiquark pairs or bag models. Hence, the origin of their formation becomes a problem separate from the description of their mechanics. An important ingredient in this type of phenomenological approach is the selection of the basic interaction Hamiltonian that describes the underlying dominant process. As noted above in (a), in the usual low-temperature superconductors, ξ varies from about 10^4 to a few hundred Å. The corresponding pairing state ϕ is too extended and ill-defined in the coordinate space; therefore, (1.1) does not play an important role. Instead, the BCS theory of superconductivity is based on the emission and absorption of phonons,

(1.2) $2e \rightarrow 2e + \text{phonons} \rightarrow 2e$.

The dependence of photons vibrations on the isotope mass and the relative weakness of the electron-phonon coupling form the basis for (b) and (c) in the usual "low temperature" superconductors, as is well-known. In contrast, reaction (1.1) is not sensitive to lattice vibrations and, because of its resonance character, the coupling would be stronger. Hence, any theory based on (1.1) can be made consistent with above (b') and (c') for the "high temperature superconductors".

In the language of particle physics, (1.1) is an s-channel process, while (1.2) is t-channel. The BCS theory may be called the t-channel theory. As we shall see, the s-channel reaction (1.1) leads to a different theory (which, however, shares many features in common with the BCS theory) of superconductivity, whose validity rests only on the localization of ϕ, and is independent of the detailed microscopic origin of the pairing mechanism; in addition, its long-range order can be represented by the macroscopic occupation number of the zero-momentum bosons, as in the Bose-Einstein condensation. Together, these two (s-channel and t-channel) formulations provide a rich body of theoretical means, which may prove useful in analysing the large variety of superconductivity and superfluidity phenomena that exist in nature.

The use of a boson field for the superfluidity of Liquid $HeII$ has had a long history. However, there are some major differences in the following application to (high temperature) superconductors:

1. The ϕ-quantum is charged, carrying $2e$, while the helium atom is neutral.

2. We assume each individual ϕ-quantum to be unstable, with 2ν as its excitation energy. (As we shall see, this assumption makes it possible for the s-channel theory to exhibit many BCS-like characteristics, yet without the isotope effect.)

In the rest frame of a single ϕ-quantum, the decay

$$(1.3) \qquad\qquad \phi \rightarrow 2e$$

occurs, in which each e carries an energy

$$\frac{k^2}{2m} = \nu.$$

Consequently, in a large system, there are macroscopic numbers of both bosons (the ϕ-quanta) and fermions (electrons or holes), distributed according to the principles of statistical mechanics.

At temperature $T < T_c$, there is always a macroscopic distribution of zero momentum bosons co-existing with a Fermi distribution of electrons (or holes). Take the simple example of *zero* temperature: Let ϵ_F be the Fermi energy. When $\epsilon_F = \nu$, the decay $\phi \rightarrow 2e$ cannot take place because of the exclusion principle; therefore, the bosons are present. Even when $\epsilon_F < \nu$, there is still a macroscopic nuber of (virtual) zero momentum boson in the form of a static coherent field amplitude whose source is the fermion pairs. This then leads to the following essential features of the s-channel model.

Below the critical temperature T_c the long range order in the boson field can always be described by its zero-momentum bosonic amplitude B, as in the Bose-Eistein condensation (and therefore similar to liquid $HeII$). Because of the transition (1.1), the zero-momentum of the boson in the condensate forces the two e to have equal and opposite momenta, forming a Cooper pair. Therefore, the same long range order also applies to the Cooper pairs of the fermions. Furthermore, as we shall see, the gap energy Δ of the fermion system is related to B by

$$(1.4) \qquad\qquad \Delta^2 = |gB|^2$$

where g is the coupling for $\phi \rightarrow 2e$.

2. - A prototype s-channel model

As a prototype of the s-channel theory of superconductivity, we assume ϕ to be of spin 0 and that the space containing ϕ is a three-dimensional homogeneous and isotropic continuum. For realistic applications, a more appropriate approximation of the latter would be the product of a two-dimensional x, y-continuum (simulating the CuO_2 plane) and a discrete lattice of spacing c along the z-direction. The two- dimensional layer character of CuO_2 planes helps in the localization of the pair state in the z-direction, making the ϕ-quantum disc-shaped. The space that ϕ moves in becomes a three-dimensional continuum when $c \rightarrow 0$, but two-dimensional when $c \rightarrow \infty$. This interesting case, plus the generalization to higher spin, will be discussed elsewhere.

Here we consider an idealized system consisting of the local scalar field ϕ and the electron (or hole) field ψ_σ where $\sigma = \uparrow$ or $\downarrow$ denotes the spin. The Hamiltonian is $(\hbar = 1)$

$$(2.1) \qquad\qquad H = H_0 + H_1$$

in which the free Hamiltonian is

$$(2.2) \qquad H_0 = \int \left[\phi^\dagger (2\nu_0 - \frac{1}{2M}\nabla^2)\phi + \psi_\sigma^\dagger (-\frac{1}{2m}\nabla^2)\psi_\sigma \right] d^3r$$

with the repeated spin index σ summed over and $\dagger$ denoting the hermitan conjugate. The interaction H_1 can be simply

$$(2.3) \qquad\qquad H_1 = g \int (\phi^\dagger \psi_\uparrow \psi_\downarrow + \text{h.c.}) d^3r.$$

Both ϕ and ψ_σ are the usual quantized field operators whose equal-time commutator and anticommutator are

$$[\phi(\vec{r}), \phi^\dagger(\vec{r}')] = \delta^3(\vec{r} - \vec{r}')$$

and

$$\{\psi_\sigma(\vec{r}), \psi_{\sigma'}^\dagger(\vec{r}')\} = \delta_{\sigma\sigma'}\delta^3(\vec{r} - \vec{r}').$$

The total particle number operator is defined to be

$$(2.4) \qquad\qquad N = \int (2\phi^\dagger \phi + \psi_\sigma^\dagger \psi_\sigma) d^3r$$

which commutes with H and is therefore conserved.

Expand the field operators in Fourier components inside a volume Ω with periodic boundary conditions:

$$\psi_\sigma(\vec{r}) = \sum_k \Omega^{-\frac{1}{2}} a_{\vec{k},\sigma} e^{i\vec{k}\cdot\vec{r}}$$

and

(2.5)
$$\phi(\vec{r}) = \sum_k \Omega^{-\frac{1}{2}} b_{\vec{k}} e^{i\vec{k}\cdot\vec{r}}$$

with $\{a_{\vec{k},\sigma}, a^\dagger_{\vec{k}',\sigma'}\} = \delta_{\vec{k},\vec{k}'}\delta_{\sigma\sigma'}, [b_{\vec{k}}, b^\dagger_{\vec{k}'}] = \delta_{\vec{k},\vec{k}'}$, etc. Equation (2.3) can then be written as

(2.6)
$$H_1 = \frac{g}{\sqrt{\Omega}} \sum_{p,k} \left[b^\dagger_p a_{\frac{p}{2}+\vec{k},\uparrow}\, a_{\frac{p}{2}-\vec{k},\downarrow} + \text{h.c.} \right].$$

In (2.2), $2\nu_0$ is the "bare" excitation energy of ϕ. Because of the interaction, the "physical" (i.e., renormalized) excitation energy 2ν in reaction (1.3) is given by

(2.7)
$$2\nu = 2\nu_0 + \frac{g^2}{2\Omega} \sum_k P \frac{1}{\nu - \omega_k}$$

where P denotes the principal value and

(2.8)
$$\omega_k = \frac{k^2}{2m}.$$

The decay width Γ of a ϕ-quantum (in vacuum) is given by

(2.9)
$$\Gamma = (g^2/\pi) m^{\frac{3}{2}} \sqrt{\frac{\nu}{2}}.$$

3. - Gap Energy

For $T < T_c$, the zero-momentum occupation number $b^\dagger_0 b_0$ of the boson field ϕ becomes macroscopic; hence, we may replace the operator b_0 by a macroscopic constant. Set in (2.5)

$$\Omega^{-\frac{1}{2}} b_0 = B = c. \text{ number}$$

and write

(3.1)
$$\phi = B + \phi_1$$

where

(3.2)
$$\phi_1 = \sum_{k\neq 0} \Omega^{-\frac{1}{2}} b_{\vec{k}} e^{i\vec{k}\cdot\vec{r}}.$$

In the following, we shall treat the effects of ϕ_1 perturbatively. Let μ be the chemical potential. Introduce

(3.3)
$$\mathcal{H} \equiv H - \mu N = \mathcal{H}_0 + \mathcal{H}_1$$

where

(3.4)
$$\mathcal{H}_0 = \sum_k \left\{ \left[\frac{k^2}{2M} + 2(\nu_0 - \mu) \right] b_{\vec{k}}^\dagger b_{\vec{k}} + (\omega_k - \mu) a_{\vec{k},\sigma}^\dagger a_{\vec{k},\sigma} \right.$$
$$\left. + g[B^* a_{\vec{k},\uparrow} a_{-\vec{k},\downarrow} + B a_{-\vec{k},\uparrow}^\dagger a_{\vec{k},\downarrow}^\dagger] \right\}$$

and

(3.5)
$$\mathcal{H}_1 = g \int (\phi_1^\dagger \psi_\uparrow \psi_\downarrow + \text{h.c.}) d^3 r$$

is the perturbation.

The zeroth-order Hamiltonian $\mathcal{H}_0$ is quadratic in field operators, and can therefore be readily diagonalized. Its fermion-dependent part can be written as a sum of matrix products, each of the form:

$$(a_{\vec{k},\uparrow}, a_{-\vec{k},\downarrow}^\dagger) \cdot A \cdot \begin{pmatrix} a_{\vec{k},\uparrow}^\dagger \\ a_{-\vec{k},\downarrow} \end{pmatrix}$$

where A is a 2×2 matrix

$$A = \begin{pmatrix} -(\omega_k - \mu) & gB^* \\ gB & \omega_k - \mu \end{pmatrix}.$$

Because of Fermi statistics, the two diagonal elements of A have opposite signs. The eigenvalue E_k of A is determined by

$$E_k^2 - (\omega_k - \mu)^2 = g^2 |B|^2;$$

i.e.,

(3.6)
$$E_k = [(\omega_k - \mu)^2 + g^2 |B|^2]^{\frac{1}{2}}.$$

Thus, we establish the formula for the gap energy (1.4):

$$\Delta^2 = g^2 |B|^2.$$

4. - Thermodynamics

The zeroth-order grand canonical partition function is

$$\mathcal{Q} = \text{trace } e^{-\beta \mathcal{H}_0},$$

whose logarithm is $p\Omega/\kappa T$, where $\beta = (\kappa T)^{-1}$, $\kappa = $ the Boltzmann constant, and p is the pressure. By using the diagonalized form of $\mathcal{H}_0$, we find

(4.1)
$$
\begin{aligned}
p = & - 2(\nu_0 - \mu)|B|^2 + \Omega^{-1} \sum_k (E_k + \mu - \omega_k) \\
& + 2(\beta\Omega)^{-1} \sum_k \ln(1 + e^{-\beta E_k}) \\
& - (\beta\Omega)^{-1} \sum_k \ln\{1 - \exp \beta[2\mu - 2\nu - (k^2/2M)]\}.
\end{aligned}
$$

In accordance with the general thermodynamical principle, at constant T and μ, the function p should be a maximum with respect to any internal parameter, sich as $|B|$. Setting $(\partial p/\partial |B|)_{\mu,T} = 0$, we have

$$\nu_0 - \mu - \Omega^{-1}\frac{g^2}{4} \sum_k \frac{1}{E_k} \tanh \frac{1}{2}\beta E_k = 0.$$

By using (2.7), we may express the above formula in terms of the physical excitation energy 2ν of the ϕ-quantum:

(4.2)
$$\nu - \mu = \Omega^{-1}\frac{g^2}{4} \sum_k \left[\frac{1}{E_k} \tanh \frac{1}{2}\beta E_k + P\frac{1}{\nu - \omega_k} \right],$$

where P denotes the principal value, as before. The right-hand side is convergent in the.ultra-violet region since the theory is renormalizable. The particle density ρ is given by $(\partial p/\partial\mu)_{T,B}$, which yields

(4.3)
$$
\begin{aligned}
\rho = & 2|B|^2 + 2\Omega^{-1} \sum_k [e^{\beta(2\nu + (k^2/2M) - 2\mu)} - 1]^{-1} \\
& + \Omega^{-1} \sum_k [E_k(1 + e^{-\beta E_k})]^{-1}[E_k + \mu - \omega_k + (E_k - \mu + \omega_k)e^{-\beta E_k}].
\end{aligned}
$$

From (4.2) and (4.3), μ and $|B|^2$ can be determined as functions of ρ and T. (Equation (4.2) is similar to the gap equation in the BCS theory, and Eq. (4.3) is the generalization of the density equation in the Bose-Einstein condensation.)

Regarding (4.1)-(4.3) as the zeroth approximation, one can develop a systematic expansion using $\mathcal{H}_1$ of (3.5) as the perturbation.

It is useful to introduce

$$(4.4) \qquad \rho_\nu \equiv (3\pi^2)^{-1}(2m\nu)^{\frac{3}{2}},$$

the fermionic density when the Fermi-energy equals ν, with the excitation energy of the ϕ-quantum $= 2\nu$. For $\rho << \rho_\nu$, one finds that the gap energy Δ_0 at zero temperature is related to the critical temperature T_c by, as in the BCS theory,

$$(4.5) \qquad \frac{\Delta_0}{\kappa T_c} = \pi e^{-\gamma} = 1.7639$$

where $\gamma = $ Euler's constant $= 0.5772$. For $\rho >> \rho_\nu$

$$(4.6) \qquad \Delta_0^2 = (2.612)g^2 \left(\frac{M\kappa T_c}{2\pi}\right)^{\frac{3}{2}}$$

and (at any temperature $T < T_c$)

$$(4.7) \qquad \Delta^2(T) = g^2|B(T)|^2 = \Delta_0^2 \left[1 - \left(\frac{T}{T_c}\right)^{\frac{3}{2}}\right],$$

as in the Bose-Einstein condensation.

A detailed study of (4.2) and (4.3) shows that typical BCS and Bose-Einstein formulas can be analytically connected within one single expression. In this way, these two approaches become closely unified. The s-channel theory has an intrinsically simpler structure than the t-channel theory; this makes it possible to take a deductive approach, thereby rendering the analysis attractive on the pedagogical level.

REFERENCES

[1]   J. C. BEDNORZ - K. A. MÜLLER, Z. Phys. **B64**, 189 (1986).

[2]   M. K. WU - J. R. ASHBURN - C. J. TORNG - P. H. HOR - R. L. MENG - L. GAO - Z. J. HUANG - Y. Q. WANG - C. W. CHU, Phys. Rev. Lett. **58**, 908 (1987). Z. X. ZHAO - L. CHEN - Q. YANG - Y. HUANG - G. CHEN - R. TANG - G. LIU - C. CUI - L. WANG - S. GUO - S. LIN - J. BI, Kexue Tongbao **6**, 412 (1987).

[3]   P. CHAUDHARI et al., Phys. Rev. **B36**, 8903 (1987).

[4]   T. K. WORTHINGTON - W. J. GALLAGHER - T. R. DINGER, Phys. Rev. Lett. **59**, 1160 (1987).

[5]   R. FRIEDBERG - T. D. LEE, Phys. Lett. **A138**, 423 (1989); "Gap Energy and Long-Range Order in the Boson-Fermion Model of Superconductivity", Preprint CU-TP-431, to be published in Phys. Rev. B.

Is CP Violation a General Property
of Weak Interactions?

I. MANNELLI (*)

Shortly after Luigi Radicati came to the Physics Institute of the University of Pisa, T.D. Lee and C.N. Yang published their famous paper on non conservation of parity in weak interactions. The flurry of experimental activity which followed showed very quickly that indeed space reflection symmetry was broken in weak transitions involving neutrinos. When, a few months later, I started working on my thesis there was still a crucial question to be answered: i.e. is parity violation a general property of weak interactions or is it exclusively due to neutrinos possessing a definite helicity? From the analysis of the production and subsequent hadronic decay of Λ hyperons in bubble chamber it was possible to give a clear answer. To find myself in the position to be able to contribute to elucidate such a question was a very exciting early experience. It is perhaps therefore not surprising that, coming back after about 30 years to the study of symmetry properties of weak interactions, I have concentrated on trying to solve experimentally a similar issue, this time related to their behaviour under CP, the combined operation of charge conjugation and parity.

It is a pleasure to dedicate to Luigi Radicati the description which follows of the principle of the $NA31$ [9] experiment at CERN and of its present results, in appreciation, as a previous student and later on as colleague and friend, of his outstanding intellectual and human qualities.

Direct CP Violation in $K \rightarrow 2\pi$

As it is well known, all experimental results concerning CP violation up to 1987, which followed the original discovery [1] that both K_S and K_L decay into the same two pion $CP = +$ final state, can be reconduced to a single fact, i.e. the amplitude for the transition $K^0 \rightarrow \overline{K}^0$ is different from the amplitude for $\overline{K}^0 \rightarrow K^0$. In the superweak model [2], linking CP violation to $|\Delta S| = 2$, transition the K_L decay to two pions is due only to the small

(*)   Scuola Normale Superiore, Pisa (Italy)

impurity ε of the eigenstate K_1 with $CP = +$ in the composition of the $K_L = (K_2 + \varepsilon K_1) \cdot (1 + |\varepsilon|^2)^{-1/2}$ itself (K_2 is the eigenstate with $CP = -$), and A_0 and A_2, the weak amplitudes for K^0 decay to two pions in isotopic spin $I = 0$ and $I = 2$, are relatively real. The ratios of the amplitudes for K_L and K_S decay to two pions

$$\eta_{+-} = \langle \pi^+ \pi^- |T| K_L \rangle / \langle \pi^+ \pi^- |T| K_S \rangle = \varepsilon$$

and

$$\eta_{00} = \langle \pi^0 \pi^0 |T| K_L \rangle / \langle \pi^0 \pi^0 |T| K_S \rangle = \varepsilon$$

are hence both equal to ε, and $\eta_{+-}/\eta_{00} = 1$. The phase of ε is determined via unitarity from the mass and total decay rate of K_S and K_L. To a good approximation $\arg(\varepsilon) = \tan^{-1} 2(m_L - m_S)/(\Gamma_S - \Gamma_L) = 43.7°$.

In more general case, where the so called direct - CP-violating amplitudes could be present in the K two-pion decay, again to a good approximation:

$$\eta_{00} = \varepsilon - 2\varepsilon', \qquad \eta_{+-} = \varepsilon - \varepsilon'$$

with $\varepsilon' = i/\sqrt{2} \operatorname{Im}(A_2/A_0) \exp[i(\delta_2 - \delta_0)]$, where δ_0 and δ_2 are the $\pi\pi$ phase shifts for $I = 0$ and $I = 2$, at $\sqrt{s} = m_K$. Experimentally $\arg(\varepsilon') = \delta_2 - \delta_0 + \pi/2$ differs from $\arg(\varepsilon)$ by no more than $20°$ [3]. Therefore taking ε' and ε as relatively real and for $|\varepsilon'/\varepsilon| \ll 1$

$$|\eta_{00}/\eta_{+-}|^2 = 1 - 6 \, \varepsilon'/\varepsilon.$$

It was first realized in 1973 [4] that it is possible to incorporate CP violation into the standard electroweak gauge theory if (at least) three generations of quarks exist, as it now appears very plausible. From this point of view only an accidental cancellation could produce $\varepsilon' = 0$. Also, stemming from the single CP violating phase appearing in the quark mixing matrix, other CP-violation phenomena could be expected, in particular for the neutral B mesons, for which $B^0 \to \overline{B}^0$ mixing has recently been observed by the $UA1$ Collaboration at the CERN Collider [5] and by ARGUS at DORIS [6].

The results of the first attempts to measure $|\eta_{00}/\eta_{+-}|$ are reported in Table 1 together with more accurate results published in 1985 [7].

The main features of the 1985 experiments are summarized in Table 2.

TABLE 1

Measurements of $\varepsilon'/\varepsilon = [1 - (|\eta_{00}/\eta_{+-}|^2)]/6$

Year	Authors	$|\eta_{00}/\eta_{+-}|$
1972	Holder et al.	1.00 ± 0.06
1972	Banner et al.	1.03 ± 0.07
1979	Christenson et al.	1.00 ± 0.09
Average		1.01 ± 0.04
$R = |\eta_{00}/\eta_{+-}|^2 = 1.02 \pm 0.08$		
Year	Authors	ε'/ϵ
1985	Black et al.	0.002 ± 0.007 ± 0.004
1985	Bernstein et al.	− 0.005 ± 0.005 ± 0.002
Average		− 0.003 ± 0.005
$R = |\eta_{00}/\eta_{+-}|^2 = 1.015 \pm 0.03$		

TABLE 2a

BNL-*Yale Experiment* (*Black et al., 1985*)

	No. of events	Background (%)
$K_L \rightarrow 2\pi^0$	1122	17.5 ± 3
$K_S \rightarrow 2\pi^0$	3317	1.2 ± 0.2
$K_L \rightarrow \pi^+\pi^-$	8506	2.0 ± 1
$K_S \rightarrow \pi^+\pi^-$	20960	

$\epsilon'/\epsilon = 0.0017 \pm 0.0082$

$|\eta_{00}/\eta_{+-}|^2 = R = 0.990 \pm 0.043 \pm 0.026$

Backgrounds: $K_L \rightarrow 3\pi^0 \rightarrow 6\gamma$

n interactions in He

K_L semileptonic decays

Incoherent K_S: 1.5 ± 0.5% of $K_S \rightarrow 2\pi^0$

0.2% of $K_S \rightarrow \pi^+\pi^-$

TABLE 2b

Chicago-Saclay Experiment (*Bernstein et al., 1985*)

	No. of events	Background (%)	Inelastics (%)	Systematics (%)
$K_L \rightarrow 2\pi^0$	3152	8.4 ± 0.4	2.9 ± 0.3	±0.6
$K_S \rightarrow 2\pi^0$	5663	0.6 ± 0.1	14.9 ± 0.5	±1.0
$K_L \rightarrow \pi^+\pi^-$	10638	3.1 ± 0.2	0.4 ± 0.1	±0.2
$K_S \rightarrow \pi^+\pi^-$	25751	0.2	1.7 ± 0.1	±0.1

Acceptance correction K_L/K_S for $2\pi^0$ ±0.5

for $\pi^+\pi^-$ ±0.2

Total systematic error   ±1.3

$\epsilon'/\epsilon = -0.0046 \pm 0.0053 \pm 0.0024$

$R = 1.028 \pm 0.032 \pm 0.014$

Main systematics:

uncertainties in background ($K_L \rightarrow 3\pi^0$, n interaction)

inelastic regeneration of K_S

acceptance correction (K_L upstream of K_S regenerator)

More recently [8] the Chicago-Fermilab-Princeton-Saclay Collaboration (experiment E731 at Fermilab) has given a result,

$$\varepsilon'/\varepsilon = 0.0032 \pm 0.0028 (\text{stat.}) \pm 0.0012 (\text{syst.}),$$

based on a sample of 6747 $2\pi^0$ decays, obtained from an improved set-up at Fermilab which is ultimately expected to yield some $100,000$ $2\pi^0$ decays and reach an accuracy of $\pm 1 \times 10^{-3}$ for ε'/ε.

At CERN the $NA31$ experiment has now produced the first evidence for direct CP violation [9]

$$\varepsilon'/\varepsilon = (3.3 \pm 1.1) \times 10^{-3}$$

and most of the remainder of this paper will be concerned with this experiment.

The principle is to measure

$$\varepsilon'/\varepsilon = (1 - R)/6,$$

where R is the double ratio of the decay rates

$$R = [\Gamma(K_L \to 2\pi^0)/\Gamma(K_L \to \pi^+\pi^-)]/\Gamma(K_S \to 2\pi^0)/\Gamma(K_S \to \pi^+\pi^-)].$$

Charged and neutral decays are detected at the same time, alternatively for K_L and K_S. Switching from K_L to K_S was done with a periodicity of about a day. The overall stability of the detection system was checked in several ways. In particular, the ratio of charged to neutral decays for K_S changed by less than $\pm 1.5\%$ over a time equal to 20 so-called miniperiods, i.e. sets of K_L and K_S data takings, each one sufficient to give a value of ε'/ε independently of any change of acceptance between miniperiods. Binning the events in momentum p_K and decay vertex longitudinal position Z_V, for small enough bins (10 GeV/c was used for p_K and 1.2 m for Z_V), the relative acceptances cancel out in the double ratio to better than 0.5%, so the ratio of the number of reconstructed events needs only a small correction, which can be reliably computed from geometrical features of the set-up, in order to represent the ratio of decay rates.

The experimental set-up [10] is sketched in Fig. 1, where transverse dimensions are enhanced by a factor of 50 relative to the longitudinal ones.

The K_L beam is produced by 450 GeV/c protons at 3.6 $mrad$. The K_L intensity is such that in total about 10^5 decays (to any final state) occur per SPS pulse, over the 120 m following the beginning of the useful decay region. Once collimated, **48 m** after the production target, the neutral beam remains in vacuum up to the dump which follows the whole set-up.

For K_S, an attenuated proton beam is brought to a target, located on a train which carries also the last beam deflecting magnets, the proton beam dump, and the K_S beam-defining collimator. The train can be moved in steps of 1.2 m over 48 m so as to imitate the flat K_L longitudinal decay distribution.

An anticoincidence counter, preceded by 7 *mm* of *Pb*, is normally inserted in the K_S beam to define precisely the beginning of the decay region.

As shown in Fig. 2 the evacuated decay volume is terminated by a thin $[4 \times 10^{-3}$ radiation lengths (r.l.)] Kevlar window and followed by a helium tank, at the

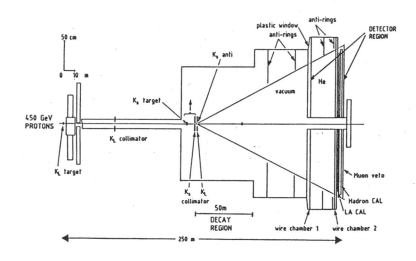

FIGURE 1

Overall schematic view of the N A31 experimental set-up

extremities of which are minidrift chambers to detect the trajectories of charged particles with 0.5 *mm* space resolution. The second set of chambers is followed by a suitably segmented scintillation-counter hodoscope.

The basic detector for photons is a liquid-argon calorimeter which represents a major improvement with respect to previous experiments. Its total thickness is 27 r.l.. The sampling is done in strips of 1.25×120 *cm* connected in such a way as to integrate the charge collected in the two halves of the calorimeter independently. The point of impact of a photon with energy greater than 5 *GeV* can be reconstructed to better than 0.5 *mm*. The measured energy resolution and linearity of response are shown in Figs. 3a and b.

In order to get a well-timed pretrigger for natural decays, a segmented plane of plastic scintillator in inserted between the two longitudinal halves of the calorimeter. Its output is read out, via wavelength shifters and light guides, by external photomultipliers; the scintillator and wavelength shifter work immersed in the liquid argon.

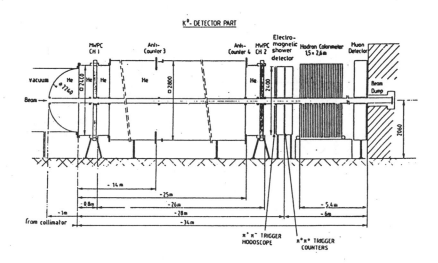

FIGURE 2

The detector region of the N A31 set-up

The invariant mass M of a system of n photons - originating from a common vertex at a distance Z_v from the detector, at which the energy E_i and the transverse coordinates x_i, y_i of each photon are measured - is given to a very good approximation, in our geometrical situation, by

$$M = \left[\sum_i^n E_i/Z_v \right] [\langle x^2 \rangle - \langle x \rangle^2 + \langle y^2 \rangle - \langle y \rangle^2]^{1/2}$$

with

$$\langle x^2 \rangle = \sum_1^n E_i x_i / \sum_1^n E_i, \quad \langle x \rangle = \sum_1^n E_i x_i / \sum_1^n E_i$$

and similarly for $\langle y^2 \rangle$ and $\langle y \rangle$.

Assuming that a K decay into $2\pi^0$ produced four detected photons, its longitudinal decay vertex position is obtained from

$$Z_v = [\sum_{j=1}^4 E_i/m_k] \cdot [\langle x^2 \rangle - \langle x \rangle^2 + \langle y^2 \rangle - \langle y \rangle^2]^{1/2}$$

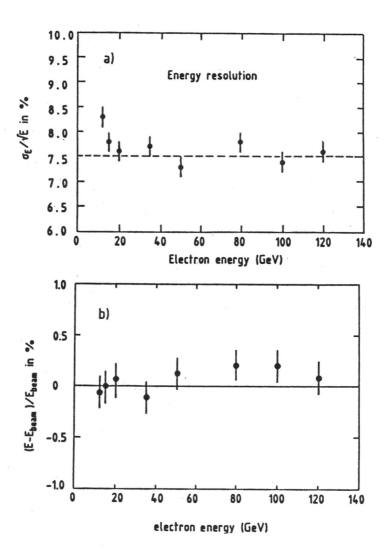

FIGURE 3

a)   *Measured energy resolution for electrons in the NA31 liquid-argon calorimeter,*

b)   *Linearity of calorimeter response*

Figures 4a and b show the precision with which the beginning of the decay region, typically 120 m in front of the calorimeter, is reconstructed for K_S decays to $\pi^0\pi^0$ and $\pi^+\pi^-$.

The curves are simple Monte Carlo predictions incorporating the measured resolution. It should be noticed that, for example, a shift of 0.8% in the absolute energy calibration of the calorimeter would result in a shift of 1 m in Z_v. In fact, knowing the longitudinal geometry of the detector with about 1 cm accuracy, checking that the same position of the K_S anticounter is reconstructed from the neutral and from the charged decays, and that the transverse geometrical scale is the same for the chambers and the liquid-argon calorimeter, the momentum scale for p_K as reconstructed for $\pi^0\pi^0$ and $\pi^+\pi^-$ (see later the method of measurement for $\pi^+\pi^-$, which depends essentially on geometry and known values of masses) can be fixed to better than 0.4×10^{-3}.

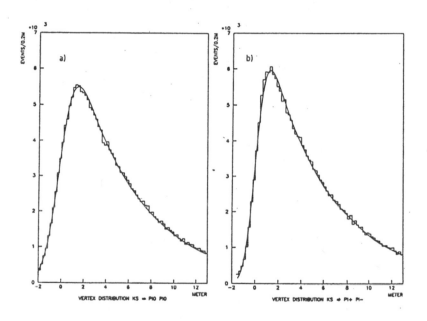

FIGURE 4

Distribution of the distance of the reconstructed decay vertex from the position of the counter defining the beginning of the accepted decay region: a) for $K_S \to \pi^0\pi^0$ and b) for $K_S \to \pi^+\pi^-$.

Of course, out of the three possible pairings of the four photons one can choose the one which is the best fit to the assumption that the photons come from $2\pi^0$. Within the resolution of the system the invariant mass of the two pairs would then coincide with the π^0 mass for true $K \to 2\pi^0$ decays.

This appears to be precisely the case for K_S (Fig. 5). For K_L the plot of the masses of two pairs shows some background (Fig. 6). The curves traced in the figure are elliptical rings of fixed area. The ratio of the axes of the ellipses is determined by the different

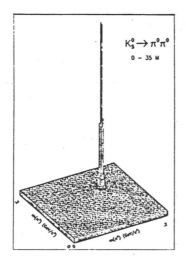

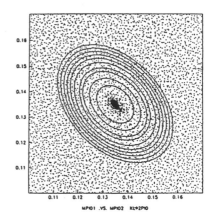

FIGURE 5

Lego plot for the invariant mass of the 2γ pairs for $K_s \to \pi^0 \pi^0$

FIGURE 6

Density distribution for $K_L \to \pi^0 \pi^0$, as a function of the mass of the 2γ pairs

resolution in the sum and in the difference of the masses of the pairs, due to the constraint of the K mass imposed on the four photons.

Except for the first rings, the population is almost exactly constant in each ring (Fig. 7a), a fact expected from Monte Carlo for $K_L \to 3\pi^0$ events for which two photons have escaped detenction. The comparison with K_S, where no such background is visible (Fig. 7b), and the flat dependence with respect to the number of the ring, allow for a precise background subtraction. It should be noticed in Fig. 8 that the background is a strong function of the distance of the decay vertex from the beginning of the K_L decay region, which coincides with the end of the K_L cleaning collimator. In fact, the reconstructed distance of the K decay vertex from the detector is equal to the true distance multiplied by the ratio of the invariant mass of the photons detected to the mass of the K. This is the reason for the shift in Z_v for the background, which is of the order of 25 m for the case when one of the π^0's from $K_L \to 3\pi^0$ is missed by the detector.

For the energy determination of the π^+ and π^-, the experiment again employs calorimetry rather than magnetic deflection. This improves the acceptance while, in the 70 to 170 GeV/cK momentum range used in the experiment, it is quite possible, as we will show, to obtain good K momentum determination and backgroup rejection.

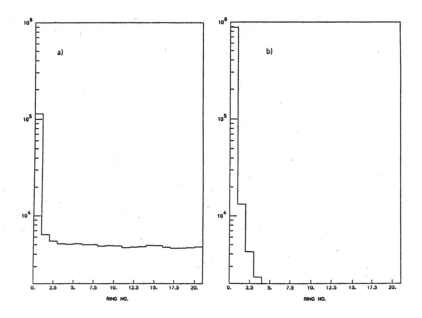

FIGURE 7

a) Number of events $K_L \rightarrow 4\gamma$ as a function of the number of the elliptical ring in which they fall according to the values of the invariant mass of 2γ pairs. b) Same for $K_S \rightarrow 4\gamma$.

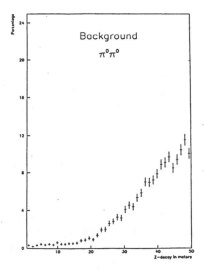

FIGURE 8

Dependence of the backgroup from $K_L \rightarrow 3\pi^0$ decays on the distance of the decay vertex from the end of the K_L cleaning collimator

For this purpose, the liquid-argon calorimeter is followed by an iron-plastic scintillator sampling calorimeter which, in conjunction with it, gives an energy resolution of $\simeq 65\%/\sqrt{E}(GeV)$ as shown in Fig. 9.

The position of the decay vertex is found directly from extrapolation of the tracks in the chambers; the momentum of K is determined using the relation

$$p_K = (1 + 1/R_E)[R_E m_K^2 - (1 + R_E)^2 m^2 \pi]^{1/2}/\theta,$$

where θ is the angle between the two pions and R_E is the ratio of the higher to the lower of the two pion energies. The above relation is remarkably accurate in our range of application. It is interesting to point out that p_K changes by only 5% when $1 < R_E < 5.6$ and we in fact (also to reject $\Lambda \rightarrow p\pi^-$ decays) only accept $1 < R_E < 2.5$. In this way the resolution in p_K is typically 1%.

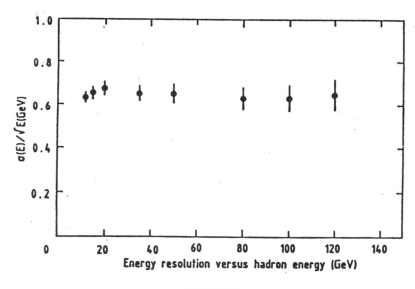

FIGURE 9

Energy resolution versus hadron energy for charged pions in the $NA31$ calorimetric measurement

In order to reject background it is necessary to compute the $\pi^+\pi^-$ invariant mass, using the calorimetric measured energy for each pion. Figure 10a shows the K_S reconstructed mass. An analogue plot for K_L, Fig. 10b, shows a very similar peak, with however a secondary peak at its left, interpreted as misidentified $K_L \rightarrow \pi^+\pi^-\pi^0$ decays with the π^0 missed by the detector.

In order to appreciate more generally the problem of backgrounds, Table 3 summarizes the size of the two-pion signal for K_L and the main background sources, the importance of the latter relative to the signal, and the amount of background we finally had to subtract, after employing all rejection criteria that we found useful.

We have already briefly discussed the criteria used for subtraction of the background induced by $K_L \rightarrow 3\pi^0$. For the charged decay modes we illustrate the main points in the following.

For $\pi^+\pi^-\pi^0$, with both photons from the π^0 escaping the detector, the reconstructed invariant mass is too low to contribute to the accepted $K_L \rightarrow \pi^+\pi^-$ region. Occasionally one of the photons could, however, closely overlap with the π^+ or π^- and be added to it, giving a higher invariant mass. The number of these events is estimated by looking at the photon density in events detected as $\pi^+\pi^-\gamma$ and normally rejected.

For $K \rightarrow \pi e\nu$ decays, the rejection comes mostly from the fact that an electron produces a shower which is almost always fully absorbed in the 27 r.l. thick liquid-argon calorimeter. That rejection criteria, based only on shower development, still leave a small residue of electrons is shown in Figs. 11a and b, where we compare the distributions of the events for K_S and K_L in the distance, d_{target}, from the K production target to the reconstructed decay plane, suitably normalized to take into account the different extrapolation distances from the wire chambers to the K_S and K_L targets. The rather flat tail for K_L is interpreted as residual $K \rightarrow \pi e\nu$, as supported by a detailed examination of the events and Monte Carlo simulation for the shape of the distribution.

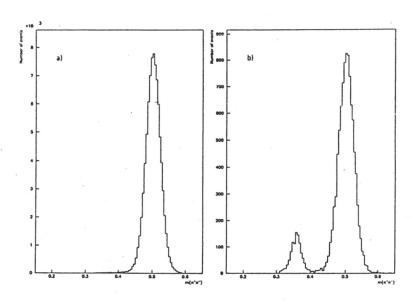

FIGURE 10

a) Invariant mass distribution for $\pi^+\pi^-$ pairs in K_S decays. b) Same in K_L decays.

Looking at the distribution of large-d_{target} events as a function of the distance Z_v of the decay vertex from the end of the K_L cleaning collimator (Fig. 12) a peak is seen at small Z_v, which, after detailed study, can be attributed to inelastic regeneration in the collimator itself. Although we can calculate, from this regenerated K_S, the number of scattered K_L, which turns out to be negligible, we decided for this preliminary analysis simply to reject both $\pi^+\pi^-$ and $\pi^0\pi^0$ decays with $Z_v < 10.5$ m, because it is rather difficult to ensure the same dependence of the acceptance for the two decay modes as a function of the scattering angle.

The decays $K \rightarrow \pi\mu\nu$ are rejected, at trigger level, by the veto counters following the iron walls behind the calorimeters; at the reconstruction stage, they are rejected by the requirements on the energy ratio of the two tracks, on their invariant mass, and on other kinematical quantities.

TABLE 3

K_L signal and background sources

Decay modes	Branching ratios (%)	Ratios relative to $\pi\pi$	Background to signal ratio (%)
$K_L \rightarrow 2\pi^0$	0.094		
$3\pi^0$	21.5	230	4
$K_L \rightarrow \pi^+\pi^-$	0.203		
$\pi e\nu$	38.7	190	
$\pi\mu\nu$	27.1	130	} 0.5
$\pi^+\pi^-\pi^0$	12.4	61	0.1
$\pi e\nu\gamma$	1.3	6	~ 0
$\pi\pi\gamma$	4.4×10^{-3}	0.02	~ 0

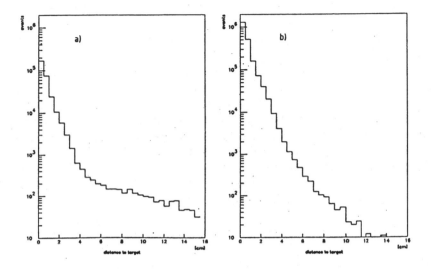

FIGURE 11

a) *Distribution of* $K_{I.} \rightarrow \pi^+\pi^-$ *reconstructed events as a function of the distance,* d_{target}, *of the K-production target from the reconstructed decay plane.*
b) *Same for* $K_S \rightarrow \pi^+\pi^-$.

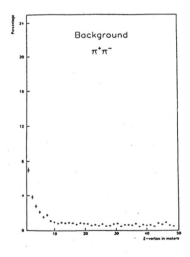

FIGURE 12

Distribution of large-d_{target} evetns as a function of the decay vertex distance from the end of the K_L cleaning collimator.

The background subtraction has been carried out bin by bin. The data analysed consist of 32 self-contained K_S and K_L data sets, or miniperiods, during which the experimental set-up and running conditions were kept as stable as possible. The double ratio R of the final number was calculated miniperiod by miniperiod for each of 10×32 bins of p_K and Z_v for $70 < p_K < 170 \; GeV/c$ and $10.5 < Z_v < 48.9 \; m$.

Given the measured resolution in p_K and Z_v, and the chosen bin sizes, the effects of the different beam sizes, different beam divergences, and momentum spectra between K_L and K_S are such that the double ratio of the numbers of reconstructed events, as shown by Monte Carlo studies, cannot possibly differ by more than 0.5% from the double ratio of the rates of decay, implying a total Monta Carlo correction on ε'/ε of less than 1×10^{-3}.

About $200,000 K_L \rightarrow \pi^0\pi^0$, $10^6 K_L \rightarrow \pi^+\pi^-$, and $12 \times 10^6 K_S \rightarrow 2\pi$, i.e. more than 20 times the integrated number of $K_L \rightarrow \pi^0\pi^0$ events of all previous experiments, were collected in 1986 and a large fraction of them are used for the analysis.

The high statistics allow us to study the dependence of the result on various parameters. As an illustration, the double ratio R is given in Figs. 13a, b, and c as a function of Z_v, of p_K, and of the chronological number of the miniperiods.

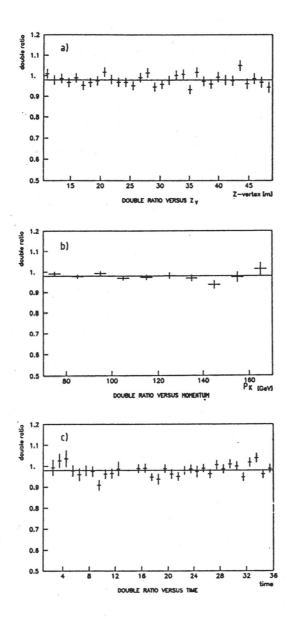

FIGURE 13

a) Double ratio: $[N(K_L \to 2\pi^0)/N(K_L \to \pi^+\pi^-)]/[N(K_S \to 2\pi^0)/N(K_S \to \pi^+\pi^-)]$ as a function of decay vertex position; b) Same as a function of momentum; c) Same as a function of time.

Of the outmost importance is the study of possible systematic distorsions. We have evaluated the differential effect on the four tupes of events selected due to all sources we could think to be relevant. The most important of them are briefly described below:

a) Trigger efficiency

By looking at a sample of events recorded with a much looser trigger requirement than the standard one, we have continuously monitored the product of the pretrigger and trigger efficiencies and found it to be, on an average: in the $2\pi^0$ mode, $(99.74 \pm 0.12)\%$ for the K_L trigger and $(99.86 \pm 0.04)\%$ for the K_S one; in the $\pi^+\pi^-$ mode $(99.58 \pm 0.10)\%$ for the K_L trigger and $(99.51 \pm 0.03)\%$ for the K_S one. This implies $\Delta R = -(0.15 \pm 0.16)\%$ for the double ratio.

b) Event losses by accidental events

By overlapping with normal events the content of events taken in correspondence with a random trigger, adjusted to occur during the burst at a frequency proportional to the instantaneous beam intensity, we have determined the fraction of events for each of the four categories which have been lost because of the disturbance induced by an accidental event. The net effect on R is $-(0.34 \pm 0.1)\%$.

c) Relative energy scale for $\pi^+\pi^-$ and $\pi^0\pi^0$

This could, in principle be an important effect, but a good understanding of the geometry allows good precision to be achieved. We estimate nevertheless a resulting for uncertainty on R of $\pm.3\%$.

d) Uncertainty in background subtraction

The largest background subtracted is in $K_L \rightarrow 2\pi^0$, where it amounts to 4%. The procedure to estimate it is, however, straightforward and we feel confident that we have determined it to an accuracy of $\pm 5\%$. The subtraction of the background for $\pi^+\pi^-$ is much smaller (0.6%), but only known with ~30% uncertainty.

The summary of the best estimates for the systematic uncertainties and our result for ε'/ε are given in Table 4 and shown in Fig. 14.

In Fig. 14, in addition to comparing our results with all previous determinations, we indicate [11] a recent best guess on the value of ε'/ε as $(2 \pm 1) \times 10^{-3}$, according to the Standard Model.

Combining quadratically the statistical error with the present limit on the systematic uncertainties, the result for ε'/ε is 3.0 standard deviations from zero.

TABLE 4

Summary of systematic uncertaines

$2\pi^0$	energy scale	$\pm 0.5 \times 10^{-3}$
	background	$\pm 0.3 \times 10^{-3}$
$\pi^+ \pi^-$	background	$\pm 0.3 \times 10^{-3}$
K_S/K_L	beam divergence	$\pm 0.2 \times 10^{-3}$
	accidentals	$\pm 0.3 \times 10^{-3}$
K_S	scattering	$\pm 0.2 \times 10^{-3}$
	anticounter inefficiency	$\pm 0.2 \times 10^{-3}$
Total systematic error		$\pm 0.8 \times 10^{-3}$

$$\varepsilon'/\varepsilon = 3.3 \pm 0.7 \quad \text{statistical error}$$
$$\pm 0.8 \quad \text{systematic error}$$
$$\times 10^{-3}$$

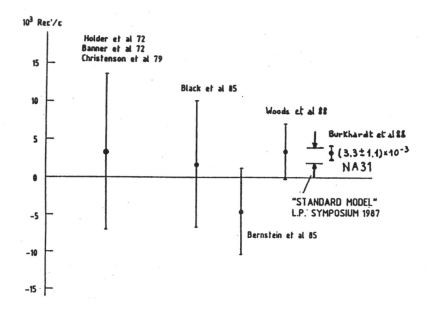

FIGURE 14

Result of the NA31 experiment for ε'/ε. In the error indicated, we have combined quadratically statistical and systematic errors.

Given the crucial importance of ε'/ε being differnt from zero from the point of view of the capability of the standard electroweak theory to incorporate CP violation, independent confirmation of this results is eagerly awaited. As already mentioned, the experiment E_{731} at Fermilab has taken data which should allow to obtain a comparable accuracy, once the necessary careful analysis will be completed. The $NA31$ collaboration has collected new data in 1988 with the same basic set-up, improved with the addition of a transition radiation detector for additional electron identification, more anticounters, a different proton momentum for producing K_S and other changes suggested by the analysis of the first run in order to reduce systematic uncertainties and improve statistics. The evaluation of the result of this second run is presently under way.

A final data taking period will take place during the summer and fall of this year.

Test of CPT Invariance in $K \to 2\pi$

Meanwhile, exploiting the same basic approach of using ratios between number of events taken under different conditions, but always originating from the same small bin of kaon momentum and longitudinal vertex position in such a way as to be (at first order) independent of the acceptance, a measurement of the difference of the phases Φ^{+-} and Φ^{00} of η^{+-} and η^{00} has been carried out by the $NA31$ collaboration. The relevance of this quantity had been pointed out among others by Barmin et al. [12], who showed that, given the very small difference between the magnitudes of η^{+-} and η^{00}

$$R = |\eta^{00}|^2/|\eta^{+-}|^2 = 1 - 6\frac{\varepsilon'}{\varepsilon} = 1 - 0.022 \pm 0.007,$$

any sizable difference between the their phases would imply CPT violation in the K to two-pion decay amplitude.

This because, assuming CPT invariance for the decay amplitude, the phase of ε' is firmly predicted to be close to the phase of ε while under the above circumstances ε' would be essentially perpendicular to ε. Only one previous significant measurement [12] exists which had given $\Phi^{00} - \Phi^{+-} = 12^0 \pm 6.2$. For this kind of experiment an interference term needs to be measured. At a distance z from a target, where an incoherent mixture of K_0 and $\overline{K}_0$ is produced, the number of decays with momentum p to a final state for which the ratio of the K_L to the K_S decay amplitude is $|\eta|e^{i\phi}$, is in fact proportional to

$$e^{-\Gamma_S t} + |\eta|^2 \cdot e^{-\Gamma_L t} + 2 \cdot |\eta| D \cdot e^{-\frac{\Gamma_S + \Gamma_L}{2}t} \cdot \cos\left(\frac{\Delta m t}{\hbar} - \Phi\right)$$

where $t = \frac{z m_K}{P c}$, $\Delta m = m_{K_L} - m_{K_S}$ and $D = \frac{N_{K_0} - N_{\overline{K}_0}}{N_{K_0} + N_{\overline{K}_0}}$ (at the momentum P).

$NA31$ has taken data, alternating every few hours, with two identical target stations and beam defining collimation systems, with targets located respectively 48 m (far) and 33.6 m (near) before the end of the final collimator, from which we measure the position of the decay vertex in order to group the events in bins. The detectors have been kept in the same conditions and the events have been reconstructed with the some basic method employed for the ε'/ε measurement. Particular attention has been devoted, using for this calibration K_S produced on the mobile train, to establish, to an accuracy better than $\pm 0.1\%$, the equality of the momentum scale for charged and neutral two pion decays. To avoid possible serious systematic uncertainties in extracting Φ from the data we have performed fits only of the ratio of events originating from the two target posission and reconstructed in the same bin of momentum and longitudinal vertex position, using of course the same value of $\Gamma_S, \Gamma_L, D(P), \Delta m$ for both the $\pi^+\pi^-$ and $\pi^0\pi^0$ decays. In this way the acceptance disappears in first order and the ratio of reconstructed events equals, to the same accuracy, the ratio of decays which have really occurred in the same bin. It has been found sufficient to parametrize

D as a quadratic function of P. It has also been checked that over a 48 m range for the vertex position and 80 GeV/c (from 70 to 150 GeV/c) for the kaon momentum, the result for $\Phi^{+-} - \Phi^{00}$ was stable with respect to any reasonable choice of bin sizes and we have finally used for fitting bins 2.4 $m \times 5$ GeV/c, i.e. in total 640 ratios. It should be noted that, for the average momentum, the distance between the two targets is such that e.g. if in correspondance to a given bin the term containing Φ reaches its relative maximum for the far target, than it is about zero for the near target position, the whole pattern for $\pi^+\pi^-$ being shifted with respect to that for $\pi^0\pi^0$ in case of a non zero difference between Φ^{+-} and Φ^{00}. An elaborate Monte Carlo calculation including resolution effects on the ratio between the number of events decaying in a bin and those expected to be reconstructed in the same bin has been performed for both charged and neutral decay modes. This ratio remains everywhere quite close to 1 and a fit with an acceptable ratio of χ^2 to number of degrees of freedom is in fact obtained with and without correcting for acceptance. The overall effect of the Monte Carlo correction on $\Phi^{+-} - \Phi^{00}$ turns out to be smaller than 1^0.

The single rates in all detectors elements are very similar for the two target stations assuring a relative immunity from accidentals. We have anyhow checked, using as for ε'/ε, special events triggered at random and overlapping the response of the detectors for them with regular events, that no distorsion at the $0.^06$ level can possibly results from this potential source of error.

Background subtraction has been performed bin by bin and has practically no effect on the results. The difference in the values obtained with and without any background subtraction is $.08^0$ for Φ^{+-} and entirely negligible for $\Phi^{00} - \Phi^{+-}$.

The best estimate of the overall possible systematic uncertainty at the present level of analysis, excluding the effects in the difference between the true values and the values according to the Particle Date Group [3] for $1/\Gamma_S = .8923 \times 10^{-10} sec$, $|\eta_{+-}| = 2.275 \times 10^{-3}$, and $\Delta m/\hbar = .534 \times 10^{10} sec^{-1}$ used in the fit is $0^0.7$ for Φ^{+-} and $1^0.1$ for $\Phi^{00} - \Phi^{+-}$. The result is

$$\Phi^{+-} = 47^0.1 \pm 1^0.4$$
$$\Phi^{00} = 0^0.8 \pm 2^0.6$$

where the error indicated is purely statistical.

Increasing by an amount equal to 1 standard deviation the values of $1/\Gamma_S$, $|\eta_{+-}|$ and Δm, the change in the fitted value of Φ^{+-} is respectively $+0^0.5, +0^0.22, +1^0.24$ and in $\Phi^{00} - \Phi^{+-} - 0^0.22, +0^0.1, 0^0$.

This result is clearly compatible with CPT invariance and fully excludes for $\Phi^{00} - \Phi^{+-}$ a value close to the nominal 12^0 of the previous experiment [12].

REFERENCES

[1] J. H. CHRISTENSON - J. W. CRONIN - V. L. FITCH - R. TURLAY, Phys. Rev. Lett. **13** (1964) 138.

[2] L. WOLFENSTEIN, Phys. Rev. Lett. **13** (1964) 569.

[3] Rewiew of Particle Properties, Phys. Lett. **170B** (1986) 1.

[4] M. KOBAYASHI - T. MASKAWA, Prog. Thoer. Phys. **49** (1973) 652.

[5] C. ALBAJAR et al., Phys. Lett. **186B** (1987) 247.

[6] H. ALBRECHT et al., Phys. Lett. **192B** (1987) 245.

[7] M. HOLDER et al., Phys. Lett. **40B** (1972) 141.
     M. BANNER et al., Phys. Rev. Lett. **28** (1972) 1597.
     J. A. CHRISTENSON et al., Phys. Rev. Lett. **43** (1979) 1209.
     J. K. BLACK et al., Phys. Rev. Lett **54** (1985) 1628.
     R. H. BERNSTEIN et al., Phys. Rev. Lett. **54** (1985) 1631.

[8] M. WOODS et al., Phys. Rev. Lett. **60** (1988) 1695.

[9] H. BURKHARDT et al., Phys. Lett. **206B** (1988) 169.

[10] H. BURKHARDT et al., Nucl. Instrum. Methods **A268** (1988) 116.

[11] M. SHIFMAN, Theoretical status of weak decays, Proceedings 1987 Symposium on Lepton and Photon Interactions at High Energies, Ed. W. Bartel and R. Rückl, North Holland.

[12] J. A. CHRISTENSON et al., Phys. Rev. Lett. **43** (1979) 1212.

Broken Symmetry in Rotating Baryons

ANDRÉ MARTIN (*)

When I was asked to write an article for Luigi Radicati's 70th birthday, I accepted with great pleasure. Then I realized that there should be some connection with symmetry or broken symmetry and this was a difficulty for me. Though some of my best friends are "symmetry" men, like Murray Gell Mann, Féza Gürsey, Louis Michel and Luigi himself, my own work has seldom had anything to do with symmetry, except for using from time to time the Wigner-Eckart theorem or some Clebsch-Gordan coefficients. Legendre polynomials which I studied and used intensively from 1962 to 1975 are, for me, above all, polynomials, and have very little do to with the group of rotations. So it looked to me difficult to find an appropriate subject... until I realized that in one problem connected with hadron structure I had exhbited without even paying attention to it, a situation of broken symmetry - like Monsieur Jourdain "qui fait de la prose sans le savoir". It is the problem of the structure of rotating baryons made of three quarks, or if you prefer the problem of the Regge trajectories of baryons. What I found is that at least for large angular momentum the three quarks prefer to dispose themselves in a quark-diquark structure instead of three equally spaced quarks. This situation is somewhat analogous to that of a rotating mass of fluid to which, as we shall see, Luigi has paid attention some years ago.

What happens to a rotating mass of fluid submitted to its own gravity, as described for instance in the book of Chandrasekhar [1], is that first, for small angular velocities, it has the shape of an ellipsoid symmetrical around the rotation axis, like the earth, for instance, but there is a bifurcation for a certain, well-defined angular momentum, and beyond this point the stable figure is an ellipsoid elongated along an axis perpendicular to the rotation axis. This is a spontaneous symmetry breaking. As you increase the angular momentum further there are more bifurcations appearing. Michel, Radicati and Constantinescu [2] have made an exhaustive list of these bifurcations towards other ellipsoid shapes, but there are also other figures appearing, like the "Poincaré pears".

If we take now a system of three quarks, interacting by two-body potentials rotating around an axis, we can try to find what is the configuration of minimum

(*)   CERN - Geneva

energy, for a given angular momentum. This is meaningful only if the two-body potential has a lower bound. Otherwise, if $V(r_{12}) \to -\infty$ for $r_{12} \to 0$ quarks will just stick together and the energy will be $-\infty$. This will not happen, fortunately, in the quantum case.

It is very easy to see that, both with relativistic and non-relativistic kinematics, the motion of the three quarks will take place in a plane perpendicular to the angular momentum axis, if the two-body potential V is an increasing function of the distance. Indeed, projecting the motion on this plane reduces the momenta of the particles and the interparticle distances, while leaving the angular momentum with respect to the axis invariant. Once this is done one finds that there are essentially two candidates to minimize the energy: a rotating equilateral triangle, which respects the symmetries of the problem, or a quark-diquark structure. If the two-body potential grows at most linearly, one finds that it is the symmetry breaking structure, the quark-diquark system, which minimizes the energy, both with relativistic and non-relativistic kinematics. In the non-relativistic case the borderline between the two situations is the case of harmonic oscillator forces.

Let us look specifically at the case of a linear potential in the extreme relativistic case (zero-mass particles). Let $V = \lambda r$ be the two-body potential.

- In the case of the equilateral triangle, let R be the distance of the particles to the centre. The angular momentum is $J = 3pR$, where p is the impulsion of one of the quarks. The interparticle distance is $\sqrt{3}R$ and hence the energy is:

$$(1) \qquad\qquad 3p + 3\sqrt{3}\lambda R = \frac{J}{R} + 3\sqrt{3}\lambda R.$$

Minimizing with respect to R we get

$$(2) \qquad\qquad E \simeq 2 \times 3^{3/4} J^{1/2} \lambda^{1/2}$$

In the quark-diquark case, we take the diquark and the quark having momentum $\vec{p}$ and $-\vec{p}$ and the distance ρ. Then $J = p\rho$ and

$$(3) \qquad\qquad E = 2p + 2\lambda\rho = 2\left(\frac{J}{\rho} + \lambda\rho\right),$$

and minimizing in ρ

$$(4) \qquad\qquad E \simeq 4J^{1/2}\lambda^{1/2}.$$

Since 2 is less than $3^{3/4} \simeq 2.279$, it is the quark-diquark configuration which is favoured.

Now you will ask me what this has to do with baryons, which are objects obeying quantum mechanics. The fact is that we shall show that for large angular momenta, and provided the interparticle potential is lower bounded, the minimum classical energy for a given angular momentum gives an estimate of the quantum ground state energy. The ratio of the two quantities approaches unity for $J \to \infty$.

This is particularly easy to see for a two-body, non-relativistic system. Take for instance a linear potential. The Hamiltonian reads, if $h = c = 2\mu = 1$, when μ is the reduced mass

$$(5) \qquad -\frac{d^2}{dr^2} + \frac{\ell(\ell+1)}{r^2} + \lambda r.$$

The expectation value of $-d^2/dr^2$ is positive. It is the radial kinetic energy. In fact, we have the operator inequality

$$(6) \qquad -\frac{d^2}{dr^2} > \frac{1}{4r^2}.$$

Hence a lower bound of the energy will be obtained by minimizing

$$\frac{(\ell+1/2)^2}{r^2} + \lambda r.$$

The minimum is

$$(7) \qquad E_{\text{inf}} = 3 \left(\frac{\lambda}{2}\right)^{2/3} (\ell+1/2)^{2/3}.$$

On the other hand, we can get an upper bound of the energy by noticing that

$$(8) \qquad r < \frac{1}{2}\left(A + \frac{r^2}{A}\right).$$

Then (5) is replaced by an exactly soluble harmonic oscillator potential and we get

$$(9) \qquad E < \frac{A}{2} + \frac{1}{\sqrt{2A}}(2\ell+3)$$

and, minimizing with respect to A

$$(10) \qquad E < E_{\text{sup}} = 3 \left(\frac{\lambda}{2}\right)^{2/3} (\ell+3/2)^{2/3}.$$

As for the classical energy it is $p^2 + \lambda r = \ell^2/r^2 + \lambda r$ with a minimum

$$(11) \qquad E_{\text{classical}} = 3 \left(\frac{\lambda}{2}\right)^{2/3} \ell^{2/3}.$$

Comparing (7), (10) and (11) we see that what we said is true. It is of course more complicated to generalize this to the three-body case and to include relativistic kinematics, but this can be done [3] and again the conclusion is the same, and therefore, from what we said at the beginning, a baryon made of three quarks interacting with linear two-body potential will eventually have for large angular momentum a quark-diquark structure.

One may ask if this conclusion will be altered if the potential is singular attractive at short distances. Because of quantum effects it is not really changed if the two-body potential is less singular than $1/r^2$ in the non-relativistic case and less singular than $1/r$ in the relativistic case. Then, if this condition is fulfilled, the recipe is the following: replace V by V_B,

$$V_B(r) = \mathrm{Sup}(V(r), -B)$$

and minimize the classical energy for fixed-angular momentum.

Now the remaining question is that of what happens in real life. Real life is:

1)   a two-body potential which is not purely linear, for which the minimization is less easy;

2)   a *finite*, *small* angular momentum. Regge trajectories are seen to be linear, i.e., $E^2 \sim J$ like in Eq. (4) already for small J. Baryon and meson trajectories are also seen to be parallel, a fact explained by the quark-diquark structure at large J, but this occurs already at small J;

3)   the three-quark wave function should be completely antisymmetric including colour. If we take three identical quarks, say uuu, i.e., the Δ trajectory, or sss, the wave function shoud be completely symmetric in space.

All these problems cannot be solved by purely analytical considerations and some courage and computer abilities are needed to investigate them. Low angular momenta, where the diquark effect is negligible [remember that the ratio between (2) and (4) is only 1.14!] have been investigated by Basdevant and Boukhraa [4]. The most interesting work, for large angular momenta, has been done by Fleck, Richard and Silvestre-Brac [5]. In spite of the obligatory symmetrization of the wave function, they have been able to uncover the diquark effect by the following procedure: you fix the distance of two of the quarks and look at the density distribution of the third quark and you find that for sufficiently large angular momenta $(L = 8)$ the third quark likes to be close to one of the other quarks. For L small this effect disappears.

We see that this effect is rather striking! We conclude that once more in physics, an asymptotic property $(J \to \infty)$ turns out to hold for a rather small value of the parameter. It would certainly be desirable to get a more physical understanding of this fact. To the question: is there a critical angular momentum beyond which the diquark appears?, the answer is that, classically,

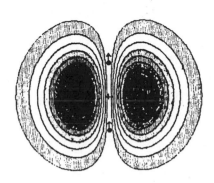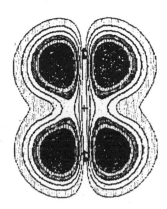

Density distributions for the third quark, the first two being held fixed (points with arrows) left L = 0, right L = 8.

there is no such transition for purely linear two-body potentials. In the quantum case, I have a suspicion that the diqaurk effect appears smoothly as the angular momentum is increased, but this is only a suspicion.

REFERENCES

[1]   S. CHANDRASEKHAR, *Ellipsoidal figures of equilibrium*, New Haven and London, Yale University Press (1969).

[2]   D.H. COSTANTINESCU - L. MICHEL - L. RADICATI, J. Phys. **40** (1979) 147.

[3]   A. MARTIN, Z. Phys. **C32** (1986) 359.

[4]   J.L. BASDEVANT - S. BOUKRAA, Z. Phys. **C28** (1985) 413.

[5]   S. FLECK - J.M. RICHARD - B. SILVESTRE-BRAC, Phys. Rev. **D38** (1988) 1519.

Reflection Symmetry and Reflection Positivity in Euclidean Quantum Field Theory

PIETRO MENOTTI (*)

1. The idea that quantum field theory formulated in euclidean space instead of Minkowski space, might be relevant for the comprehension of our world is due to Schwinger [1]. However it took some time mainly through the work of Wick [2], Symanzik [3] and Osterwalder and Schrader [4] to reach a complete understanding of the problem. Formally the point is that as noted originally by Minkowski, by taking an imaginary time one goes back to the euclidean metric. This however turned out to be something more than a formal trick. In fact it has been shown that all the physically relevant properties can be extracted from a euclidean formulation. Technically, given a euclidean theory which satifies the axioms of euclidean invariance, symmetry in the permutation of the field arguments and "positivity" it is possible to construct from it a theory in Minkowski space, satisfying the Wightman axioms.

The above mentioned results are of general nature and simply put a relationship between two different formulations. In practice one has to construct a euclidean theory by giving a euclidean lagrangian. In the same way as not all lagrangians in Minkowski space satisfy the correct physical requirements (like hermiticity), so not all euclidean lagrangians generate euclidean Green's functions satisfying the correct euclidean axioms. Of these the least obvious to satisfy is that of positivity. It is the analogue of hermiticity in the Minkowski formulation and from it the unitary and the spectrum properties of the related Minkowski theory follow. In order to have such positivity requirement the lagrangian has to be reflection symmetric, (similar to a sort of time reversal invariance) but still this is not enough. In fact the presence in the lagrangian of time derivatives (as they always appear) makes the problem very delicate and it has to be considered carefully.

In sect. 2 we formulate reflection positivity; in order to give a rigorous definition of the functional integral we shall go over to a discretized version of the theory (lattice theory) where all problems come to surface and then we shall examine how reflection positivity can be satisfied in a variety of scalar,

(*)   Dipartimento di Fisica della Università, Pisa and INFN Sezione di Pisa

vector, fermion and gauge theories. In sect. 3 we address the case of higher spin fields and finally that of the gravitational field.

2. The reflection positivity axiom states that given a function F of the field variables with support in the positive euclidean time it is possible to define an antilinear map Θ of F into the functions of the fields at negative times so that the following expectation value is positive.

$$(1) \qquad\qquad \langle F\Theta F \rangle \geq 0.$$

The abstract expression (1) is given in a lagrangian theory by the functional integral

$$(2) \qquad\qquad \langle F\Theta F \rangle = \int d[\phi] e^A F\Theta F / Z$$

with

$$(3) \qquad\qquad Z = \int d[\phi] e^A \text{ and } A = \int \mathcal{L} dx.$$

Eq. (3) has the typical structure of the functional integral appearing in statistical mechanics with Bolzmann factor e^A, but the requirement of reflection positivity does not appear in statistical mechanics even if some interesting models satisfy it.

In order to define the problem let us consider the simplest example of the scalar complex field in which the antilinear map is provided by

$$(4) \qquad\qquad \Theta[\varphi(x)] = \varphi^*(\theta x) \text{ with } \theta x = (\mathbf{x}, -x_4)$$

and extended by antilinearity to all functions of the fields φ and φ^* by

$$(5) \qquad\qquad \Theta[F(\varphi, \varphi^*)] = F^*(\Theta\varphi, \Theta\varphi^*)$$

where F^* is defined by $F^*(\varphi^*) = [F(\varphi)]^*$.

Obviously $\Theta[\partial\varphi(x)/\partial x_4] = -\partial\varphi^*(\theta x)/\partial x_4$ and the usual lagrangian of the $g(\phi^*\phi)^2$ theory satisfies

$$(6) \qquad\qquad \Theta[\mathcal{L}(x)] = \mathcal{L}(\theta x).$$

We shall call a lagrangian with the property (6) reflection symmetric. At first sight it would appear that in presence of a real reflection symmetric lagrangian the theory should be always reflection positive. In fact denoting with ϕ_+ the fields at positive times and with ϕ_- the fields at negative times one is tempted to write

$$(7) \qquad Z\langle F\Theta F \rangle = \int d[\phi_+] d[\phi_-] e^{A_+[\phi_+] + A_-[\phi_-]} F(\phi_+)\Theta[F(\phi_+)]$$

where

(8)
$$A_+ = \int_{x_4>0} \mathcal{L}(x)dx, \quad A_- = \int_{x_4<0} \mathcal{L}(x)dx$$

which due to (6) equals

(9)
$$\int d[\phi_+]d[\phi_-]e^{A_+[\phi_+]+A_+[\phi_-]}F(\phi_+)[F(\phi_-)]^*$$

which is obviously ≥ 0. However reflection symmetry, though necessary is not sufficient for having reflection positivity. In fact the writing of the action A as the sum of a function A_+ of fields at positive times and a function A_- of fields at negative times is not correct because in presence of time derivatives there are terms connecting points with infinitesimal positive time with points with infinitesimal negative time. To appreciate this we shall turn to a discretized version of space time (lattice discretization) which allows a rigorous definition of the functional integral and brings to surface the problems which occur at the reflection plane $x_4 = 0$.

The replacement of the continuous space time by a discrete set of elements (points, links, etc..) is by now a common device to give the functional integral a well defined meaning [5]. The derivatives in our example of the scalar field, become finite differences and there in not a unique way to perform such a transition. One coult replace $\partial_\mu(\phi)(x)$ by

(10)
$$\frac{1}{a}[\phi(x+a\hat{\mu}) - \phi(x)]$$

as well as by the symmetric difference

(11)
$$\Delta_\mu\phi(x) = \frac{1}{2a}[\phi(x+a\hat{\mu}) - \phi(x-a\hat{\mu})]$$

or even by more complex structures which formally go over to the derivative for $a \to 0$.

Replacing the derivative by (10) the elementary contribution to the kinetic part of the lagrangian sits on the one dimensional simplex $(x, x+a\hat{\mu})$, the link, and we have

(12)
$$\Theta(\mathcal{L}((x, x+a\hat{\mu}))] = \mathcal{L}(\theta(x, x+a\hat{\mu}))$$

i.e. $\mathcal{L}$ is reflection symmetric.

If we reflect with respect to sites (i.e. with respect to a three dimensional hyperplane containing sites) we have $A = A_+ + A_- + A_0$ with $\exp(A_0) > 0$, with A_0 function only of the fields at $x_4 = 0$ and $A_+(A_-)$ function only of the fields at $x_4 \geq 0(x_4 \leq 0)$. Then we get immediately reflection positivity. If we reflect with respect to links (i.e. with respect to a three dimensional hyperplane cutting in two the time links) we have $A = A_+ + A_- + \Delta A$ with A_+ function only of the fields at $x_4 > 0$ and ΔA being the part which connects positive with negative times. It is simple to prove that $\exp(\Delta A)$ is a positive kernel (not simply a positive function). Thus we have reflection positivity both reflecting with respect to sites and with respect to links. If instead of (10) we adopt (11) $\mathcal{L}$ is still reflection symmetric. With regard to reflection positivity however, it holds if we reflect with respect to sites but it is violated if we reflect with respect to links [6].

A more interesting example is provided by the Dirac theory. In Minkowski space the lagrangian

$$(13) \qquad \mathcal{L} = -\frac{1}{2}\overline{\psi}\left(\gamma_\mu \frac{\overleftrightarrow{\partial}}{\partial x^\mu}\right)\psi - m\overline{\psi}\psi$$

($\mu = 1\ldots 4$, $x_4 = it$, γ_μ all hermitean and $\overline{\psi} = \psi^+\gamma_4$) is hermitean. On the other hand the euclidean version of $\mathcal{L}$ is not hermitean, even at the formal level. It is however reflection symmetric. Let us consider in fact the antilinear map on the anticommuting Grassmann variables $\overline{\psi}, \psi$

$$(14) \qquad \Theta[\psi(x)] = \overline{\psi}(\theta x)\gamma_4$$

$$(15) \qquad \Theta[\overline{\psi}(x)] = \gamma_4\psi(\theta x)$$

which is extended by antilinearity to an arbitrary polynomial of ψ and $\overline{\psi}$, reversing in addition the order of the anticommuting fields. We have

$$(16) \qquad \Theta[\overline{\psi}(x)\gamma_\mu \frac{\partial}{\partial x_\mu}\psi(x)] = \Theta[\psi(x)]\gamma_\mu\left(\frac{\overleftarrow{\partial}}{\partial x_\mu}\right)\Theta[\overline{\psi}(x)]$$

i.e. $\Theta[\mathcal{L}(x)] = \mathcal{L}(\theta(x))$.

We have learned however that reflection symmetry is not sufficient; moreover Grassmann integration here occurs, for which an integral of the structure (9) is not assured to be positive. The simple minded discretization of $\mathcal{L}$ which is reflection symmetric is given by

$$(17) \qquad -\frac{1}{2}\overline{\psi}(x)\gamma_\mu \overleftrightarrow{\Delta}_\mu \psi(x) - m\overline{\psi}(x)\psi(x).$$

Lagrangian (17) possesses the well known disease of giving rise to a "doubling" of fermion species. It was the the price of getting cheeply reflection symmetry by adopting the symmetric derivative (11). One can prove [7] that $\mathcal{L}$ is reflection positive when reflecting with respect to links, but not when reflecting with respect to sites [9]. The way out of the doubling problem was shown by Wilson [10] and amounts to replacing $\mathcal{L}((x, x + a\hat{\mu}))$ by

(18)
$$\mathcal{L}((x, x + a\hat{\mu})) = -\frac{1}{a}[\overline{\psi}(x)(\gamma_\mu - 1)\psi(x + a\hat{\mu}) - \overline{\psi}(x + a\hat{\mu})(\gamma_\mu + 1)\psi(x)]$$
$$-\frac{1}{8K}[\overline{\psi}(x)\psi(x) + \overline{\psi}(x + a\hat{\mu})\psi(x + a\hat{\mu})].$$

It is immediately verified that (18) is reflection symmetric. In addition it implies reflection positivity [8] on all possible choice of the observable F [9], both reflecting with respect to sites and with respect to links. The Wilson projectors $(1 \pm \gamma_\mu)$ cannot be changed into $(1 \pm r\gamma_\mu)$ with $r \neq 1$ without violating positivity at least for some observables [9]. The proof goes over without difficulties in presence of interaction with gauge (abelian and not abelian) fields [7-9], and the situation is thus completely satisfactory.

Going over now to the vector field, the standard massive vector lagrangian

(19)
$$\mathcal{L} = -\frac{1}{4}(\partial_\mu A_\nu - \partial_\nu A_\mu)^2 - \frac{1}{2}m^2 A_\mu^2$$

can easily be discretized by assigning the vector variables to the links, $A(x, x + a\hat{\mu}) = -A(x + a\hat{\mu}, x)$ and replacing $\partial_\mu A_\nu - \partial_\nu A_\mu$ with the discrete external derivative (boundary operation) $A(x, x + a\hat{\nu}) + A(x + a\hat{\nu}, x + a\hat{\nu} + a\hat{\mu}) + A(x + a\hat{\nu} + a\hat{\mu}, x + a\hat{\mu}) + A(x + a\hat{\mu}, x)$. The elementary lagrangian now sits on a two dimensional simplex, the so called plaquette and reflection symmetry $\Theta[\mathcal{L}(P)] = \mathcal{L}(\theta P)$ is immediately checked. The action A breaks down in $A = A_+ + A_- + A_0$, being A_0 the contribution of the space-space plaquettes at $x_4 = 0$, and thus reflection positivity can be proved under reflection with respect to sites, which is sufficient if we are at the end interested in the continuum limit. What permits the proof of positivity is the absence in the lagrangian of time derivatives of the field A_4 which would give rise to reflection properties on the hyperplane $x_4 = 0$ conflicting with the positivity of the kernel $\exp(\Delta A)$ arising in the decomposition of the action A.

The Maxwell field can be treated as a special case $(m = 0)$ of the above theory in the so called non compact formulation; the freedom implied by gauge invariance allows also, by a proper choice of the gauge in the neighbourhood of $x_4 = 0$ (the $A_4 = 0$ gauge), to extend the proof of reflection positivity to reflection with respect to links. A more profound approach [5], which extends also to non abelian fields, is to introduce as elementary variable the finite connection along the path $(x, x + a\hat{\mu})$; the complete proof of positivity goes through rather easily [7-9].

3. Extension of the above ideas to higher $(\geq 3/2)$ spin fields poses some problems. The Rarita - Schwinger lagrangian [11] is given by

(20)
$$\mathcal{L} = \frac{1}{4}\epsilon^{\mu\nu\rho\sigma}\overline{\psi}_\mu\gamma_5\gamma_\nu\overleftrightarrow{\partial}\psi_\sigma - \frac{m}{4}\overline{\psi}_\mu[\gamma_\mu, \gamma_\nu]\psi_\nu.$$

It is hermitean in Minkowski space and reflection symmetric in the euclidean under the antilinear transformation

(21)
$$\psi_\mu(x) \to \eta_\mu \overline{\psi}_\mu(\theta x) \gamma_4$$

(22)
$$\overline{\psi}_\mu(x) \to \eta_\mu \gamma_4 \psi_\mu(\theta x)$$

where $\eta_l = 1$ for $l = 1, 2, 3$ and $\eta_4 = -1$. Due to the vector character of ψ_μ in the index μ, it is natural to assign ψ_μ to the links i.e. to write $\psi(x, x + \hat{\mu})$. This however gives serious problems in getting a reflection symmetric discretized lagrangian, unless one uses the symmetric derivative (11) for $\partial_\rho \psi_\sigma$ in eq. (20) and averages over the mid point in $\overline{\psi}_\mu(x)$. Such a procedure gives rise to a doubling phenomenon for the spin 3/2 particle. Despite the paid price the proof of reflection positivity does not succeed. In fact if we reflect with respect to the links we have an ambiguous behavior of the cut link under reflection, while if we reflect with respect to the sites we have a non positive definite $\exp(A_0)$.

Coming now to the spin 2 field we can start from the general Pauli - Fierz lagrangian [12]

(23)
$$\mathcal{L} = \mathcal{L}_0 + \mathcal{L}_m$$

with

(24)
$$\mathcal{L}_0 = -\frac{1}{4} \partial_\lambda h_{\mu\nu} \partial_\lambda h_{\mu\nu} + \frac{1}{4} \partial_\lambda h_{\mu\mu} \partial_\lambda h_{\nu\nu}$$
$$-\frac{1}{2} \partial_\lambda h_{\lambda\mu} \partial_\mu h_{\nu\nu} + \frac{1}{2} \partial_\lambda h_{\lambda\mu} \partial_\nu h_{\nu\mu}$$

and

(25)
$$\mathcal{L}_m = -\frac{1}{4} m^2 (h_{\mu\nu} h_{\mu\nu} - h_{\mu\mu} h_{\nu\nu})$$

The symmetric tensor field $h_{\mu\nu}$ is not associated to any simple geometrical structure like the A_μ to the link and the $\partial_\mu A_\nu - \partial_\nu A_\mu$ to the plaquette etc. Thus the only possibility is to associate $h_{\mu\nu}$ to the sites and then mid point derivatives are necessary in order to have a reflection symmetric lagrangian. We have seen however, how symmetric derivatives imply doubling [6]. Moreover reflecting with respect to sites clashes against the ambiguous reflection properties of the fields h_{4i} on the hyperplane $x_4 = 0$. The massless case $m = 0$, which corresponds to the linearized gravitational field is an exception due to the invariance of the action under the linear gauge transformations

(26)
$$h_{\mu\nu} \to h'_{\mu\nu} = h_{\mu\nu} + \Delta_\mu \xi_\nu + \Delta_\nu \xi_\mu.$$

This allows to choose (it is enough in the surroundings of the hyperplane $x_4 = 0$) the gauge $\Sigma_i h_{ii} = 0, h_{4i} = 0$ under which $\mathcal{L}_0$ reduces to

(27)
$$-\frac{1}{4}\Delta_4 h_{ij}\Delta_4 h_{ij} - \frac{1}{4}\Delta_l h_{mn}\Delta_l h_{mn}$$
$$-\frac{1}{2}\Delta_l h_{lm}\Delta_m h_{44} + \frac{1}{2}\Delta_l h_{lm}\Delta_n h_{nm}.$$

But

(28)
$$\exp\left(-\frac{1}{4}\sum_{ij}(h_{ij}(\mathbf{x}, 1) - h_{ij}(\mathbf{x}, -1))^2\right)$$

is a positive definite kernel and is the only term connecting positive with negative euclidean times. This is sufficient to have reflection positivity; however the result has only a formal value due to the well known non definiteness in sign of the euclidean Einstein lagrangian (it is non definite also in the Minkowski space). Such non definiteness makes the functional integral divergent; actually in perturbation theory one performs a formal gaussian integration. One wonders what happens in the gravitational case at the exact (i.e. non linearized) level. We already pointed out the poor geometric nature of $g_{\mu\nu}$ and we know that the basic objects are the four one forms e^a (the vierbeins) and the one forms Γ^{ab} (the connections). Such one forms are naturally associated to the elementary line elements (the links) and the Riemann curvature two-form to the plaquettes. It is in fact possible [13] to give a discretized theory by associating to each link a finite element of the Poincaré group $U(x, x + a\hat{\mu})$ and defining the finite curvature along the closed elementary contour (μ, ν) by

(29)
$$R_{\mu\nu}^{ab}(J, x) = -\frac{1}{4a^2}tr[J^{ab}(U_{\mu\nu}(x) - U_{\nu\mu}(x))]$$

and the finite vierbein by

(30)
$$t^a(x, x + a\hat{\mu}) = -\frac{1}{4a}tr(K^a U(x, x + a\hat{\mu})DU(x + a\hat{\mu}, x))$$

where $U_{\mu\nu}$ is the product of the U's along the elementary contour and J^{ab}, D, K^a are respectively the generators of rotations, dilatations and special conformal transformations in a proper representation. It is easy now to write a lagrangian which in the continuum limit reduces to the Einstein - Hilbert action in the first order Palatini form [13]. The enforcement of reflection symmetry on such a lagrangian produces an action which at the linearized level gives the already discussed discretization of $\mathcal{L}_0$ of eq. (24), thus giving rise to the doubling phenomenon for the graviton [6]. The theory is exactly reflection positive when one integrates non perturbatively over all the Poincaré group. With regard to the divergence of the functional integral due to the non definiteness of the Einstein action, it can be cured at the regularized level either by going over to the Sitter model, either by properly introducing terms quadratic in the Riemann tensor i.e. going over to the so called higher derivative gravity. In the Poincaré case a special limit procedure can be envisioned and the problem is finally linked to the possibility of a non conventional renormalization of such a theory [14].

4. In conclusion reflection positivity in the euclidean formulation is completly understood for conventional (spin $\leq$ 1) theories including gauge theories. For higher spin there is only a partial understanding; in the case of gravity the problem mixes with the non definiteness in sign of the Einstein action.

REFERENCES

[1]   J. SCHWINGER, Proc. Nat. Acad. Sci. U.S. **44** 956 (1958); Phys. Rev. **115** 721 (1959).

[2]   G.C. WICK, Phys. Rev. **96** 1125 (1954).

[3]   K. SYMANZIK, in Proceedings of the International School of Physics "Enrico Fermi" course 45 ed. R. JOST, Academic Press, New York London, 1969.

[4]   K. OSTERWALDER - R. SCHRADER, Comm. Math. Phys. **31** 83 (1973); K. OSTERWALDER in Proceedings of the 1973 Erice Summer School, G. VELO - A.S. WIGHTMAN editors, Lecture Notes in Physics vol. **25** Springer 1973 Berlin, Heidelberg, New York.

[5]   K.G. WILSON, Phys. Rev. D **10** 2445 (1974); A.M. POLYAKOV, Phys. Lett. **59 B** 82 (1975).

[6]   P. MENOTTI - A. PELISSETTO, Ann. Phys. (N.Y.) **170** 287 (1986).

[7]   K. OSTERWALDER - R. SEILER, Ann. Phys. (N.Y.) **110** 440 (1978).

[8]   M. LÜSCHER, Comm. Math. Phys. **54** 283 (1977).

[9]   P. MENOTTI - A. PELISSETTO, Comm. Math. Phys. **113** 369 (1987); and in "Field Theory on the Lattice" A. BILLOIRE et al ed. Nucl. Phys. B (Proc. Suppl.) **4** 644 (1988).

[10] K.G. WILSON: Quark and Strings on the Lattice. In: New Phenomena in Subnuclear Physics. A. ZICHICHI (ed.) New York: Plenum 1977.

[11] W. RARITA - J. SCHWINGER, Phys. Rev. **60** 61 (1941).

[12] M. FIERZ - W. PAULI, Proc. Roy. Soc. Lond. **173 A** 211 (1939).

[13] P. MENOTTI - A. PELISSETTO, Phys. Rev. D **35** 1194 (1987); for a different approach based on the Regge calculus see: R. FRIEDBERG - T.D. LEE, Nucl. Phys. **B 242** 145 (1984); H. CHEEGER - W. MÜLLER - R. SCHRADER, Comm. Math. Phys. **92** 405 (1984); B. BERG, Phys. Rev. Lett. **55** 904 (1985), H. HAMBER - R. WILLIAMS, Phys. Lett. **157 B** 368 (1985).

[14] S. WEINBERG, in "General Relativity: An Einstein Centenary Survey" edited by S.W. HAWKING - W. ISRAEL (Cambridge University Press, Cambridge England 1979).

Covariant Symmetric Non-associative Algebras on Group Representations

LOUIS MICHEL (*)

1. - Introduction

It is a great pleasure to dedicate this paper to Luigi, as token of our thirty year old friedship. Our two families have been several times neighbours; with our wives, we have seen growing the Radicati and Michel children. It was always a deep joy to meet. We also wrote many letters to each other. This Festschrift is a great circumstance to thank Luigi for all that I learned from him, from his approach to life and... from his approach to physics.

We have published together six papers [MR]. Most of these papers are related to spontaneous symmetry breaking. Most of them use or sharpen a mathematical tool which is the subject of this short paper. Jordan algebras are examples of "covariant symmetric non-associative algebras on group representations" (we shall simply call them "$\vee - algebras$"); they were invented $[JO1, 2]$ for the need of physics sixty years ago. The second example of $\vee$-algebras explicitly used in physics was introduced by Gell-Mann [GE] on the octect, i.e. the adjoint representation of SU_3; he called it the "D-algebra". This algebra was introduced independently by Biedenharn [BI] for all SU_n adjoint representations. We made a systematic approach of these algebras and extended them to different representations of Lie groups. In physical applications the vectors of these algebras describe physical states; we emphasized that the idempotents of the algebras, or we can also say, the one dimensional subalgebras:

(1.1) $$v \vee v = \lambda v$$

are good candidates for states with spontaneous symmetry breaking.

(*)   I.H.E.S., F-91440 Bures-sur-Yvette.

2. - The ∨-algebras and their automorphisms

Consider a vector space $\mathcal{E}$ on the field $K = \mathbb{C}$ or $\mathbb{R}$. It becomes an algebra when we choose a homomorphism:

$$(2.1) \qquad \mathcal{E} \otimes \mathcal{E} \xrightarrow{\alpha} \mathcal{E}.$$

To have a commutative algebra, we consider only the symmetrized tensor product:

$$(2.2) \qquad \mathcal{E} \otimes_S \mathcal{E} \xrightarrow{\alpha} \mathcal{E}.$$

When there are no ambiguities about the algebra we consider, we shall often use the notation:

$$(2.3) \qquad x, y \in \mathcal{E}, \quad x \vee y \stackrel{\text{def}}{=} \alpha(x \otimes_S y).$$

Given $x \in \mathcal{E}$, the correspondence $\mathcal{E} \ni y \mapsto x \vee y$ is an endomorphism of $\mathcal{E}$ that we denote by:

$$(2.4) \qquad \forall y \in \mathcal{E}, \quad D_x y \stackrel{\text{def}}{=} x \vee y.$$

The symmetric algebras on $\mathcal{E}$ form a vector space, generally denoted by $\text{Hom}(\mathcal{E} \otimes_S \mathcal{E}, \mathcal{E})$ of dimension $n^2(n+1)/2$ where $n = \dim\mathcal{E}$.

The automorphism group of $\mathcal{E}$ is $GL_n(K)$. It acts naturally on $\text{Hom}(\mathcal{E} \otimes_S \mathcal{E}, \mathcal{E})$. Using the same letter for $g \in GL_n$ and the corresponding isomorphism $\mathcal{E} \xrightarrow{g} \mathcal{E}$, this action is:

$$(2.5) \qquad \alpha \mapsto g \circ \alpha \circ g^{-1} \otimes_S g^{-1}.$$

We denote by $(GL_n)_\alpha$ the stabilizer of α; it is the automorphism group of the algebra defined by α.

In physics, we often start from a symmetry group G and its representation on $\mathcal{E}$ of dimension n; it is given by a homomorphism $G \xrightarrow{T} GL_n$. We do not require the representation to be irreducible and we denote its character by:

$$(2.6) \qquad g \in G, \quad \chi(g) = \text{tr } T(g).$$

We recall that the determinant and the characters of the symmetrized tensor product representation satisfy:

$$(2.7) \qquad \begin{aligned} &\det(T(g) \otimes_S T(g)) = (\det T(g))^{n(n+1)/2}, \\ &tr(T(g) \otimes_S T(g)) = \chi_S(g) = \frac{1}{2}\left(\chi(g)^2 + \chi(g^2)\right) \end{aligned}$$

If T is an irreducible representation of G and if the reduction of the symmetrized tensor product representation $T \otimes_S T$, contains the representation T without multiplicity, there exists a unique (up to a scale factor) G-equivariant map α (sometimes called an intertwining operator) which defines a $\vee$-algebra which has G as group of automorphisms. If T is contained in $T \otimes_S T$ with a multiplicity ν, this is also the dimension of the vector space of the covariant algebras. It may happen that G is a strict subgroup of Aut $\mathcal{A} \subset GL_n$; this occurs more often when the representation T is reducible. Indeed the dimension of the vector space of covariant $\vee$-algebras is then larger than one and for some directions of this vector space the automorphism group of the corresponding $\vee$-algebras might be larger (this is similar to the situation, well known to physicists, of accidental degeneracy accompanied with larger symmetry). A famous mathematical example of this phenomenon is told at the end of section 5.

It may happen that G is a strict subgroup of Aut $\mathcal{A} \subset GL_n$; this occurs more often when the representation T is reducible. That G is an automorphism group of the algebra, means:

$$(2.8) \qquad (T(g)x) \vee (T(g)y) = T(g)(x \vee y).$$

This implies for D_x defined in (2.4):

$$(2.9) \qquad T(g)D_x T(g)^{-1} = D_{T(g)x}.$$

If the representation T is irreducible or, more generally, if it does not contain the trivial representation, it leaves no non trivial linear form invariant. So, by taking the trace of the preceding equation:

$$(2.10) \qquad \text{No vector} \neq 0 \text{ invariant by } T \Rightarrow tr D_x = 0.$$

Unitary group representations are very important in physics: they leave invariant a Hermitian scalar product whose real, imaginary part is a symmetric (= orthogonal), antisymmetric (= symplectic) G-invariant bilinear form. In particular, the unitary representation might be real; if not it can always be considered as an orthogonal representation of double dimension.

We end this section by recalling definitions valid for all types of algebras. Given a $\vee$-algebra on $\mathcal{E}$ and two vector subspaces $\mathcal{B}_1, \mathcal{B}_2 \subset E$, we denote $\mathcal{B}_1 \vee \mathcal{B}_2$ the vector space $\{b_1 \vee b_2, \forall b_1 \in \mathcal{B}_1, \forall b_2 \in \mathcal{B}_2\}$. The vector subspace $\mathcal{B}$ is a subalgebra, an ideal of the $\vee$-algebra $\mathcal{A}$ when:

$$(2.11) \qquad \mathcal{B} \text{ subalgebra of } \mathcal{A} \leftrightarrow \mathcal{B} \vee \mathcal{B} \subset \mathcal{B}, \quad \mathcal{B} \text{ ideal of } \mathcal{A} \leftrightarrow \mathcal{B} \vee \mathcal{A} \subset \mathcal{B}.$$

The intersection of two subalgebras (respectively, ideals) is a subalgebra (an ideal). The sum of two ideals is an ideal.

3. - General methods of construction of ∨-algebras

In this section we will consider three general methods to construct ∨-algebras without starting from a group representation, as we have done in the previous section.

As we have seen, the n-dimensional ∨-algebras form a vector space of dimension $n(n + 1)/2$, and we could determine all automorphism groups of these algebras by finding the stabilizers of GL_n on $\text{Hom}(\mathcal{E} \otimes_S \mathcal{E}, \mathcal{E})$ which is equivalent to the action on $(\mathcal{E}' \otimes_S \mathcal{E}') \otimes \mathcal{E}$ where $\mathcal{E}'$ is the dual space of $\mathcal{E}$. This representation of GL_n is the direct sum of two irreducible representations and the stabilizers are the intersections of the stabilizers of the two irreducible representations. This study could be done for each n.

Another approach is possible. The GL_2-representation $g \mapsto g \otimes g$ on the n^2 dimensional space can be realized on the space of $n \times n$ matrices by the action:

$$(3.1) \qquad\qquad m \mapsto gmg^\top.$$

There is a unique decomposition of m into a symmetric and an antisymmetric part:

$$(3.2) \qquad m = s + a, \quad s = \frac{1}{2}\,(m + m^\top), \quad a = \frac{1}{2}\,(m - m^\top)$$

which is invariant under the action (3.1). The action on the symmetric part realizes the representation $g \otimes_S g$. One must find the subgroups G of GL_n leaving stable an n-dimensional subspace of S, the set of $n \times n$ symmetrical matrices and such that the restriction of the G-representation on this invariant subspace be equivalent to the natural representation $G \subset GL_n$. We shall carry this program for $n = 2$ in the next section. Note that for $n = 1$, there is only (up to scaling) one ∨-algebra of dimension 1. The elements of $K^\times (= \mathbb{C}^\times, \mathbb{R}^\times)$ (the multiplicative group of the field) which are automorphisms of this algebra must satisfy $\lambda^2 = \lambda$, so $\lambda = 1$, the automorphism group is trivial.

The third approach starts from a *symmetric trilinear form* and a *non degenerate symmetric bilinear form*. We first recall how to build a completely symmetrical m linear form from $p_m(u)$ a homegenous polynomial of degree m defined on the n-dimensional vector space $\mathcal{E}_n$. The gradient of p_m at u along the vector $x \in \mathcal{E}_n$ is by definition:

$$(3.3) \qquad \mathcal{D}_{x,u}p_m(u) = \lim_{\theta \to 0}\ (p_m(u + \theta x) - p_m(u))$$

For a fixed u the gradient is a linear form on $\mathcal{E}_n$, for a fixed x, it is a degree $m - 1$ homogeneous polynomial. Moreover, the homogeneity of p_m implies:

$$(3.4) \qquad p_m(\lambda u) = \lambda^m p_m(u) \Rightarrow \mathcal{D}_{u,u}p_m(u) = mp_m(u).$$

Remark also that:

$$(3.5) \qquad \mathcal{D}_{x,u}\mathcal{D}_{y,u}p_m(u) = \mathcal{D}_{y,u}\mathcal{D}_{x,u}p_m(u).$$

Then:

$$(3.6) \qquad \tilde{p}_m(x_1, x_2, \ldots, x_m) = (m!)^{-1}\mathcal{D}_{x_1,u}\mathcal{D}_{x_2,u}\ldots\mathcal{D}_{x_m,u}p_m(u).$$

is a completely symmetrical m-linear form such that:

$$(3.7) \qquad \tilde{p}_m(u, u, \ldots, u) = p(u).$$

A bilinear form $\tilde{q}(x,y)$ is non degerate if, and only if, $\forall x \in \mathcal{E}_n, \tilde{q}(x,y) = 0 \Rightarrow y = 0$. This is equivalent to say for the quadratic polynomial $q(u)$ (usually called a quadratic form) that the gradients $\mathcal{D}_{x_i,u}p_m(u)$ of n linearly independent vectors x_i are linearly independent. We call this quadratic form non degenerate (in a coordinate system $q(x) = q_{\alpha\beta}x^\alpha x^\beta$ is non degenerate $\leftrightarrow$ det $q_{\alpha\beta} \neq 0$). Given a homogeneous polynomial t of degree 3 and a non degenerate quadratic form q, one defines the $\vee$-algebra $\mathcal{A}$ on $\mathcal{E}_n$ by:

$$(3.8) \qquad \forall z \in \mathcal{E}_n, \quad \tilde{q}(x \vee y, z) = \tilde{t}(x, y, z).$$

This method allows to build a $\vee$-algebra with a given group of automorphisms G when the linear representation of G on $\mathcal{E}_n$ leaves invariant a non degenerate quadratic form (e.g., it is an orthogonal representation) and also a third degree polynomial. For instance, the permutation group S_n acting by permutation of coordinates of $\mathcal{E}_n$ is represented by real orthogonal matrices (with elements 1 or 0); it leaves invariant the one dimensional subspace $\mathcal{E}'_1$ of vectors with all components equal and the orthogonal subspace $\mathcal{E}'_{n-1}$ of vectors whose sum of their components vanishes. The polynomial $t = \sum_{\alpha=1}^{n}(x^\alpha)^3$ is evidently S_n-invariant; with the invariant quadratic form $q = \sum_{\alpha=1}^{n}(x^\alpha)^2$, it defines a S_n covariant $\vee$-algebras. The vector subspaces $\mathcal{E}'_1$ and $\mathcal{E}'_{n-1}$ carry subalgebras.

This last method does not build the most general $\vee$-algebras. Indeed, given a $\vee$-algebra on $\mathcal{E}$, we say that it leaves invariant a symmetric bilinear form (denoted simply by (u, v), it can be degenerate), when:

$$(3.9) \qquad (x \vee y, z) = (x \vee z, y).$$

From the symmetry of the $\vee$ product and the symmetry of the bilinear form, (3.9) defines a completely symmetrical trilinear form that we have denoted $\{x, y, z\}$ in our papers (e.g. [MR5]). With (2.4) and (3.9) we see that D_x is a symmetric operator: $D_x = D_x^\mathsf{T}$. For any symmetric bilinear form on $\mathcal{A}$, we define:

$$(3.10) \qquad \mathcal{B}^\perp \overset{\text{def}}{=} \{x \in \mathcal{A}, \forall b \in \mathcal{B}, (x, b) = 0\},$$

LEMMA 1. *If a $\vee$-algebra $\mathcal{A}$ carries an invariant symmetric bilinear form, $\mathcal{B}$ ideal of $\mathcal{A} \Rightarrow \mathcal{B}^{\perp}$ ideal of $\mathcal{A}$. If the form is non degenerate, $\mathcal{B} \cap \mathcal{B}^{\perp}$ is a trivial $\vee$-algebra.*

Indeed, $\forall a \in \mathcal{A}$, $\forall b \in \mathcal{B}$, $\forall b' \in \mathcal{B}^{\perp}$, $0 = (b', b \vee a) = (b, b' \vee a) = (b \vee b', a)$. If the form is degenerate, $\mathcal{A}^{\perp}$ is a non trivial ideal.

4. - The 2 dimensional $\vee$-algebras and their automorphisms

It is a classic result on the diagonalization of quadratic forms on $\mathbb{C}$ that in the symmetric representation of $GL_n(C)$:

$$(4.1) \qquad g \in GL_n(C), \quad s^{\top} = s \mapsto gsg^{\top}$$

there are $n+1$ orbits; they are characterized by the rank of s. One orbit is open dense, that of the invertible matrices; I belongs to it. Its stabilizer is the group of matrices which satisfy $gIg^{\top} = I$; this is the (complex) orthogonal group O_n. The group which leaves invariant the one dimensional subspace $\{\lambda I\}$ of the $n(n+1)/2$-dimensional space S is generated by O_n and the dilations:

$$(4.2) \qquad gg^{\top} = \lambda I, \quad \text{with } \lambda^n = (\det g)^2.$$

In dimension $n = 2$ we use the three Pauli matrices:

$$(4.3) \qquad \tau_1 = \begin{pmatrix} 0 & 1 \\ 1 & 0 \end{pmatrix}, \quad \tau_2 = \begin{pmatrix} 0 & -i \\ i & 0 \end{pmatrix}, \quad \tau_3 = \begin{pmatrix} 1 & 0 \\ 0 & -1 \end{pmatrix}.$$

An easy computation yields for the matrices g which satisfy (4.2) for $n = 2$:

$$(4.4) \qquad \lambda = \det g, \; g = aI - bi\tau_2, \quad \lambda = -\det g, \; g = a\tau_3 + b\tau_1,$$
$$\text{with } a^2 + b^2 = \lambda.$$

The group O_2 and the dilations leave also invariant the 2-dimensional subspace of S spanned by the symmetric matrices $\alpha\tau_3 + \beta\tau_1$. Its representation on this space is:

$$(4.5) \qquad \lambda = \det g, \; s \mapsto gsg^{\top} \equiv (aI - bi\tau_2)(\alpha\tau_3 + \beta\tau_1)(a + bi\tau_2)$$
$$\Leftrightarrow g \mapsto (a^2 - b^2)I - 2abi\tau_2;$$

$$(4.5') \qquad \lambda = -\det g, \; s \mapsto gsg^{\top} \equiv (a\tau_3 + b\tau_1)(\alpha\tau_3 + \beta\tau_1)(a\tau_3 + b\tau_1)$$
$$\Leftrightarrow g \mapsto (a^2 - b^2)\tau_3 + 2ab\tau_1.$$

The conditions for this representation to be equivalent to that of (4.4) are:

$$(4.6) \qquad (a^2 - b^2) = a, \; 2ab = \epsilon b, \; \epsilon = \pm 1.$$

This implies either $b = 0$, so $a = 1$, $g = I$, $g = \tau_3$ or $a = \epsilon/2$ ϵ; but $\epsilon = 1$ is ruled out, because it implies $\det g = 0$. So we obtain finally for the elements g of the automorphism group G of the algebra:

$$(4.7) \qquad I, \ \frac{1}{2}\left(I \pm \sqrt{3}i\tau_2\right), \ \tau_3, \ \frac{1}{2}\left(\tau_3 \pm \sqrt{3}\tau_1\right).$$

This is the 6-element group C_{3v}, which is generated by two reflections and is subgroup of the real orthogonal group O_2; it is isomorphic to the permutation group S_3. By polarization equation (4.6) gives the algebra law; if we denote by $x, y; x', y'; \ldots$ the vector coordinates:

$$(4.8) \qquad \begin{pmatrix} x \\ y \end{pmatrix} \vee \begin{pmatrix} x' \\ y' \end{pmatrix} = \begin{pmatrix} xx' - yy' \\ -xy' - yx' \end{pmatrix}.$$

It is well known [MI3] that the invariant polynomials of the representation (3.9) of S_3 are polynomials in the two invariants $q = x^2 + y^2$, $t = \left(x^3 - 3xy^2\right)/3$. So we have obtained an example of application of the third general method of the previous section:

$$(4.9) \ \ u = \begin{pmatrix} x \\ y \end{pmatrix}, \ (u, u) = q, \ u \vee u = \frac{1}{3}\,\mathrm{grad}\ t, \ (u \vee u, u) = t, \ u \vee u \vee u = (u, u)u.$$

Remark that this algebra is simple (no ideal except itself and 0). Note also that $(u \vee u) \vee u = u \vee (u \vee u)$ since the algebra is symmetric and the polarization of the third degree invariant yields an invariant symmetric triple product of vectors (= symmetric trilinear form):

$$(4.10) \qquad \left(\begin{pmatrix} x \\ y \end{pmatrix} \vee \begin{pmatrix} x' \\ y' \end{pmatrix}, \ \begin{pmatrix} x'' \\ y'' \end{pmatrix}\right) = xx'x'' - yx'x'' - xy'x'' - xx'y''.$$

We are left to study the second non trivial orbit of $GL_2(C)$ on the symmetric quadratic forms, the orbit of those of rank one. The subgroup of $GL_2(C)$ which leaves invariant the one dimensional subspace $\begin{pmatrix} \alpha & 0 \\ 0 & 0 \end{pmatrix}$ is the subgroup G of upper triangular matrices $\begin{pmatrix} a & b \\ 0 & d \end{pmatrix}$. Indeed:

$$(4.11) \quad \begin{pmatrix} a & b \\ 0 & d \end{pmatrix} \begin{pmatrix} \alpha & \gamma \\ \gamma & \beta \end{pmatrix} \begin{pmatrix} a & 0 \\ b & d \end{pmatrix} = \begin{pmatrix} a^2\alpha + 2ab\gamma + bd\beta & d(a\gamma + b\beta) \\ d(a\gamma + b\beta) & d^2\beta \end{pmatrix}.$$

The verification is done for $\beta = \gamma = 0$. But the reducible representation of G is not decomposable: it does not leave invariant a two dimensional subspace. For this we have to impose $b = 0$. Then the group of diagonal matrices $\begin{pmatrix} a & 0 \\ 0 & d \end{pmatrix}$ is represented on the subspace of quadratic forms of coordinates γ, β by $\begin{pmatrix} ad & 0 \\ 0 & d^2 \end{pmatrix}$. The two representations are equivalent if, and only if, $d = 1$.

So the symmetry group is isomorphic to $\mathbb{C}^\times$ and its representation is reducible. The corresponding algebra law is:

$$(4.12) \qquad \begin{pmatrix} x \\ y \end{pmatrix} \vee \begin{pmatrix} x' \\ y' \end{pmatrix} = \begin{pmatrix} 0 \\ yy' \end{pmatrix}.$$

The linear forms invariant by the group are, including a factor β:

$$(4.13) \qquad \begin{pmatrix} 0 & 0 \\ 0 & \beta \end{pmatrix}, \text{ i.e. } \left(\begin{pmatrix} x \\ y \end{pmatrix}, \begin{pmatrix} x' \\ y' \end{pmatrix} \right) = \beta yy'.$$

while the quadratic forms invariant by the algebra depend on two parameters:

$$(4.14) \qquad \begin{pmatrix} \alpha & 0 \\ 0 & \beta \end{pmatrix}, \text{ i.e. } \left(\begin{pmatrix} x \\ y \end{pmatrix}, \begin{pmatrix} x' \\ y' \end{pmatrix} \right) = \alpha xx' + \beta yy'.$$

So (4.13) shows that the one dimensional subspace of $\begin{pmatrix} x \\ 0 \end{pmatrix}$ is an ideal, while (4.14) shows, for $\beta = 0$ that the one dimensional subspace of $\begin{pmatrix} 0 \\ y \end{pmatrix}$ is also an ideal.

Any two dimensional subalgebra of a $\vee$-algebra is a (sub)algebra of either of the two forms found here.

5. - Other examples of $\vee$-algebras

The SU_n covariant $\vee$-algebra carried by the SU_n adjoint representation (of dimension $n^2 - 1$) [BI] [GE] [MR5] has been used in physics. The vector space of the SU_n Lie algebra can be represented by the $n \times n$ traceless Hermitean matrices $x = x^*$; the invariant orthogonal scalar product, the Lie and $\vee$ algebra laws are normalized to:

$$(5.1) \qquad (x, y) = \text{tr} xy, \quad x \wedge y = -\frac{i}{2} [x, y], \quad x \vee y = \frac{1}{2} \{x, y\} - \frac{1}{n} (x, y) I$$

where [ ] and { } indicate the commutator and the anticommutator. The $\vee$-algebra is trivial for $n = 2$. The roots r of the algebra have square length 2 and satisfy the characteristic equation:

$$(5.2) \qquad (r, r) = 2, \quad r^n - \frac{1}{2} (r, r) r^{n-2} = 0.$$

So:

$$(5.3) \qquad r \vee r \vee r = \frac{n-2}{n} r$$

Radicati and I have introduced the pseudo roots q:

(5.4)
$$q \stackrel{\text{def}}{=} r \vee r, \quad (q,q) = \frac{2(n-2)}{n}, \quad q \vee q = \frac{n-4}{n} q,$$
$$(q,r) \equiv (r \vee r, r) = 0, \quad q \vee r \equiv r \vee r \vee r = \frac{n-2}{n} r$$

A Cartan subalgebra $\mathcal{C}$ is a maximal Abelian Lie subalgebra: it has dimension $n-1$. It contains $n(n-1)$ roots that we label r_k, $1 \leq k \leq n(n-1)$. The Cartan subspace $\mathcal{C}$ is also a $\vee$ subalgebra, isomorphic to the $\vee$-algebra on the space $\mathcal{E}'_{n-1}$ we have studied in 3. after (3.8); it has S_n as automorphism group: this is here the Weyl group of SU_n. For $a \in \mathcal{C}$, the spectra of the operators ad a and D_a on the $n(n-1)$ dimensional space $\mathcal{C}^{\perp}$ orthogonal to the Cartan subspace are respectively:

(5.5) Spectra on $\mathcal{C}^{\perp}$:   ad $a = \{i(a, r_k)\}$, $D_a = \{(a, q_k)\}$ with $q_k = r_k \vee r_k$

Remark that with the interplay of these two algebras one makes a Z_2-graded SU_n Lie algebra. Another example of $\vee$-algebra for the 18-dimensional real irreducible representation of $SU_3 \times SU_3$ is given in [MR1].

When we restrict SU_n to SO_n the $n^2 - 1$ dimensional adjoint representation of SU_n decomposes into the direct sum of the adjoint representation of SO_n and its symmetric traceless rank 2 tensor representation of dimension $n(n+1)/2$. The latter carries a SO_n covariant $\vee$ algebra (see e.g. [MI2]). This is also the case of the SO_n representation of symmetric traceless tensors of rank $2k$; (e.g. for $k = 2$ [MI6,8,9]).

Let us give now an example of $\vee$ algebra from the mathematical literature. In 1955, in a famous paper [CH], Chevalley found that all known simple (non Abelian) finite groups are simple Lie groups on finite fields except the five Mathieu groups discovered between 1861 and 1873. From 1965 to 1975, 21 other so-called sporadic groups were discovered and it is now a theorem that there are no more. The largest sporadic group is called indifferently "Friendly giant" or "Monster". It was defined [GR] as the automorphism group of a $\vee$-algebra of dimension 196883. The construction of this $\vee$-algebra is clarified in [TI]. In a n-dimensional orthogonal space, a lattice is said to be even if the norm (x, x) of each vector is even. The Grammian Γ (= matrix of the scalar products of basis vectors) of an even lattice is a $n \times n$ matrix with integer elements. The lattice is self-dual if $|\det \Gamma| = 1$. Even self-dual lattices exist only in dimension multiple of 8; one for $n = 8$, the root lattice E_8; two for $n = 16$, $E_8 \oplus E_8$ and a lattice of D_{16} (well known in string theory); twenty four for $n = 24$, among them there is a unique one with shorter vectors of norm 4: the Leech lattice (found in 1965). The quotient of its symmetry group G, divided by its center $\{I, -I\}$ is a sporadic group Co_1 found by J. Conway (1968). The norm 4 and 6 vectors form two orbits of G whose stabilizers are two other sporadic groups, respectively Co_2 (of index 98280) and Co_3. The smallest irreducible representation of G are orthogonal and of dimension $24, 276, 299, \ldots$.

One builds naturally a larger group $C \supset G$, but I cannot give here the details; on its representation (direct sum of 3 irreducible representations) of dimension $299 + 98280 + 98304$, the C covariant $\vee$ algebras form a 6 dimensional vector space. One (up to a scale factor) of these $\vee$-algebras has an exceptionally large automorphism group: the Friendly Giant.

6. - The meaning of idempotent of $\vee$-algebra in physics

Radicati and I showed that in the space of internal symmetry (essentially flavours at that time) of the fundamental interactions, the direction of spontaneous symmetry breaking are idempotents of $\vee$ algebras [MR1,2,3,4]. This was extended by some of our students [PE] [DA]. It is true that from these algebras one can build G-invariant degree four polynomial (bounded below) similar to those proposed sixty years ago by Landau as mathematical model for second order phase transitions, and, more recently by Higgs (in the Lagrangian of the scalar Higgs field) for the spontaneous breaking of symmetry in gauge theories [MI1,2,4,5,7]. There is a difference with the Landau polynomial; for $\vee$-algebras with an invariant polynomial $(x \vee x, x)$ of degree three, the "Landau" polynomial has a degree three term which excludes second order phase transition, but describes first order phase transitions "not far from second order" as they occur sometimes in crystals and often in liquid crystals. It is true that for an orthogonal representation of a symmetry group G without fixed vectors $\neq 0$, the invariant polynomial of degree 3 are of the form $p(x) = (x \vee x, x) - \frac{3}{2} \lambda(x, x)$, so their extremas satisfy: $x \vee x = \lambda x$. However, for the dimension $n \geq 1$, they are all saddle points: indeed the Hessian is $2D_x - \lambda I$; at an extremum x, $(x, H_x x) = \lambda(x, x)$, tr $H_x = -n\lambda$ (see 2.10).

The situation is different if we consider a general bifurcation problem with a symmetry group G. At a bifurcation point, the solutions are tangent to the space of an irreducible (orthogonal, if the problem is on the real) representation of G; with some analyticity hypothesis D. Sattinger [SA] (and his papers quoted there) has shown that bifurcations generally occur for irreducible representations with a G-covariant $\vee$-algebra, in the direction of idempotents.

I had the occasion to verify these properties [MI6,8,9] for the renormalization of the Landau-Higgs model. The symmetry group G acts on an orthogonal n-dimensional representation. The physics must be independent from the basis chosen for this representation; indeed the renormalization equation is O_n covariant and the critical exponents are O_n invariants. At that time the bifurcation equation was written [BR] in the $\varepsilon = d - 4$ expansion. It was convenient to consider all Lagrangians with quartic polynomials and to study the renormalization flow as a vector field u in the vector space P of quartic polynomials; the dimension of P is $1 + ((n + 6)(n + 1)n(n - 1)/24)$ since we assume that the n-dimensional representation of G is irreducible on the real. There is a O_n covariant $\vee$-algebra on P. The renormalization flow u of a G invariant Lagrangian $\mathcal{L}$ stays tangent to the space P^G of G invariant quartic polynomials. In the neighbourhood of small ϵ the number and the type of stability of the renormalization fixed point u^* depend only on the leading term of the renormalization equation, i.e. $u^* \vee u^* = \frac{2}{3} \varepsilon u^*$ if no further degeneracy appears (this is not the case for $n = 4$; this exceptional case has been completely treated in [TO]). Let $\tilde{G}$ the stabilizer of $\mathcal{L}$ in the O_n action; it might be strictly larger than G and it is the true symmetry group of the problem. If there are solutions u_i^*, we can form others by action of the stabilizer in O_n of $\mathcal{P}^G$. It can be shown that the latter is equal to the normalizer $N_{O_n}(\tilde{G})$. I could prove the theorem: if there is a stable fixed point, it is unique. This shows that the symmetry of a stable fixed point satisfies $\tilde{G} = N_{O_n}(\tilde{G})$. With J.C. Toledano, we have studied the physical implications of these results [MI10].

However there is much more to say on the use of covariant $\vee$-algebras in physics; I hope one day to work again on this subject with Luigi!

REFERENCES

[BI] L.C. BIEDENHARN, *On the representations of the semi-simple Lie groups*. I. *The explicit construction of invariants for the unimodular unitary group in N-dimensions*. J. Math. Phys., **4** (1963) 436-445.

[BR] E. BREZIN - J.C. LE GUILLOU - J. ZINN JUSTIN, *Discussion of critical phenomena for the n-vector model*. Phys. Rev. B10 (1974) 892-900.

[CH] C. CHEVALLEY, *Sur certains groupes simples*. Tohoku Math. J. (2) **7** (1955) 44-66.

[DA] C. DARZENS, *On chiral symmetry breaking II*. Ann. Phys., **76** (1973) 236-249.

[GE] M. GELL-MANN, *Symmetries of Baryons and Mesons*. Phys. Rev., **125** (1962) 1067-1084.

[GR] R.L. GRIESS Jr., *The Friendly Giant*, Inventiones Math., **69** (1982) 1-102.

[JO] P. JORDAN

1. *Über Verallgemeinerungsmöglichkeiten der Formalismus der Quantenmechanik*, Gött Nach. (1933) 209-214.

2. *Über die Multiplikation quantenmechanischer Grössen*, Z. Phys. **80** (1933) 209-214.

[MI] L. MICHEL

1. *Simple mathematical models of symmetry breaking. Application to particle physics*. p. 251-262 in *"Mathematical Physics and Physical Mathematics"*, R. Maurin and R. Rączka edit., D. Reidel PubL. Co, Dordrecht 1976.

2. *Les brisures spontanées de symétrie en physique*. p. 234-245 Colloque C7, supp. J. Physique **36** (1975) C7 41-51.

3. *Invariants Polynomiaux des groupes de symétrie moléculaire et crystallographique*. p. 75-91 in Proc. 5th Intern. Colloq. Group Theoretical Methods in Physics, Academic Press 1977.

4. *Minima of Higgs-Landau potentials*. p. 157-203 in *Regards sur la physique contemporaine*, Editions CNRS, Paris 1979.

5. *The description of symmetry of physical states and spontaneous symmetry breaking*. p. 21-28 in *"Symmetries and broken symmetries in condensed matter physics"* Editor N. Boccara, IDSET Paris 1981.

6.   *Symmetry and the renormalization group fixed points of quartic Hamiltonians*, p. 63-82 in *Symmetry in particle physics*, I. Bars et al. edit. Plenum Press, New-York 1984.

7.   *Landau theory of second order phase transition and invariant theory.* p. 763-773, in *"Group Theoretical method in Physics"*, edit Markov, Man'ko, Shabad, Harwood Academic Pub. 1987.

8.   *Renormalization group fixed points of general n-vector model.* Phys. Rev. B29 (1984) 2777-2783.

9.   *Covariant vector fields and renormalization groups.* p. 162-184 in Proc. 13th International Colloquium in Group Theoretical Methods in Physics. World Scientific, Singapore, 1984.

10.   with J.C. TOLEDANO, *Symmetry criterion to the lack of a stable fixed point in the renormalization group recursion relations.* Phys. Rev. Lett. **54** (1985) 1832-1833.

[MR] L. MICHEL - L. RADICATI

1.   *On the dynamical breaking of $SU(3)$.* p. 19-34 in *"Symmetry principles at high energy"*. (Proc. 5th Coral Gables conf.), Benjamin, New-York 1968.

2.   *Breaking of the $SU(3) \times SU(3)$ Symmetry in Hadronic Physics*, p. 191-203 in *"Evolution of Particle Physics"* Academic Press, New-York 1970.

3.   *Geometrical properties of the fundamental interactions* in *"Mendeleev Symposium"*, Atti Accad. Sci. Torino, Sci. Fis. Mat. Natur., (1971) 377-389.

4.   *Properties of the breaking of Hadronic internal symmetry.* Ann. of Phys., **66** (1971) 758-783.

5.   *The geometry of the octet.* Ann. Inst. Henri Poincaré, **18** (1973) 185-214.

6.   with D. CONSTANTINESCU, *Spontaneous symmetry breaking and bifurcation from the Mac Laurin and Jacobi sequence.* J. Physique, **40** (1979) 147-159.

[PE] F. PEGORARO - J. SUBBA RAO, *Effect of weak interaction in the breaking of hadronic internal symmetry.* Nucl. Phys., B44 (1972) 221-235.

[SA] D.H. SATTINGER, *Bifurcation and symmetry breaking in applied mathematics.* Bull. Am. Math. Soc., **3** (1980) 779-819.

[TI] J. TITS, *Le Monstre*, Séminaire Bourbaki, 620 (1983).

[TO] J.C. TOLEDANO - L. MICHEL - P. TOLEDANO - E. BREZIN, *Fixed points and stability for anisotropic system with 4-component order parameters.* Phys. Rev. B31 (1985) 7171-7196.

The Baryon Magnetic Moments, Semileptonic Matrix Elements and Masses: A Field Theoretical Parametrization and the non Relativistic Quark Model

G. MORPURGO (*)

1. - The problem

This article deals with the relationship between the non relativistic quark model ($NRQM$) and the relativistic field theory[†]. I will show that one can look to the $NRQM$ in a way rather different from the customary one. It will emerge that for the validity of many results of the $NRQM$ it is not compulsory that the motion of the quarks inside the hadrons is slow; the existence of a correspondence between the usual simple states of the $NRQM$ and the exact states plus some rather general property of the underlying field theory (leading to the dominance of the additive quark terms) can be sufficient; so that - in principle - one can understand why the $NRQM$ works well also for the hadrons composed of "light" constituent quarks.

I will exemplify this in three cases, the magnetic moments of the baryons, their semileptonic decays and the octet and decuplet baryon mass formulas; but the conclusion is more general. In essence we show the following: 1) each of the quantities mentioned above (magnetic moments, semileptonic matrix elements, masses), calculated with a complete relativistic field theory (e.g. QCD) can be parametrized as a sum of many terms with different spin-flavour structures, each term being multiplied by some coefficient; 2) the $NRQM$ calculations can be interpreted as a selection of certain terms among the many ones just mentioned;

(*) Istituto di Fisica dell'Università and INFN-Genova (Italy).

[†] I am doubly indebted to Luigi Radicati in connection with my work on the non relativistic quark model ($NRQM$). First, his (and Gürsey's) proposal [1] of SU_6 contributed [2] to stimulate my first paper [3] - 24 years ago - on the $NRQM$. The second reason for gratitude refers precisely to the present article; although I noted long ago [4] the general ideas to be developed here, it has been only in thinking to this paper for Luigi's seventieth birthday that I have straightened them out and seen that they might be, perhaps, of some interest. I am therefore glad to dedicate this paper to him.

in other words among the many terms of the complete parametrization obtained from field theory, the $NRQM$ selects a few and neglects many others; 3) for the baryon magnetic moments, semileptonic matrix elements, masses, it is possible (if we consider flavour breaking up to a certain order only) to determine the coefficients of all the terms in the complete parametrization by fitting the experimental data. One can then verify that the few terms selected in the $NRQM$ have coefficients appreciably larger than those of the others. However the mechanism through which the underlying relativistic field theory (QCD?) operates so that the terms selected by the $NRQM$ are larger than the others must still be clarified.

Before entering in the details, it is appropriate to summarize the main steps of the procedure to be used in the parametrization of the field theoretical results.

a) We start with the introduction of a set of model states. These model states $|\phi_B >$ are, for the baryons, three quark states with the usual "naive" wave functions of the $NRQM$ (baryon B at rest, $L = 0$ and total spin 1/2 or 3/2 respectively for the lowest octet and decuplet). We consider these model states as the eigenstates of some simple flavour symmetric non relativistic model hamiltonian $\mathcal{H}$; we also introduce the exact states $|\psi_B >$, the (baryon) eigenstates of the hamiltonian H of the complete relativistic field theory (say the hamiltonian of QCD, if QCD is the correct theory).

b) We then introduce, and show how to construct, a unitary operator V that, when applied to the model states, transforms them into the exact eigenstates $|\psi_B >$: $|\psi_B > = V|\phi_B >$.

c) Consider now the calculation of, say, the magnetic moment of a baryon $< \psi_B|\underline{M}|\psi_B >$, where $\underline{M} = (1/2) \int [\underline{j}(\underline{r}) \times \underline{r}]d^3\underline{r}$ in an obvious notation. Using the operator V, this amounts to calculate the expectation value of the operator $V^+\underline{M}V$ in the state $|\phi_B >$. The operator $V^+\underline{M}V$ is a very complicated field operator, enormously more complicated, of course, than $\underline{M}$; but, because we calculate only its expectation value on a three quark state $|\phi_B >$, the only part of $V^+\underline{M}V$ that intervenes is its projection on the three quark states. It is evident that this projection is equivalent to some three quark operator $\underline{\tilde{M}}$ constructed in terms of the spin-flavour-space variables of the three quarks intervening in ϕ_B; thus the problem has been reduced to that of constructing the most general 3-quark axial vector operator $\underline{\tilde{M}}$ of the variables $\underline{\sigma}_i$'s, λ_i's and $\underline{r}_i$'s of the three quarks ($i = 1, 2, 3$). Because the magnetic moment is an axial vector, the three body operator $\underline{\tilde{M}}$ to be constructed must be an axial vector. [When we will parametrize the masses, the quantity to be constructed will be V^+HV, a scalar under space rotations].

d) We stress at this point that the simplicity of the model states ϕ_B plays a vital role. Let us continue to fix the attention on the magnetic moments. Because the orbital angular momentum L of ϕ_B is zero, the axial vector $\tilde{M}$ has no term proportional to $\underline{L}$; because ϕ_B can be factorized in the product of a space, spin-flavour, colour factor, it becomes possible to eliminate (by integration) all reference to the space coordinates and colour. Because the factor referring to the spin-flavour variables of the three quarks is symmetric, due to the antisymmetry of the space-colour factor, it is necessary that $\tilde{M}$ be a symmetric spin-flavour operator of the three quarks.

e) The problem is thus to classify all the symmetric spin-flavour operators of the three quarks. For the spin structure a great simplification is achieved using a property of the Pauli $\underline{\sigma}$'s: the expectation value of $(\underline{\sigma}_i \times \underline{\sigma}_k)$ $(i \neq k)$ on any real wave function $R(1, 2, 3)$ of the spin flavour variables of the three quarks vanishes. This property reduces considerably the number of the axial vectors [and also of the scalars] that can be constructed through the spin operators of the three quarks and that, therefore, appear in the final parametrization.

f) As to the flavour structure of the final parametrization, its complexity is reduced if one assumes, as we do, that in the exact field theoretical strong hamiltonian H of the quark gluon fields, the only term breaking the flavour is due to the mass difference of the strange quark λ with respect to the P and N quarks [isospin is assumed to be conserved]. If this is so, the most general flavour operator that can intervene in a spin-flavour expression operating between the initial and final quark states $|\phi_B >$ is P_i^λ, the projection operator on the λ quark of index i. Flavour breaking to successive orders corresponds to the presence of terms containing an increasing number of P^λ factors (up to a maximum of course of 3).

g) In view of e) and f) the number of the spin-flavour structures that intervene in the parametrization turns out to be sufficiently small if we confine, as we shall do, to first order in flavour breaking.

h) I repeat that the method of parametrization to be developed here is founded largely on the simple choice of the model hamiltonian $\mathcal{H}$ (that implies simple model states $|\phi_B >$). These states are the original $NRQM$ states with $L = 0$, that, as far as their spin-flavour structure is concerned, coincide [5] with the SU_6 states of Gursey and Radicati [1]. It appears that the role of these model states does not depend on whether SU_6 is or is not an approximate symmetry; their importance is due to the fact that they are a set of simple factorizable states that can be put in correspondence with the exact states of the system; their magic properties are factorizability, given values of L and S, and for the lowest baryons and mesons, $L = 0$.

i) In the following we will proceed in the order indicated above to the general parametrization of the baryon magnetic moments. We then examine how this parametrization compares with experiment and with the $NRQM$. Then we will do the same for the semileptonic decay matrix elements and the masses.

2. - The operator V transforming the model states into the exact ones

We start from the exact field theoretical Hamiltonian H of the strong interactions, involving the quark and gluon fields. H may be the Hamiltonian of QCD or may be different; it is, anyway, the Hamiltonian of a relativistic field theory of quarks and gluons. We will assume only two properties for H: a) that the only flavour breaking in H is due to the mass term; b) that the electromagnetic and the weak currents are the conventional ones expressed only in terms of the quark fields. QCD has these properties.

Consider a baryon B at rest (belonging to the lowest octet or decuplet) and call $|\psi_B >$ its state:

$$(1) \qquad\qquad H\ |\psi_B > = M_B\ |\psi_B >$$

To simplify the presentation we fix the attention until further notice on the calculation of the magnetic moment M_z of an octet baryon; this is the expectation value of the operator $M_z = (1/2) \int (\underline{j}(\underline{r}) \times \underline{r})_z d^3\underline{r}$ in the state ψ_B:

$$(2) \qquad\qquad M_z = < \psi_B|M_z|\psi_B >$$

where the current $j_\mu(x)$ that appears in M_z:

$$(3) \qquad j_\mu(x) = e\ \left[\frac{2}{3}\ \bar{u}_R \gamma_\mu u_R - \frac{1}{3}\ \bar{d}_R \gamma_\mu d_R - \frac{1}{3}\ \bar{s}_R \gamma_\mu s_R \right]$$

is expressed in terms of the renormalized quark fields $u_R(x), d_R(x), s_R(x)$ (the index R stays for renormalized). We assume that at small momentum transfer the propagators of the renormalized quark fields are governed by masses in the range of a few hundred MeV (the so called "constituent" quark masses) and not of the order of a few MeV (the bare quark masses). In fact we will use the notation P, N, λ for the renormalized quark fields:

$$P(x) \equiv u_R(x), \quad N(x) \equiv d_R(x), \quad \lambda(x) \equiv s_R(x)$$

and identify these fields with the "constituent quarks". Note, however, that the masses of P, N, λ will not intervene explicitly in what follows except for the parameter $(m_\lambda - m_P)/m_\lambda \equiv \Delta m/m_\lambda$ that characterizes the flavour breaking.

In the following we will often abbreviate with $\Psi(x)$ the quark fields:

$$(4) \qquad\qquad \Psi(x) = \begin{vmatrix} P(x) \\ N(x) \\ \lambda(x) \end{vmatrix}$$

so that the current $j_\mu(x)$ is:

$$(5) \qquad\qquad j_\mu(x) = e\ \bar{\Psi}(x) P^q \gamma_\mu \Psi(x)$$

where:

(6)
$$P^q = \frac{2}{3} P^P - \frac{1}{3} P^N - \frac{1}{3} P^\lambda$$

and P^P, P^N, P^λ are the projection operators that select the P, N, λ fields. In terms of the Gell Mann matrices, where $\lambda_3 = \text{Diag} [1, -1, 0]$ and $\lambda_8 = \text{Diag} [1, 1, -2]$:

(7) $\quad P^P = \frac{1}{6} (2 + 3\lambda_3 + \lambda_8), \quad P^N = \frac{1}{6} (2 - 3\lambda_3 + \lambda_8), \quad P^\lambda = \frac{1}{3} (1 - \lambda_8)$

In this notation the mass term in the hamiltonian H is:

(8)
$$m \int \overline{\Psi}(\underline{r})\Psi(\underline{r})d^3\underline{r} + \Delta m \int \overline{\Psi}(\underline{r})P^\lambda \Psi(\underline{r})d^3\underline{r}$$

In addition to the exact Hamiltonian H of the complete relativistic field theory introduced above, we now define an auxiliary Hamiltonian $\mathcal{H}$ operating only (if we are interested in the baryons) in the three quark (no antiquark, no gluon) sector. This auxiliary Hamiltonian $\mathcal{H}$ is *our choice*; we select it as the simplest, most naive non relativistic quark model Hamiltonian. We call it the model Hamiltonian; its eigenstates, the model eigenstates, will be indicated as $|\phi_B >$:

(9)
$$\mathcal{H} |\phi_B > = M_0|\phi_B >$$

The model Hamiltonian is assumed to be flavour independent and to conserve separately the total spin and the total orbital angular momentum of the three quarks; its eigenstates $|\phi_B >$, corresponding to the lowest octet and decuplet baryons, are assumed to be the usual $L = 0$ baryon wave functions:

(10)
$$\phi_B = X_{L=0}(\underline{r}_1, \underline{r}_2, \underline{r}_3) \cdot W_B(1, 2, 3) \cdot S_c(1, 2, 3)$$

where $X_{L=0}$ is the space symmetrical factor, W_B is the spin flavour symmetrical factor and S_c is the singlet colour antisymmetrical factor. I insist on $L = 0$ because this feature will be essential to simplify things in the following. Note that, as well known, $L = 0$ plus space-colour antisymmetry [5] implies that the spin flavour part of the wave function is symmetrical and has the SU_6 structure. The flavour independence of the auxiliary Hamiltonian $\mathcal{H}$ implies that the space part $X_{L=0}$ of the wave function is the same for all the octet baryon states and that $\mathcal{H}$ has, as indicated in (9) a degenerate eigenvalue M_0 corresponding to all the octet baryons (this is why M_0 in (10) does not carry the baryon index B). We do not assume that the space part $X_{L=0}$ and the eigenvalue M_0 for the decuplet baryons are the same as for the octet.

To introduce the operator V we now decompose the exact Hamiltonian H as:

(11)
$$H = H_0 + H_1$$

where H_0 is the free Hamiltonian and H_1 is the interaction between quarks and gluons, everything being expressed in terms of the renormalized fields as stated before. Calling η a projection operator on the states of three quarks (no antiquark-no gluon), we rewrite H as:

$$(12) \qquad\qquad\qquad H = K_0 + K_1$$

where:

$$(13) \qquad\qquad\qquad K_0 = H_0 + \eta \mathcal{X} \eta - \eta H_0 \eta$$

$$(14) \qquad\qquad\qquad K_1 = H_1 - \eta \mathcal{X} \eta + \eta H_0 \eta$$

Writing $H_0 \equiv \eta H \eta + (1 - \eta) H (1 - \eta)$ and $H_1 \equiv (1 - \eta) H \eta + \eta H (1 - \eta)$, we can also express K_0 and K_1 as:

$$(15) \qquad\qquad\qquad K_0 = \eta \mathcal{X} \eta + (1 - \eta) H (1 - \eta)$$

$$(16) \qquad\qquad K_1 = (1 - \eta) H \eta + \eta H (1 - \eta) - \eta \mathcal{X} \eta + \eta H \eta$$

Because in the 3-quark sector K_0 and $\mathcal{X}$ are the same, the states $|\phi_B > |0$ gluons$>$ are eigenstates of K_0 belonging to the eigenvalue M_0 (omitting the gluon zero point energy):

$$(17) \qquad\qquad K_0 |\phi_B > |0 \text{ gluons} > = M_0 |\phi_B > |0 \text{ gluons} >$$

The unitary operator V is now defined as the operator that, acting on the eigenstates $|\phi_B >$ of K_0 (that is of $\mathcal{X}$), transforms them into the eigenstates of $K_0 + K_1$, that is of H. This can be done, in principle, by the adiabatic method, that is imagining (compare [6] for some details) an adiabatic insertion of the "perturbation" K_1 in (12). We write (abbreviating $|\phi_B > |0$ gluons$>$ in $|\phi_B >$):

$$(18) \qquad\qquad\qquad |\psi_B > = V |\phi_B >$$

It is evident that the structure of the field operator V is very complicated; it transforms a simple 3-quark (0 antiquark, 0 gluon) state (the eigenstate of the simple model Hamiltonian) into the exact state $|\psi_B >$. The complexity of V is clear if we imagine the state $|\psi_B >$ expanded in Fock states of quarks and gluons; this expansion is a superposition of all the infinite Fock states with the correct (exact) quantum numbers of B; schematically:

$$(19) \qquad |\psi_B > = |qqq > + |qqq\bar{q}q > + |qqqG > + |qqqGG > + \ldots$$

where we have left understood the amplitudes that should appear in front of the various states; note incidentally that the 3-quark state in the Fock expansion

(19) of $|\psi_B >$ is a superposition of 3-quark states with different values of L (configuration mixing), whereas the 3-quark model state $|\phi_B >$ has, by construction, only $L = 0$.

3. - The effective three quark magnetic moment operator $\tilde{\underline{M}}$

Consider the expression (2) for the magnetic moment; in it we replace $|\psi_B >$ with $V|\phi_B >$; where V is the operator introduced above. Thus we have:

$$(20) \qquad M_z = \; < \psi_B | M_z | \psi_B > \; = \; < \phi_B | V^+ M_z V | \phi_B >$$

where:

$$(21) \qquad V^+ \underline{M} \; V = V^+ \int [(\underline{j}(\underline{r}) \times \underline{r})/2] d^3 \underline{r} \; V$$

The operator (21) is a complicated one containing all sorts of creation and destruction operators of quarks, antiquarks and gluons. We are interested, however, only in the expectation value of $V^+ \underline{M} \; V$ in the state $|\phi_B >$. Because the state $|\phi_B >$ contains only three quarks, we are interested only in the projection $\tilde{M}$ of $V^+ \underline{M} \; V$ in the subspace of the 3-quark, no antiquark no gluon states $|3q >$:

$$(22) \qquad \tilde{\underline{M}} = \sum_{3q, 3q'} \sum |3q > < 3q|V^+ \underline{M} \; V|3q' > < 3q'|$$

This projection $\tilde{\underline{M}}$ is necessarily a function of the three quark variables $1, 2, 3$ in ϕ_B; stated differently, the evaluation of the expectation value of the field operator $V^+ \underline{M} \; V$ in the state $|\phi_B >$ reduces to the calculation of the expectation value of some three body quantum mechanical operator $\tilde{\underline{M}}$ in ordinary non relativistic three body quantum mechanics; this is a trivial but important simplification.

We now proceed to determine the most general form of the three body axial vector operator $\tilde{\underline{M}}$. Before showing that the complexity of $\tilde{\underline{M}}$ is greatly reduced by the choice of the wave functions (10) it is convenient, however, to clarify two points. The first is the following: the current $\underline{j}(\underline{r})$ is a field operator expressed in terms of creation and destruction operators of quark Dirac fields; these operators act on four component spinors. On the other hand the non relativistic model Hamiltonian $\mathcal{H}$ and its eigenstates $|\phi_B >$ are expressed in terms of two component spinors. What is then the meaning of (20)? The answer is simple: the two component spinors in ϕ_B should be interpreted as four component spinors with two zero's in the lower components. This amounts to extend formally the model Hamiltonian $\mathcal{H}$ to a 4×4 matrix that operates on four component spinors but does not connect the spaces of the upper and lower components and is zero on the latter.

The second point refers to the colour: we have written the current (3) without specifying the colour index of the quarks and without summing over it. This can be done because all the ensuing applications refer to expectation values or matrix elements between colour singlets; in such calculations the colour plays no role, except for guaranteing a symmetric space factor $X_{L=0}$ in (10); to simplify the notation we will suppress the colour degree of freedom in what follows.

4. - The elimination of the space variables and the spin structure of $\underset{\sim}{M}$

It is easy to see that the most general form of $\underset{\sim}{M}$ to be inserted between $L = 0$ states can be written:

$$(23) \qquad \underset{\sim}{M} = \sum_{\nu} Z_{\nu}(\underline{r}, \underline{p}) \underline{G}_{\nu}(\underline{\sigma}, f)$$

where Z_{ν} is an operator depending on the $\underline{r}$'s $(\equiv \underline{r}_1, \underline{r}_2, \underline{r}_3)$ and $\underline{p}$'s $(\equiv \underline{p}_1, \underline{p}_2, \underline{p}_3)$ of the three quarks and $\underline{G}_{\nu}$ is an axial vector operator depending on their spin $\underline{\sigma}(\equiv \underline{\sigma}_1, \underline{\sigma}_2, \underline{\sigma}_3)$ and flavour f $(\equiv f_1, f_2, f_3)$ variables. The expression (23) contains the fact already mentioned that the axial vector property of $\underset{\sim}{M}$ is carried only by the spin factors $\underline{G}_{\nu}$. The factorization property of ϕ_B allows to integrate over the $\underline{r}, \underline{p}$ variables; after this integration the operator $\underset{\sim}{M}$ - a function of the spin-flavour variables only - is written as:

$$(24) \qquad \underset{\sim}{M} = \sum_{\nu} g_{\nu} \underline{G}_{\nu}(\underline{\sigma}, f)$$

where g_{ν} are numerical real coefficients. The sum over ν in (23) or (24) is extended to all the independent hermitian axial vectors $\underline{G}_{\nu}(\underline{\sigma}, f)$ of the three quarks.

We have analyzed in [6] the expression of the most general hermitian axial vector formed with the spins $\underline{\sigma}_1, \underline{\sigma}_2, \underline{\sigma}_3$ of three spin 1/2 particles, and we refer to that analysis. We only mention the main simplification that emerges in constructing these axial vectors and give the conclusion. The main simplification is this: if $c(f)$ is any hermitian real flavour dependent operator of the three quarks (by "real" we mean that its matrix elements between real functions are real) and if R is any real spin flavour wave function of the three quarks it is:

$$(25) \qquad < R|(\underline{\sigma}_1 \times \underline{\sigma}_2)c(f)|R > = 0$$

Due to this property the most general $\underline{G}(\underline{\sigma}, f)$ can only be - when operating on states ϕ_B with a given J -:

$$(26) \qquad \underline{G}(\underline{\sigma}, f) = \underline{\sigma}_1 \Gamma_1^J(f) \quad \text{or} \quad = \underline{\sigma}_2 \Gamma_2^J(f) \quad \text{or} \quad = \underline{\sigma}_3 \Gamma_3^J(f)$$

where the $\Gamma_i^J(f)$ are real flavour operators. Any other spin-flavour structure can be reduced to a combination of the expressions (26). Note finally the following: a) the property (25) holds only for a diagonal matrix element, not necessarily for a transition one; b) in addition to the expectation value (25) also the expectation value of the scalar:

$$(27) \qquad\qquad (\underline{\sigma}_1 \times \underline{\sigma}_2) \cdot \underline{\sigma}_3$$

times any hermitian real flavour dependent operator vanishes on any real spin flavour state of three particles. This property will be important in the analysis of the masses (Sect. 10).

5. - The flavour structure of $\underline{\tilde{M}}$

Consider the three body operator $\underline{\tilde{M}}$ (22); $\underline{\tilde{M}}$ is linear in the electromagnetic current and the latter is linear in P^q [compare the equations (6) and (7)]; because the only flavour dependence of V is through the operator P^λ (that appears in the flavour breaking term (8)) and because P^λ commutes with P^q, it follows that also $\underline{\tilde{M}}$ is linear in P^q. Moreover because each order of flavour breaking implies the presence of a P^λ, the flavour factor $\Gamma^J(f)$ (equation (26)) of $\underline{\tilde{M}}$ can only be one of the following:

$$(28) \qquad\qquad P_i^q, \quad P_i^q P_k^\lambda, \quad P_i^q P_k^\lambda P_j^\lambda, \quad P_i^q P_1^\lambda P_2^\lambda P_3^\lambda$$

where the indices i, k, j (each of which can have the three values $1, 2, 3$) refer to the three quarks; i, k, j can be equal or different. Note that for equal values of the index it is:

$$(29) \qquad\qquad P_i^q P_i^\lambda = -\frac{1}{3} P_i^\lambda$$

If we include the flavour breaking at no more than first order, only the first two structures on the left among those listed above (28) intervene.

Thus all the spin flavour structures breaking the flavour up to first order are obtained, according to (26), multiplying $\underline{\sigma}_1, \underline{\sigma}_2$ or $\underline{\sigma}_3$ either by P_i^q or by $P_i^q P_k^\lambda$. Since the spin-flavour factor of ϕ_B is symmetric in the variables of the three quarks, the effective magnetic moment operator $\underline{\tilde{M}}$ must also be symmetric in $1, 2, 3$; it follows that, to first order in flavour breaking, $\underline{\tilde{M}}$ is a linear combination of the following seven expressions (written without using the simplification (29); we will exploit it in a moment):

$$(30) \qquad \underline{G}_1 = \sum_i P_i^q \underline{\sigma}_i \qquad\qquad \underline{G}_2 = \sum_i P_i^q P_i^\lambda \underline{\sigma}_i$$

$$\underline{G}_3 = \sum_{i \neq k} P_i^q \underline{\sigma}_k, \quad \underline{G}_4 = \sum_{i \neq k} P_i^q P_i^\lambda \underline{\sigma}_k, \quad \underline{G}_5 = \sum_{i \neq k} P_i^q P_k^\lambda \underline{\sigma}_k,$$

(31)
$$\underline{G}_6 = \sum_{i \neq k} P_i^q \underline{\sigma}_i P_k^\lambda, \quad \underline{G}_7 = \sum_{i \neq k \neq j} P_i^q P_k^\lambda \underline{\sigma}_j$$

[If we had included all the possible flavour breaking terms, that is also containing 2 and 3 P^λ's factors, the number of possible expressions would have been 11 instead of 7-compare [6] for their complete list]. The expressions (30) (31) simplify introducing the abbreviations:

(32)
$$\underline{\Sigma}^q \equiv \sum_i P_i^q \underline{\sigma}_i, \qquad \underline{\Sigma}^\lambda \equiv \frac{1}{3} \sum_i P_i^\lambda \underline{\sigma}_i$$

as well as the charge Q, the strangeness S and the total spin $2\underline{J}$:

(33)
$$Q \equiv \Sigma_i \, P_i^q, \quad S \equiv -\Sigma_i P_i^\lambda, \quad 2\underline{J} \equiv \Sigma_i \underline{\sigma}_i$$

The seven $\underline{G}_\nu$'s $(\nu = 1, \ldots 7)$ (30) (31) are written below in terms of the quantities (32) (33); we have now used the simplification (29):

$$\underline{G}_1 = \underline{\Sigma}^q$$
$$\underline{G}_2 = -\underline{\Sigma}^\lambda$$
$$\underline{G}_3 = Q \cdot (2\underline{J}) - \underline{\Sigma}^q$$
$$\underline{G}_4 = \frac{1}{3} S \cdot (2\underline{J}) + \underline{\Sigma}^\lambda$$
$$\underline{G}_5 = 3Q \cdot \underline{\Sigma}^\lambda + \underline{\Sigma}^\lambda$$
$$\underline{G}_6 = -S \cdot \underline{\Sigma}^q + \underline{\Sigma}^\lambda$$
$$\underline{G}_7 = -Q \cdot S \cdot (2\underline{J}) - \frac{1}{3} S \cdot (2\underline{J}) - 3Q \cdot \underline{\Sigma}^\lambda + S \cdot \underline{\Sigma}^q - 2 \underline{\Sigma}^\lambda$$

6. - The baryon magnetic moments; a comparison with the $NRQM$

Each of the seven $\underline{G}_\nu$'s linear in the flavour breaking is a linear combination of products of the quantities (32) and (33) containing at most the first power of S or of $\underline{\Sigma}^\lambda$. Therefore we can write the most general form for the magnetic moment $\underline{M}$ of a baryon to first order in the flavour perturbation as the expectation value over the model states of:

(35)
$$\underline{M} = \sum_1^7 {}_\nu g_\nu \underline{G}_\nu \equiv \mu \, \underline{\Sigma}^q + A \, \underline{\Sigma}^\lambda + F \, Q \cdot (2\underline{J}) + H \, S \cdot (2\underline{J})$$
$$+ L \, Q \cdot \underline{\Sigma}^\lambda + K \, S \cdot \underline{\Sigma}^q + G \, Q \cdot S \cdot (2\underline{J})$$

where $g_1, g_2, g_3 \ldots g_7$ or, equivalently, μ, A, F, H, L, K, G are seven real parameters. Because the magnetic moments of the seven octet baryons

$P, N, \Lambda, \Sigma^{+,-}, \Xi^{0,-}$ have been measured, it is possible to determine the seven coefficients in (35); the expectation values of Σ_z^q and Σ_z^λ for the baryons (necessary to calculate the expectation value of the right hand side of (35) on the baryon states) are listed below:

	$< \Sigma_z^q >$	$< \Sigma_z^\lambda >$
P	$+1$	0
N	$-2/3$	0
Λ	$-1/3$	$+1/3$
Σ^+	$+1$	$-1/9$
Σ^-	$-1/3$	$-1/9$
Ξ^0	$-2/3$	$+4/9$
Ξ^-	$-1/3$	$+4/9$
$< \Sigma^0 \uparrow |\Sigma_z^q|\Lambda \uparrow> = -1/\sqrt{3}$		$< \Sigma^0 \uparrow |\Sigma_z^\lambda|\Lambda \uparrow> = 0$

The last line gives the matrix elements (to be used later) of Σ_z^q and Σ_z^λ for the transition $\Sigma^0 \rightarrow \Lambda\gamma$.

The equation (35) and the values of $< \Sigma_z^q >$, $< \Sigma_z^\lambda >$ given above lead to the following expressions for the seven magnetic moments (we indicate by the baryon symbol the magnetic moment of the baryon in proton magnetons):

$$P = \mu + F \qquad\qquad N = -\frac{2\mu}{3}$$

(36)
$$\Lambda = -\frac{\mu}{3} + \frac{A}{3} - H + \frac{K}{3}$$

$$\Sigma^+ = \mu - \frac{A}{9} + F - H - \frac{L}{9} - K - G$$

$$\Sigma^- = -\frac{\mu}{3} - \frac{A}{9} - F - H + \frac{L}{9} + \frac{K}{3} + G$$

$$\Xi^0 = -\frac{2\mu}{3} + \frac{4A}{9} - 2H + \frac{4K}{3}$$

$$\Xi^- = -\frac{\mu}{3} + \frac{4A}{9} - F - 2H - \frac{4L}{9} + \frac{2K}{3} + 2G$$

The P and N magnetic moments ($P = 2.793$, $N = -1.913$) depend only on the two parameters μ and F [this remains true also if all the flavour breaking terms are included]. It follows (in proton magnetons):

(37)
$$\mu = 2.869 \qquad\qquad F = -0.076$$

The $\Lambda, \Sigma^{+,-}, \Xi^{0,-}$ determine the remaining five parameters; solving (36) with

[7]:

$$\Sigma^+ = 2.48, \quad \Sigma^- = -1.16, \quad \Lambda = -0.61, \quad \Xi^0 = -1.25, \quad \Xi^- = -0.65$$

one finds:

(38) $A = +1.005, \quad K = +0.289, \quad H = +0.086, \quad G = -0.155, \quad L = -0.175$

[on choosing $\Sigma^+ = 2.38$ (and the other magnetic moments as above) one obtains $A = 1.08, \; K = 0.31, \; H = 0.12, \; G = 0.13, \; L = -0.10$].

It should be enphasized that (38) are, to first order in flavour breaking, the exact values of the parameters A, K, H, G, L; that is the values that the exact relativistic field theory must produce, whatever the theory is, provided only that the electromagnetic current is given by (5) and the flavour breaking is represented only by the mass term (8).

Consider now the $NRQM$; one of its sucesses has been the simple description of the magnetic moments. The $NRQM$ gives for the magnetic moments in a self explanatory notation the two parameter (μ', a) formula (we write μ' instead of μ and in (40) A' instead of A because the values of μ', A' in the 2-parameter $NRQM$ fit differ slightly from those μ, A of the complete fit, of course):

(39) $$\underline{M} = \mu' \sum \left[(2/3)\underline{\sigma}^P - (1/3)\underline{\sigma}^N - (a/3)\underline{\sigma}^\lambda \right]$$

In the present notation (39) is rewritten:

(40) $$\underline{M} = \mu'\underline{\Sigma}^q + A'\underline{\Sigma}^\lambda$$

with:

(41) $$A' = \mu'(1 - a)$$

Note that in the above $NRQM$ formulas $a = m_P/m_\lambda$; $1 - a = \Delta m/m_\lambda$, the quantity that characterizes the flavour breaking. The formula (39) [or (40)] fits the data fairly well with $\mu' = 2.79, \; a = 0.65$ [$a = 0.65, \; \mu' = 2.79$ implies $A' = 0.98$].

Comparing (40) with (35) one sees the reason of the success of the $NRQM$ description; the two parameters μ', A' that the model introduces are indeed the largest ones; moreover the value of the quantity a from the fit of the $NRQM$ and that from the exact (first order flavour breaking) solution are almost identical, in spite of the presence in the latter solution of 5 additional parameters F, H, K, G, L. Among these K is the largest ($|K/A| \doteq 0.29$); this is not introduced, usually, in the $NRQM$ description; its meaning is a correction to the magnetic moment of the individual quark linear in the strangeness, that is, essentially, linear in the mass of the baryon in which the quarks are. The remaining parameters are rather smaller than A [$|F/A| \doteq 0.075$, $|G/A| \doteq 0.15$, $|L/A| \doteq 0.17$, $|H/A| \doteq 0.085$], but of course in some cases they have an appreciable cumulative effect; for instance the value of the combination:

$$\Lambda + 2\Sigma^- + \Sigma^+ = G - 4H + (L/9) - F \doteq -0.44$$

is due essentially to the small value of H; this remark clarifies perhaps many minor "controversies" in the past on the "quality" of the fit of the $NRQM$.

To summarize, it is true that, though affected by many corrections, the simple $NRQM$ parametrization approximates reasonably the exact (first order flavour breaking) parametrization: μ and A, the only parameters in the $NRQM$, are indeed the largest ones; the remaining parameters are at most 15% of μ [in fact 10%; but if we express the equation (35) in terms of $g_1, g_2 \ldots g_7$, the largest parameter among the "small" ones is g_4; its ratio to g_1 is almost 15%].

This implies that we have now a detailed check that the assumption of additivity, the main assumption of the $NRQM$ [the property that distinguishes the "large" terms G_1, G_2 in (35) with respect to the remaining ones], is fulfilled to $\approx$ 15%; but the question is now: why is this so? Why does the correct relativistic theory (say QCD) lead to approximate additivity? Take, for instance the P and N magnetic moments: in this case, independently of any flavour breaking aproximation, only the terms μ and F in (35) intervene; why does the correct theory predict a value of F so small as it is? By looking, say in QCD, to the diagrams that contribute to the various structures G_ν it will perhaps be possible to answer these questions; for the moment they are still unanswered; the answer does not necessarily imply a slow motion of the quarks inside the baryons, thus confirming the statement at the beginning.

7. - The $\Sigma^0 \to \Lambda\gamma$ transition

Another application of the parametrization of the magnetic moments introduced in the past sections is to the $\Sigma^0 \to \Lambda\gamma$ transition. We are again dealing with two baryons with $J = 1/2$ of the same octet so that the equation (35) with the same values of the parameters $\mu, A, K \ldots L$ can be used; however now we have to calculate a transition matrix element, not a diagonal one, so that the vanishing of the terms proportional to $(\underline{\sigma}_i \times \underline{\sigma}_k)$ is no more necessarily true. However if we are interested in the $\Sigma^0 \to \Lambda\gamma$ *rate* correct to first order only in flavour breaking, it can be shown [6] that no contribution arises from terms containing $(\underline{\sigma}_i \times \underline{\sigma}_k)$ and therefore no new parameter intervenes in addition to the 7 parameters $\mu, A, K \ldots L$, present in (35). We get [6]:

$$(42) \qquad < \Sigma^0 \uparrow |M_z| \Lambda^0 \uparrow> = (\mu + KS) \; < \Sigma^0 \uparrow |\Sigma_z^q| \Lambda^0 \uparrow>$$

$$+ A < \Sigma^0 \uparrow |\Sigma_z^\lambda| \Lambda^0 \uparrow> = -(\mu - K) \; \frac{1}{\sqrt{3}}$$

$$= -\frac{(2.869 - 0.289)}{\sqrt{3}} = -1.49 \quad (|\text{Exp}| = 1.61 \pm 0.09)$$

in agreement with the experiment inside (approximately) one standard deviation.

Note finally that the eq. (42) $[< \Sigma^0 \uparrow |M_z| \Lambda^0 \uparrow> = -(\mu - K)/\sqrt{3}]$ is identical - when μ and K are expressed, via (36), in terms of the baryon magnetic moments - to a relation obtained by Okubo (Phys. Letters 4 (1963) 14, eq. (8b)) using only flavour invariance plus first order T_3^3 violation. This must be so of course; what is new and unexpected in this paper is that the number of parameters appearing in the full spin-flavour parametrization of the magnetic moments is the same as that in the Okubo's parametrization that refers only to the flavour space [compare Okubo's equations 7, 8 and the footnote † on page 161 of his paper]. It is this simplicity of the full spin-flavour parametrization that allows to clarify the connection of the $NRQM$ to the complete field theory.

8. - The $\Delta \to P\gamma$ $M1$ transition

The matrix element of the $\Delta \to P\gamma$ transition predicted by the $NRQM$ is ≈ 1.45 times below the truth (the rate, a factor ≈ 2.1); the calculation includes a form factor effect $1 - (1/6)k^2 <r^2> \doteq 0.82$ in the matrix element. This discrepancy is one of the largest in the $NRQM$ predictions. We will show below how it can be qualitatively understood via the non additive terms present in the general parametrization; it will not be necessary to assume that such terms are abnormally large.

We repeat, first of all, that the expression (35) is a parametrization of the magnetic moments to be used only in the expectation values of the baryon octet states with $J = 1/2$. The extension of (35), valid both for $J = 1/2$ and $J = 3/2$, can be derived easily, however. For the baryons of strangeness zero, the only ones to be considered in this section, this extension is:

$$(43) \quad \underline{M} = (\alpha - \delta)\underline{\Sigma}^q + (\beta - \gamma) \left[\frac{1}{4} (4|\underline{J}|^2 - 7)\underline{\Sigma}^q + \frac{1}{4} \underline{\Sigma}^q (4|\underline{J}|^2 - 7) \right]$$
$$+ \left[\delta - \beta + \frac{1}{2} \gamma (4|\underline{J}|^2 - 7) \right] \underline{Q} \cdot (2\underline{J})$$

where $\alpha, \beta, \gamma, \delta$ are four real parameters that replace μ, Λ in (35); note that, among these, only $\alpha - \delta$ multiplies an expression additive in the quarks.

In calculating the expectation value of (43) over an octet state with $J = 1/2$, it becomes:

$$(44) \quad \underline{M}(J = 1/2) = [\alpha - \delta - 2(\beta - \gamma)] \underline{\Sigma}^q + [\delta - \beta + 2\gamma] \underline{Q}(2\underline{J})$$

while for a $J = 3/2 \to J = 1/2$ transition (43) it is:

$$(45) \quad \underline{M}(3/2 \to 1/2) = [\alpha - \delta + \beta - \gamma] \underline{\Sigma}^q$$

Comparing (44) with (35) we have:

$$(46) \qquad \mu = \alpha - \delta - 2(\beta - \gamma) \qquad F = \delta - \beta + 2\gamma$$

whereas the operator $\underline{M}(3/2 \to 1/2)$ that intervenes in the $3/2 \to 1/2$ transition is multiplied by:

$$(47) \qquad \hat{\mu} = \alpha - \delta + \beta - \gamma = \mu + 3(\beta - \gamma)$$

Thus the transition magnetic moment operator $\underline{\hat{M}}$ responsible for the $\Delta \to P\gamma$ transition:

$$(48) \qquad \underline{\hat{M}} = \hat{\mu} \, \underline{\Sigma}^q$$

differs from the operator $\mu \underline{\Sigma}^q$ that appears [eq. (35)] in the calculation of the magnetic moments of the baryon octet. If we take into account that, in addition to the above difference, the transition $\Delta \to P\gamma$ magnetic moment can have a contribution of the form $\underline{M}^\eta$:

$$(49) \qquad \underline{M}^\eta = \frac{\eta}{2} \sum_{i,k} (\underline{\sigma}_i \times \underline{\sigma}_k) \, (P_i^q - P_k^q)$$

(where η is some real parameter) the ratio between the rate $\Gamma(\Delta \to P\gamma)$ for the $\Delta \to P\gamma$ transition and that calculated for the same transition in terms of the $NRQM$ is seen to be (compare [6] for more details):

$$(50) \qquad \frac{\Gamma(\Delta \to P\gamma)}{\Gamma_{NRQM}(\Delta \to P\gamma)} = 1.05 \, \frac{\hat{\mu}^2 + 9\eta^2}{\mu^2}$$

The experimental value of (50) is, as stated, ≈ 2.1; if η is not much larger than 15% μ the term (49) with coefficient η is largely insufficient to produce this; but because $\hat{\mu}/\mu = 1 + [3(\beta - \gamma)]/\mu$ (eq. (47)) it is sufficient to assume that $(\beta - \gamma)/\mu = 0.13$ to have the right hand side of (50) equal to 2.1. Thus, because of the factor 3 in (47), the non additive terms with a coefficient of the order 13% of μ can explain the discrepancy between the naive $NRQM$ and the data.

We conclude this section with a few words on the electric quadrupole ($E2$) forbiddenness in the $\Delta \to P\gamma$ transition. As shown long ago [8] the usual additive quadrupole operator Q_M of the $NRQM$, having the structure:

$$(51) \qquad Q_M = \sum_i P_i^q \, \Omega_M(\underline{r}_i) + \sum_i P_i^q \, (\underline{\sigma}_i * \underline{V}(\underline{r}_i))_M$$

($\Omega_M(\underline{r}_i)$ is a quadrupole operator with spherical component M and $\underline{V}(\underline{r}_i)$ an axial vector operator, both functions of the space coordinates $\underline{r}_i$) implies the vanishing of the $E2$ amplitude in the $\Delta \rightarrow P\gamma$ decay. The presence in addition to (51) of small non additive terms in the $E2$ operator such as:

$$K \sum_{i>k} (P_i^q + P_k^q) \; [\sigma_{iz}\sigma_{kz} - (\underline{\sigma}_i \cdot \underline{\sigma}_k)/3] \; S(\underline{r}_1, \underline{r}_2, \underline{r}_3)$$

where K is a coefficient and $S(\underline{r}_1, \underline{r}_2, \underline{r}_3)$ is a scalar symmetrical function of the coordinates (and momenta), can produce the transition. The experiment shows that the ratio between the amplitudes $E2/M1$ is only 5% to 13% of the value expected if the $E2$ transition were not inhibited [9]; therefore it appears once more that the additivity of the $NRQM$ is a good approximation.

9. - The semileptonic matrix elements of the baryon octet

The parametrization of the matrix elements of the electromagnetic current described in the past sections can be extended to the matrix elements of the weak current. We write again the transformation between the model and exact states of the baryons A and B : $|\psi_A> \; = V|\phi_A>$, $|\psi_B> \; = V|\phi_B>$ and introduce the vector $<1>$ and axial vector $<\sigma_z>$ matrix elements with $\Delta S = 0$ and $\Delta S = 1$ ($S =$ strangeness) in the limit of four momentum transfer $\Delta P_\mu = 0$.

Proceeding exactly as for the magnetic moments (compare [10] for more details) the most general parametrization for *the case of no flavour breaking* has the following form [we write $\tau^+ = \lambda_1 + i\lambda_2$ and $\lambda^+ = \lambda_4 + i\lambda_5$]:

$$(52) \qquad <1>_{\Delta S=0} = \; < \phi_B \uparrow | \sum_i^3 \tau_i^+ |\phi_A \uparrow>;$$

$$<1>_{\Delta S=1} = \; < \phi_B \uparrow | \sum_i^3 \lambda_i^+ |\phi_A \uparrow>$$

$$<\sigma_z>_{\Delta S=0} = \; < \phi_B \uparrow | \; a \sum_i^3 \tau_i^+ \sigma_{zi} + b \sum_{i\neq k}^3 \tau_i^+ \sigma_{zk} \; |\phi_A \uparrow>$$

$$(53) \qquad \equiv \; < \phi_B \uparrow |(a-b) \sum_i^3 \tau_i^+ \sigma_{zi} + b \sum_i^3 \tau_i^+ \; (2J_z)|\phi_A \uparrow>$$

$$<\sigma_z>_{\Delta S=1} = \; < \phi_B \uparrow | \; a \sum_i^3 \lambda_i^+ \sigma_{zi} + b \sum_{i\neq k}^3 \lambda_i^+ \sigma_{zk} \; |\phi_A \uparrow>$$

$$(54) \qquad \equiv \; < \phi_B \uparrow |(a-b) \sum_i^3 \lambda_i^+ \sigma_{zi} + b \sum_i^3 \lambda_i^+ \; (2J_z) \; |\phi_A \uparrow>$$

Note that in the no flavour breaking case the axial vector matrix elements

$< \sigma_z >$ are completely characterized by two real parameters a and b. [In the second form of (53) (54), $(2J_z)$ has to be replaced by 1, because $J_z = 1/2$; we have left it indicated as $(2J_z)$ to make more clear the passage from the first to the second form in the above equations].

Note that, a priori, the possible structures that might intervene in the parametrization of the no flavour breaking axial matrix elements are:

1) for $< \sigma_z >_{\Delta S=0}$,

$$\sum_1^3 {}_i \tau_i^+ \underline{\sigma}_i, \quad \sum_1^3 {}_{i \neq k} \tau_i^+ \underline{\sigma}_k, \quad \sum_1^3 {}_{i \neq k} (\tau_i^+ - \tau_k^+)(\underline{\sigma}_i \times \underline{\sigma}_k), \quad \sum_1^3 {}_{i \neq k \neq j} \tau_j^+ (\underline{\sigma}_i \times \underline{\sigma}_k)$$

2) for $< \sigma_z >_{\Delta S=1}$,

$$\sum_1^3 {}_i \lambda_i^+ \underline{\sigma}_i, \quad \sum_1^3 {}_{i \neq k} \lambda_i^+ \underline{\sigma}_k, \quad \sum_1^3 {}_{i \neq k} (\lambda_i^+ - \lambda_k^+)(\underline{\sigma}_i \times \underline{\sigma}_k), \quad \sum_1^3 {}_{i \neq k \neq j} \lambda_j^+ (\underline{\sigma}_i \times \underline{\sigma}_k)$$

However it can be shown [10] that also in this case operators containing $(\underline{\sigma}_i \times \underline{\sigma}_k)$ do not contribute to the matrix elements between ϕ_A and ϕ_B; and therefore the most general expressions reduce to (53) and (54) [note that here we have not to deal with diagonal matrix elements as for the magnetic moments].

The two parameters a and b can be linearly related to the Cabibbo's parameters D and F. To find the relationship it is sufficient to consider two independent matrix elements and express them in terms of a, b or D, F. For the $N \to P$ matrix elements it is:

(55) $< 1 >_{NP} = 1$ $< \sigma_z >_{NP} = \dfrac{5}{3}(a - b) + b = \dfrac{5}{3}a - \dfrac{2}{3}b = D + F$

For $\Sigma^- \to \Lambda$ we have:

(56) $< 1 >_{\Sigma-\Lambda} = 0$ $< \sigma_z >_{\Sigma-\Lambda} = \sqrt{2/3}\,(a - b) = \sqrt{2/3}\,D$

and thus:

(57) $a = \dfrac{1}{3}D + F, \qquad b = F - \dfrac{2}{3}D$

We thus find that a and b have a very direct physical meaning: a is the coefficient of an expression additive in the quarks, whereas b is the coefficient of an expression containing at the same time the variables of two quarks. As we have already seen for the magnetic moments, the $NRQM$ (in its simplest form) amounts to selecting only the terms additive in the individual quarks; here this implies to set $b = 0$, which indeed corresponds, by (57) to the well known $NRQM$ result $D/F = 3/2$. Another point worth noting is that the $NRQM$ approximation ($b = 0$) implies, by (55), $(g_A/g_V)_{NP} = (5/3)a$; the frequent assertion that the $NRQM$ predicts $(g_A/g_V)_{NP} = (5/3)$ is incorrect, as already remarked long ago; this wrong (time honoured!) assertion is due to the identification of abstract SU_6 with the $NRQM$ and of the current quarks with the constituent quarks.

As to the experimental value of $b/a = [1 - (2D/3F)]/[1 + (D/3F)]$, it depends at the moment appreciably on the way how the experimental data are analyzed. Fixing the Cabibbo angle through the semileptonic decays ($\sin\theta_C = 0.231 \pm 0.003$), one gets [11] $D/F = 1.58 \pm 0.04$ and thus $b/a \doteq$ 3%; if the Cabibbo angle is determined differently (e.g. like in [12] from the $0^+ \rightarrow 0^+$ β transitions: $\sin\theta_C = 0.225 \pm 0.002$) one gets $D/F = 1.74$ and $b/a \doteq 0.1$.

Let us now add a few words on the parametrization of the semileptonic matrix elements *including the flavour breaking up to first order* [10]. In this case the number of intervening parameters is too large to allow, without additional assumptions, their determination from the present data. In this respect the situation is different from that of the magnetic moments; there, as we saw (sect. 6), the seven parameters appearing in the parametrization with flavour breaking to the first order could be determined from the seven magnetic moments of the octet baryons and the result used (sect. 7) to determine the rate of the $\Sigma^0 \rightarrow \Lambda\gamma$ to the first order in the flavour breaking.

Here to proceed one might keep only those flavour breaking terms that are simply additive in the quarks, as suggested by the $NRQM$. Then only one first order flavour breaking term remains. The only change to (52) (53) (54) is that, instead of (54), the $\Delta S = 1$ axial vector matrix element takes the form:

$$(58) \quad <\sigma_z>_{\Delta S=1} = <\phi_B \uparrow | \ (a + \Delta a) \sum_1^3 \lambda_i^+ \sigma_{zi} + b \sum_{\substack{1 \\ i \neq k}}^3 \lambda_i^+ \sigma_{zk} \ |\phi_A \uparrow>$$

where Δa is a new parameter; we expect the flavour breaking parameter Δa to be of the order $a(\Delta m/m_\lambda)$. However, if $b = 0$, a variation of a (by Δa) cannot produce a variation in D/F because we have just noted that (for $b = 0$) $D/F = 3/2$, independently of the value of a; because b, though not zero, is very small (a few percent as we stated) a variation of a by Δa cannot produce the large difference between the values of D/F for the $\Delta S = 0$ and $\Delta S = 1$ transitions (respectively $\doteq 1.34$ and $\doteq 1.82$) that, according to ref. [12], is required by the data and is indicative of flavour breaking. Thus the overall situation is unclear although one does not see any reason why the axial vector flavour breaking effects to first order should be negligible for the semileptonic matrix elements.

10. - Parametrization of the masses

If in the equation (20) the magnetic moment operator M_z is replaced by the exact Hamiltonian H of the system, the argument used to parametrize the magnetic moments can be used to parametrize the masses of the baryons for the octet and decuplet; instead of an axial vector we must now parametrize a scalar (under space rotations). The factorizability (10) of the model state reduces again the expectation value of the hamiltonian to a combination of spin flavour operators multiplied by some coefficients; in this case the same coefficients for all octet and decuplet states. As to the spin flavour invariants, we recall the following: 1) as stated in sect. 4 the expectation value of $\underline{\sigma}_1 \cdot (\underline{\sigma}_2 \times \underline{\sigma}_3)$ on a real spin flavour wave function vanishes; the same is true for $\underline{\sigma}_1 \cdot (\underline{\sigma}_2 \times \underline{\sigma}_3)\Gamma(f)$ where $\Gamma(f)$ is any real flavour operator. The only spin invariants in the parametrization of a scalar are thus:

$$(59) \qquad\qquad 1; \qquad (\underline{\sigma}_i \cdot \underline{\sigma}_k)$$

2) as to the flavour space, the charge operator P^q will now be absent, of course [we will not consider the electromagnetic mass differences]; because the only flavour operator in the theory is P^λ, only products of P^λ's can intervene, containing a maximum of three P^λ's.

Because each spin-flavour operator must be symmetrical in the three quarks (the spin-flavour factor of our model states is symmetrical) the following 5 independent operators can appear in the masses to the first order in flavour breaking:

$$(60) \quad 1, \ \sum_i P_i^\lambda, \ \sum_{i>k}(\underline{\sigma}_i \cdot \underline{\sigma}_k), \ \sum_{i>k}(\underline{\sigma}_i \cdot \underline{\sigma}_k)(P_i^\lambda + P_k^\lambda), \ \sum_{\substack{i \neq k \neq j \\ (i>k)}}(\underline{\sigma}_i \cdot \underline{\sigma}_k)P_j^\lambda$$

In addition to the above structures (60) the following ones (61) would be present if we include flavour breaking to all orders.

$$(61) \ \sum_{i>k} P_i^\lambda P_k^\lambda, \ \sum_{i>k}(\underline{\sigma}_i \cdot \underline{\sigma}_k)P_i^\lambda P_k^\lambda, \ \sum_{\substack{i \neq k \neq j \\ (i>k)}}(\underline{\sigma}_i \cdot \underline{\sigma}_k)(P_j^\lambda(P_i^\lambda + P_k^\lambda), \ P_1^\lambda P_2^\lambda P_3^\lambda$$

Again, as for the magnetic moments, we will not consider terms of orders higher than the first in flavour breaking. Excluding therefore all the terms (61), the most general expression for the mass, to first order in flavour breaking is:

$$
\begin{aligned}
M = M_0 &+ B\sum_i P_i^\lambda + C\sum_{i>k}(\underline{\sigma}_i \cdot \underline{\sigma}_k) + D\sum_{i>k}(\underline{\sigma}_i \cdot \underline{\sigma}_k)(P_i^\lambda + P_k^\lambda) \\
&+ E\sum_{\substack{i \neq k \neq j \\ (i>k)}}(\underline{\sigma}_i \cdot \underline{\sigma}_k)P_j^\lambda
\end{aligned}
$$

(62)

Because (62) contains five parameters (M_0, B, C, D, E) and we must fit 4 octet masses and 4 decuplet ones, we will get three (well known) mass relations.

Before doing this a remark is appropriate: the same expressions (60) (61) listed above appear in the parametrization of any quantity independent of the charge P^q, scalar under rotations; in obtaining the expressions (60) (61) we used, in fact, only the factorizability of the states. Thus we can parametrize in terms of the nine quantities listed in (60), (61) the hamiltonian H, or, equally well, its square H^2, or H^3 or more generally any function $F(H)$ of H. This has two consequences:

1) It is a proof (general, not only for the case of the masses) that the number of independent invariants [for the masses the 9 listed in (60) + (61)] must be always larger than or equal to the number of the states, as expected [we can have nothing out of nothing]. If it were not so, we would find, for the masses, rigorous relations that would be true simultaneously for the masses M, their squares or any function $F(M)$ of the masses; this is clearly impossible, except for the uninteresting case of complete mass degeneracy.

2) It is only neglecting invariants that correspond to flavour breaking of some order, that we can obtain approximate relations between the masses (the Gell Mann Okubo mass formula will be seen to be one of them); but these relations can equally be derived for any power or function of the masses. Of course these relations do not have all the same accuracy: if, say, it is a fair approximation to include the flavour breaking only to first order in dealing with H, it is certainly a worse approximation to do the same dealing with $H^{10} = (H_0 + H_1)^{10}$. What is, in this sense, the best function of H, the one that minimizes the effect of the neglected flavour breaking terms, is hard to say, a priori [but it is hard to see a reason why the quadratic mass formulas should be better than the linear ones, a long debated question; this provides some justification for the linear mass formulas]. Of course all the above remarks apply also to the conventional group theoretical derivation of, say, the Gell Mann-Okubo mass formula.

Similar remarks apply to the magnetic moments; we might have parametrized $H \underline{M} H$ instead of $\underline{M}$, for instance, in a way formally identical to that used for $\underline{M}$. We did not raise this point when speaking of the magnetic moments only because we preferred to raise it for the masses, where the situation is much more familiar.

11. - The mass formulas and their comparison with those obtained with the $NRQM$

With some algebra the masses of the octet and decuplet baryons can be written in terms of the coefficients $M_0 \ldots E$ of the parametrization (62). Writing again $S = \text{strangeness} = -\Sigma_i P_i^\lambda$ and defining:

$$(63) \quad \tilde{M}_0 \equiv M_0 - \frac{9}{2} C, \quad \beta = 3D + \frac{3}{2} E - B, \quad \gamma = \frac{C}{2}, \quad \delta = D - E, \quad \epsilon = -\frac{E}{2}$$

the expression (62) can be rewritten:

$$(64) \qquad M = \tilde{M}_0 + \beta S + \gamma[4J(J+1)] + \delta \sum_i (\underline{\sigma}_i P_i^\lambda \cdot 2\underline{J}) + \epsilon S[4J(J+1)]$$

It is:

$$(65) \qquad \langle\uparrow | \sum_i (\underline{\sigma}_i P_i^\lambda \cdot 2\underline{J}) | \uparrow\rangle_{J=1/2}^{J=3/2} = \sum_3^5 \cdot \langle\uparrow | \sum_i \sigma_{iz} P_i^\lambda | \uparrow\rangle$$

where $| \uparrow\rangle$ means a state with J_z respectively $1/2$ and $3/2$, and the factors 3 and 5 multiplying the right hand side of (65) refer respectively to $J = 1/2$ and $3/2$. The values of $\langle\uparrow | \sum_i \sigma_{iz} P_i^\lambda | \uparrow\rangle$ are listed below for the $\underline{8}$ and $\underline{10}$ baryons:

N	Λ	Σ	Ξ	Δ	Σ^*	Ξ^*	Ω
0	1	$-1/3$	$4/3$	0	1	2	3

From (64) and (65) we get, therefore (baryon symbols stay for baryon masses):

$$N = \tilde{M}_0 + 3\gamma$$
$$\Lambda = \tilde{M}_0 - \beta + 3\gamma + 3\delta - 3\epsilon$$
$$(66)$$
$$\Sigma = \tilde{M}_0 - \beta + 3\gamma - \delta - 3\epsilon$$
$$\Xi = \tilde{M}_0 - 2\beta + 3\gamma + 4\delta - 6\epsilon$$

and:

$$\Delta = \tilde{M}_0 + 15\gamma$$
$$\Sigma^* = \tilde{M}_0 - \beta + 15\gamma + 5\delta - 15\epsilon$$
$$(67)$$
$$\Xi^* = \tilde{M}_0 - 2\beta + 15\gamma + 10\delta - 30\epsilon$$
$$\Omega = \tilde{M}_0 - 3\beta + 15\gamma + 15\delta - 45\epsilon$$

From (66), (67) we get 3 mass relations: the Gell Mann Okubo formula for the octet:

$$(68) \qquad \frac{N+\Xi}{2} = \frac{3\Lambda+\Sigma}{4}$$
$$\qquad\quad\; (1129) \qquad (1135)$$

and the equal spacing formula for the decuplet:

$$(69) \qquad \Omega^- \; \Xi^* = \Xi^* - \Sigma^* = \Sigma^* - \Delta$$
$$\qquad\quad\; 139 \qquad\quad 149 \qquad\quad\; 152$$

The values (in MeV) of the parameters $\tilde{M}_0, \beta, \gamma, \delta, \epsilon$ in (64) obtained from (66) (67) and also the parameters M_0, B, C, D, E, in (62) are displayed below [the values for the masses used are from [13]]

(70) $\tilde{M}_0 = 865.2, \ \beta = -228.6, \ \gamma = 24.5, \ \delta = -19.4, \ \epsilon = -1.5$

or, alternatively:

(71) $M_0 = 1086, \ B = 188.4, \ C = 49.2, \ D = -15.4, \ E = 3.0$

As is apparent already from the formulas (68) (69), the equation (62) [or (64)] represents the data to a few percent, in spite of the fact that the flavour breaking has been taken into account only to the first order. Note that the values listed in (70) (71) are not a best fit; to mention only the parameters that will play a role in the following, δ comes simply from $\Sigma - \Lambda = -4\delta$; γ is obtained from $\Delta - N = 12\gamma$; $\epsilon = -1.5$ is the average of two determinations: a) $\Sigma^* - \Sigma - [\Xi^* - \Xi] = 12\epsilon$, giving $\epsilon = -2$ and b) $\Sigma^* - \Sigma - 6\delta - 12\gamma = -12\epsilon$, giving $\epsilon = -1$.

   We now compare these results with those from the version of the $NRQM$ due to De Rujula, Georgi and Glashow [14] where the potential between the quarks (and, in particular, the spin dependent one) is taken as the Fermi Breit approximation to the QCD one gluon exchange potential. In ref. [14] the mass formula for the baryons, to the first order in the flavour breaking, is given by the eq. (6). Comparing that formula with our mass formula in its first form (62), the two are seen to be equal (to order $(\Delta m/m_\lambda)^2$) if we suppress in (62) the parameter E:

(72) $E = 0$

and if D and C are related by:

(73) $\dfrac{D}{C} = -\dfrac{m_\lambda - m_p}{m_\lambda} \equiv -\dfrac{\Delta m}{m_\lambda}$

where the m's are the quark masses. [In particular, using (63), the value (73) of D/C and (72) one can verify that our formulas for the masses (66), (67) satisfy the equations (5) and (11) of ref. [14]].

   We thus see from (71) that E, the only parameter outside of the $NRQM$ description (with two body potentials), is indeed small, with respect to the others; moreover $(-D/C)$ is $(15.4/49.2) = 0.31$ from (71); and this number is in good agreement with $(1 - a) = 0.35$ obtained (eq. (41)) from the magnetic moments, which value, in a $NRQM$ description, is also interpreted as $\Delta m/m_\lambda$.

12. - The exotic states

The reinterpretation of the $NRQM$ presented here implies some change in the way of looking at the exotic states, those states with quantum numbers incompatible with $3q$ for the baryons or with $q\bar{q}$ for the mesons. The usual statement is that the discovery of such exotic states would seriously contradict the $NRQM$. It seems that this is no more true; we say a few words on how the situation is qualitatively changed in this respect.

The exact states are constructed from the model states by application of the "adiabatic" operator V. The non exotic states, say for the mesons, are schematically $V|q\bar{q}>$; we are however fully allowed to consider also exotic states, the states with exotic quantum numbers coming for instance from $V|q\bar{q}q\bar{q}>$ (or "glueballs" possibly coming from $V|GG>$ or $V|GGG>$). One expects that the majority of these exotic states are high in mass and that, therefore, their decay widths for decay into non exotic states are so large to make them unobservable. However it may happen (or it may not) that a few of these states are low in mass, widely separated from the majority; in such case they might be narrow. It could be of some interest to examine specific models of theories to see if and under which circumstances the latter situation (a few exotic states low in mass) can in fact occur.

13. - Concluding remarks

There is no point in summarizing again the results of the above analysis; we will only add a few words on problems still to be examined. First, we hope to look from this point of view at the other classical problems of the $NRQM$: for instance the vector meson → pseudoscalar meson $+\gamma$ transitions, the e.m. mass formulas, the leptonic decays of mesons etc. The second problem has been noted repeatedly: one should try to understand which are the features of a field theory such that the terms suggested by the $NRQM$ turn out to be more important than the others in the parametrizations; in other words why the terms additive in the variables of the individual quarks have coefficients larger than those carrying the variables of two quarks and the latter are larger (at least so it appears from the masses) than those related to the variables of three quarks. Thirdly, as to the flavour breaking, one might ask if the condition that we have imposed, essentially that flavour breaking is associated only to the projection operator P^λ, is a necessary one for getting useful parametrizations [15].

FOOTNOTES and REFERENCES

[1]   F. Gürsey - L.A. Radicati, Phys. Rev. Lett. **13**, 173 (1964).

[2]   Compare C. Pickering, *"Constructing quarks"* (Edinburgh Univ. Press, 1984 p. 118).

[3]   G. Morpurgo, *Physics* **2**, 95 (1965).

[4]   G. Morpurgo, *Rapporteur talk at the XIV Int. conf. on high energy physics*, Vienna 1968 [proceedings, CERN sci. publ. ed. by J. Prentki and J. Steinberger, p. 225, compare sect. 5.1].

[5]   Compare ref. 3; if the space-colour wave function is antisymmetric with $L = 0$, the spin-flavour factor of a proton is necessarily: a normalization factor times $\text{Symm}[\alpha_1(\alpha_2\beta_3 - \alpha_3\beta_2)P_1P_2N_3]$; the states for the other octet baryons are constructed similarly; they coincide with the SU_6 states.

[6]   G. Morpurgo, *Field theory and the non relativistic quark model: a parametrization of the baryon magnetic moments and masses* - INFN/GE, March 15, 1989 (to be published).

[7]   The values of the magnetic moments and the predictions of specific models have been discussed by J. Franklin in two talks presented at the 8-th Int'l Symposium on high energy spin physics, Univ. of Minnesota Sept. 12-17, 1988 (AIP Conf. Proc. - to be published).

[8]   C. Becchi - G. Morpurgo, Phys. Letters **17**, 352 (1965).

[9]   R. Davidson - N.C. Mukhopadhyay - R. Wittman, Phys. Rev. Lett. **56**, 804 (1986).

[10] G. Morpurgo, *Parametrization of the semileptonic baryon matrix elements*, INFN/GE-April 24, 1989 (to be published).

[11] J.M. Gaillard - G. Sauvage, Ann. Rev. Nucl. and Particle Sci., vol. **34**, 351 (1984).

[12] J.F. Donoghue - B.R. Holstein, Phys. Rev. **D25**, 2015 (1982); see also A. Garcia - P. Kielanowsky, Phys. Letters, **110B**, 498 (1982).

[13] Summary tables of particle properties, Particle data group, Phys. Lett. 1988.

[14] A. De Rujula - H. Georgi - S. Glashow, Phys. Rev. **D12**, 147 (1975); for convenience we transcribe the eq. (6), (5) and (11) of this paper quoted in the text: eq. (6), $M = A + B\sum_i \frac{\Delta m_i}{m_p} + C\sum_{i>j} \underline{s}_i \cdot \underline{s}_j[1 - (\Delta m_i + \Delta m_j)/m_p]$ (here A, B, C are the coefficients of De Rujula et al., different from ours); eq. (5), $m_p/m_\lambda = 2(\Sigma^* - \Sigma)/(2\Sigma^* + \Sigma - 3\Lambda)$; eq. (11) $\Sigma - \Lambda = \frac{2}{3}[1 - (m_p/m_\lambda)][\Delta - N]$.

[15] A final question is not directly related to hadron physics and the $NRQM$ but, more generally, to the meaning of certain models in physics. Take the shell model of the nucleus. Can it also be seen as expressing the dominance of certain terms in a parametrization that starts from a unitary correspondence between the simple shell model product states and the exact states of the nucleus? In fact the expression "model state" was introduced long ago in nuclear physics in a study of the relationship between the shell model states and the exact states of a nucleus [compare R.J. Eden and N.C. Francis, Phys. Rev. **97**, 1366 (1955)].

The Nonlinear Hanle Effect

GIOVANNI MORUZZI - FRANCO STRUMIA (*)

1. - Introduction

The classical Hanle effect is due to the removal of the zero field degeneracy in the upper state of a resonance fluorescence transition by an external field [1,2]. In fact, because of degeneracy, any linear superposition of the Zeeman sublevels of a state with $J > 0$ is an eigenstate of the atomic Hamiltonian at zero field. As soon as the degeneracy is removed, only the pure Zeeman sublevels remain stationary states. The transition between the zero field and the high field case occurs in a field region whose extension depends on the homogeneous width of the Zeeman levels. While the external field is swept in this region, the angular distribution of the polarized components of the fluorescence radiation changes as a function of the field intensity; different behaviors are observed according to the choice of the polarization of the excitation light, and of the experimental geometry. The classical Hanle effect can be described in terms of absorption and spontaneous emission, disregarding stimulated emission [2]. However, the presence of stimulated emission becomes important at excitation light intensities high enough to effect the population distribution between the upper and lower level of the transition, thus decreasing the absorption coefficient of the vapor. This non-linear (saturation) effect, exactly like the classical Hanle effect, depends on the polarization of the exciting radiation and on the degeneracy removal in the upper (and/or lower) state of the transition. Thus, at high intensities of the excitation light, a change in the spacing between the Zeeman sublevels affects also the total absorption of the sample, and, consequently, the total intensity of the emitted fluorescence radiation, as well as other physical properties of the investigated vapor. This phenomenon is known as *non-linear Hanle effect* (*NLHE*), and has very interesting applications in laser physics.

The *NLHE*, contrary to other saturation spectroscopy experiments, requires only one laser beam, and does not require the presence of a standing wave. In these conditions, the resolution is largely independent of the distorsions in the wave front of the laser beam. A *NLHE* signal is observed when the degeneracy in either the lower or in the upper level of the saturated transition

(*) Dipartimento di Fisica dell'Università di Pisa

is removed by an external electric or magnetic field. Thus, exactly as for the classical Hanle effect, the width of the signal is independent of the Doppler line width because the detenction system is equally sensitive to all frequencies within the line profile, and, since the signal can be detected, for instance, by measuring the global fluorescence intensity, no attempt is done to resolve the shape of the optical line. The signal is also independent of the laser frequency tuning within the Doppler line width. It is interesting to note that the $NLHE$ is characterized by the occurrence of an absorption *increase*, whereas an absorption *decrease* is typical of practically all other saturation techniques.

Apart from the removal of the zero field degeneracy (stimulated Hanle effect), two further analogous nonlinear effects are possible. The first of these (stimulated level crossing) requires the presence of at least two levels whose energy is field dependent, so that they are degenerate at a particular nonzero value of the external field. The third case requires three levels connected by a two-frequency laser beam. Degeneracy between two of the levels is induced when the difference between the two laser frequencies ω_1 and ω_2 is equal to their separation: $\Omega = \omega_1 - \omega_2$. This last method is also known as *Optical-Optical Double Resonance* ($OODR$) spectroscopy and in fact, in some respects, can be considered as an extension of the optical-microwave double resonance spectroscopy. The $NLHE$ spectroscopy was first proposed by A. Javan in 1964 [3] and clearly described in successive papers [4, 5, 6].

Like the classical Hanle effect, the $NLHE$ may be detected by observing the fluorescence radiation [3, 7]. In this case the anisotropy and polarization of the fluorescence enable a deeper insight into the experiment, since the evolutions of the diagonal and off diagonal elements of the density matrix can be independently detected. This techinique is not practicable for very dense samples and in the infrared spectral region, where a transmission experiment is more suitable.

Also transmission experiments can be difficult in some physically interesting conditions. For instance, when the absorbing medium is optically thin, the absorbed power is only a small fraction of the incoming light intensity, and the effect is a small change in this small fraction. This is why the $NLHE$ was first observed either directly on laser active media [3, 4, 5, 6, 8, 9, 10, 11, 12, 13], or on a resonant absorber placed within the laser cavity (intracavity technique) [14, 15, 16].

A first general theory of the $NLHE$ and of the related stimulated level-crossing effect was developed by Feld et al. [5]. This theory provided a qualitative explanation of the physical origin of the effect, and obtained a formal expression for the absorption profile at small saturation intensities, by means of a perturbation density matrix approach, as it had been done in earlier theoretical works [17, 18, 19, 20]. At that time, however, it was practically impossible to achieve an experimental quantitative verification of the size of the effect.

In the eighties new non-conventional techniques, namely the optoacoustic and optogalvanic techniques, were applied to the $NLHE$ detection [21]. These techniques allow a simple and very sensitive detection of small changes in the absorption of a vapor sample, and have thus opened new possibilities of experimental investigations. It is worth noting that in the media in which these techniques are applied, the observation of the fluorescence radiation, and hence of the classical Hanle effect or level crossing, is often difficult, if not, as in the IR molecular optoacoustic spectroscopy, impossible.

In molecular spectroscopy, the first optoacoustic recording of the $NLHE$ was observed by inserting a microphone into an absorption cell external to the CO_2 laser cavity [22]. The signal corresponded to an absorption increase of the CO_2 laser radiation, whose intensity was mechanically chopped at an acoustical frequency, by the CH_3OH molecules when an external electric field was applied to the cell. The optoacoustic signal showed that the absorption coefficient increases significantly (by more than 20%) when the linear Stark effect (the molecule has a large permanent electrical dipole moment) removes the degeneracy between the different M components of the transition. The experimental results for CH_3OH could be statisfactorily interpreted by a rate equation approach [23, 24]. The first optoacoustic $NLHE$ signal was observed almost simultaneously to the first optoacoustic sub-Doppler saturation Lamb-dip signal [25]. It is worth nothing that the $NLHE$ signal was larger than the Lamb-dip signal in similar experimental conditions.

In the optogalvanic technique, the light absorption by an atomic vapor subject to an electrical discharge is monitored by measuring the change in the discharge impedance, due to atomic excitation. The observation of the $NLHE$ by optoacoustic methods suggested a similar possibility for the optogalvanic spectroscopy [26]. The absorption increase of an atomic vapor located in a hollow cathode discharge was reported in [27, 28, 29]. The first quantitative analysis of the $NLHE$ in the typical experimental conditions of optogalvanic spectroscopy, based on the rate equation approach, and its experimental verification was reported in [30]. More recently Sokabe et al. [31] have developed a similar rate equation theory in order to interpret their observation of $NLHE$ in the IR absorption of CH_3OH by optoacoustic technique.

2. - Saturation Intensity and Saturated Line width

The theory of the $NLHE$ is based on the concepts of saturation intensity and saturated line width. Let us consider an atomic or molecular vapor interacting with a monochromatic laser beam traveling along the z axis of a Cartesian reference frame. We shall assume that the angular frequency ω of the light beam is close to resonance with a non-degenerate atomic (or molecular) transition between a lower level $|i>$ and an upper level $|k>$, with central angular frequency ω_0.

The gain coefficient $\alpha\,(\omega, z)$ for the light beam is defined by the relation

$$(1) \qquad \frac{dI}{I} = \alpha\,(\omega, z)\,dz,$$

where I is the intensity of the light beam inside the vapor at coordinate z (invariance for x and y translations is assumed), and $I + dI$ is its intensity at coordinate $z + dz$. A semiclassical treatment shows that the small signal gain coefficient $\alpha_0(\omega, z)$, i.e. the gain coefficient when the light intensity is so low that the populations of the upper and lower levels of the transition are not sensibly perturbed, can be written

$$(2) \qquad \alpha_0(\omega, z) = \frac{\pi\omega}{\varepsilon_0 c\hbar}(N_k - N_i)|< k\,|\hat{\mathbf{e}} \cdot \mathbf{P}|\,i >|^2\,f(\omega),$$

where N_k and N_i are the population densities at z of the upper and lower state, respectively. If $N_k > N_i$, as for a lasing transition, α_0 is positive; if $N_k < N_i$, as for thermal equilibrium, α_0 is negative, and $-\alpha_0$ is the usual absorption coefficient. In this chapter we shall denote unit vectors by circumflexes, and here $\hat{\mathbf{e}}$ is the unit vector describing the light polarization. $\mathbf{P}$ is the electric dipole moment operator, and $f(\omega)$ is a normalized function ($\int f d\omega = 1$) describing the shape of the absorption line. In the absence of Doppler broadening $f(\omega)$ wuold be a Lorentzian of half width γ, γ being due to the homogeneous broadening:

$$(3) \qquad f(\omega) = \frac{1}{\pi} \cdot \frac{\gamma}{(\omega - \omega_0)^2 + \gamma^2}.$$

For a Doppler broadened line, with Doppler half width $\Delta\omega_D >> \gamma$, we have

$$(4) \qquad f(\omega) = \frac{1}{\Delta\omega_D} \cdot \frac{\sqrt{log2}}{\sqrt{\pi}}\,e^{-log2\,[(\omega - \omega_0)/\Delta\omega_D]^2}.$$

In the intermediate cases, i.e. when the orders of magnitude of γ and $\Delta\omega_D$ are comparable, $f(\omega)$ is the convolution of a Lorentzian line with a Gaussian line, which is known as the Voigt profile.

Whereas the shape, height and width of the gain profile do not depend on the intensity of the light beam as long as the intensity is low enough not to perturb the population difference $(N_k - N_i)$ in a sensible way, changes are observed at higher intensities. For an arbitrary light intensity there is no formula of general validity like (2), but each physical situation must be treated separately. The simplest case is the two-level system. This case is approximated, for instance, by an atomic vapor with a ground atomic state $|i >$ and a highly excited state $|k >$ (highly with respect to kT). We shall assume that our atom has no levels close to the ground state, so that only $|i >$ is populated at thermal equilibrium. We shall exclude the presence of pumping and collisional relaxation mechanisms. We shall assume that the frequency of the light beam is close to resonance with the $|k >\leftarrow |i >$ transition, and that the beam polarization and the selection rules for the transition are such to forbid multiple quantum transitions. A further assumption is that our transition is non-degenerate. All these conditions are verified, for instance, for a transition from a $|\mu = 0 >$ ground level to an $|m = +1 >$ excited level and σ^+ polarization of the light beam.

The time evolution of the density matrix ρ representing our vapor is

$$(5) \qquad \frac{d\rho}{dt} = -\frac{i}{\hbar}[\mathcal{H}_0 + \mathcal{V}, \rho] + \Gamma\rho,$$

where $\mathcal{H}_0$ is the unperturbed atomic Hamiltonian, $\mathcal{V}$ the perturbation Hamiltonian describing the interaction of the vapor with the light beam, and Γ is a superoperator (an operator operating on operators) describing the combined effects of relaxation and spontaneous decay (only spontaneous decay, wiht our assumptions). The matrix element of the interaction Hamiltonian for an electric dipole transition between the states $|i >$ and $|k >$ can be written

$$(6) \qquad \mathcal{V}_{ik} = -\frac{\sqrt{I}}{\sqrt{2c\varepsilon_0}} < i|[\hat{e}\, e^{-iwt} + \hat{e}^* e^{iwt}] \cdot \mathbf{P}|k >,$$

where I is the intensity of the excitation beam. Let us now write down the matrices of equation (5) explicitly

$$(7) \qquad \mu - \begin{bmatrix} \rho_{kk} & \rho_{ki} \\ \rho_{ik} & \rho_{ii} \end{bmatrix}; \quad \mathcal{H}_0 - \begin{bmatrix} \hbar\omega_0 & 0 \\ 0 & 0 \end{bmatrix};$$

$$\mathcal{V} = \begin{bmatrix} 0 & \hbar W e^{-iwt} \\ \hbar W e^{iwt} & 0 \end{bmatrix}; \quad \Gamma\rho = \begin{bmatrix} -2\gamma\rho_{kk} & -\gamma\rho_{ki} \\ -\gamma\rho_{ik} & +2\gamma\rho_{kk} \end{bmatrix};$$

where

$$(8) \qquad W = -\frac{\sqrt{I}}{\hbar\sqrt{2c\varepsilon_0}} < i|\hat{e} \cdot \mathbf{P}|k > .$$

It is always possible to choose the relative phase between $|i >$ and $|k >$ so that W is real, and we shall assume this phase choice. The matrix elements of $\Gamma\rho$ have been evaluated by assuming that there is no collisional relaxation, and that the lifetime of the excited level $|k >$, due to spontaneous decay, is τ. In these conditions the natural line width in absence of the light beam γ equals $1/2\tau$ because of the quadratic dependence of the photon number on the electric field of the emitted wave. In our case, which is formally equivalent to the Bloch equations for a spin $1/2$ in the rotating wave approximation, the steady state solution for the density matrix has constant diagonal elements and off diagonal elements oscillating at $exp(\pm iwt)$:

$$(9) \qquad \rho_{kk} = \sigma_{kk}; \quad \rho_{ii} = \sigma_{ii} = 1 - \sigma_{kk}; \quad \rho_{ik} = \sigma_{ik}e^{iwt}; \quad \rho_{ki} = \rho_{ik}^*$$

where σ is a constant Hermitian matrix.

If we substitute (7), (8) and (9) into (5) we obtain

$$(10) \qquad \sigma_{ik} = \frac{W(2\sigma_{kk} - 1)}{(\omega_0 - \omega + i\gamma)}; \quad \rho_{kk} = \sigma_{kk} = \frac{W^2}{(\omega - \omega_0)^2 + \gamma^2 + 2W^2}.$$

At equilibrium, the rate at which photons are absorbed from the light beam, $-\alpha I/\hbar\omega$, must equal the photon emission rate due to spontaneous emission $2\gamma N_k$. If we denote by N the total atomic density, then $N_k = \rho_{kk} N$. Thus we get for the gain coefficient α

$$(11) \qquad \alpha = -\frac{2\gamma N \hbar\omega}{I} \cdot \frac{W^2}{(\omega - \omega_0)^2 + \gamma^2 + 2W^2}$$

If we now define the saturation intensity I_S as

$$(12) \qquad I_S = \frac{\hbar^2 c \varepsilon_0 \gamma^2}{|<i|\hat{e}\cdot\mathbf{P}|k>|^2},$$

and the satuarted line width γ_S as

$$(13) \qquad \gamma_S = \gamma\sqrt{1 + I/I_S},$$

equation (11) finally, becomes

$$(14) \qquad \alpha = -\frac{|<i|\hat{e}\cdot\mathbf{P}|k>|^2 \pi\omega}{\varepsilon_0 c \hbar} \cdot \frac{N}{\sqrt{1 + I/I_S}} \cdot \frac{\gamma_S/\pi}{(\omega - \omega_0)^2 + \gamma_{S^2}}.$$

The term I/I_S, which in our case equals $2W^2/\gamma^2$, is called saturation parameter and is usually denoted by S. The last fraction in equation (14) is a normalized Lorentzian curve of width γ_S. Equation (14) thus turns into eq. (2), with $f(\omega)$ given by (3), at the limit $I << I_S$ (i.e. $S << 1$). The height of the Lorentzian curve (14) is seen to decrease with increasing intensities of the light beam because of the denominator of the second fraction. The saturation intensity (12) is the intensity at which the height of the gain profile is reduced by a factor $1/2$. It is important to note that, although we have got again a Lorentzian profile, the satured line width γ_S is not simply obtained from the time-energy uncertainty relation.

If we now consider a Doppler broadened gain profile, under the assumption that the natural line width γ is much smaller than the Doppler width $\Delta\omega_D$, we obtain, after some calculations:

$$(15) \qquad \alpha = -\frac{|<i|\hat{e}\cdot\mathbf{P}|k>|^2 \omega}{\varepsilon_0 c \hbar \Delta\omega_D} \cdot \frac{N\sqrt{\pi log2}}{\sqrt{1 + I/I_S}} e^{-log2[(\omega-\omega_0)/\Delta\omega_D]^2}$$

Since in our assumed conditions the width of the gain curve is practically determined by the Doppler width only, equation (15) can be rewritten

$$(16) \qquad \alpha(\omega, S) = \frac{\alpha_0(\omega)}{\sqrt{1 + S}},$$

where $\alpha_0(\omega)$ is the small signal gain coefficient obtained by inserting (4) into (2). Such a simple relation between the saturated gain and the small signal gain is, however, not valid in the general case.

At the limit $I << I_S (S << 1)$ equation 16 reduces to (2) with $f(\omega)$ given by (4). It is important to note that, while any information on the natural line width γ is lost at the $I << I_S$ limit, equation 15 retains information on γ through the presence of I_S (see eq. 12). Of course, it would be an extremely difficult experimental task to extract an accurate value of γ from measurements of the absorption coefficient of a two-level system at high intensities. The point is, however, that the saturated response is sensitive to γ even though the linear response is not [5].

3. - The Three-Level Case: Homogeneously Broadened Lines

In order to observe the actual $NLHE$ we need the presence of coherence, and therefore our laser beam must connect at least three levels, two of which are degenerate or near-degenerate. An example is given in fig. 1, which represents a case analogous to the $6^1 S_0 - 6^3 P_1$ transition of an even isotope of mercury (of course, it would not be easy to observe saturation effects on the $253.7 nm$ line!). The ground state is assumed to have $J = 0$ and the excited state $J = 1$.

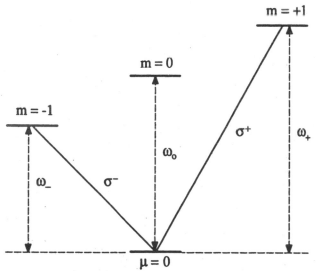

Fig. 1 *A transition from a* $J = 0$ *ground state to a* $J = 1$ *excited state. The energy of the excited state is assumed to be high enough so that the state is practically unpopulated at thermal equilibrium. A static uniform magnetic field* B_0 *parallel to the propagation direction* z *of the exciting light beam, and a linear polarization of the light beam along the* x *axis are assumed. In this conditions the radiation cannot excite the* $|\mu = 0 > \rightarrow |m = 0 >$ *transition, so that the system behaves as a three level system.*

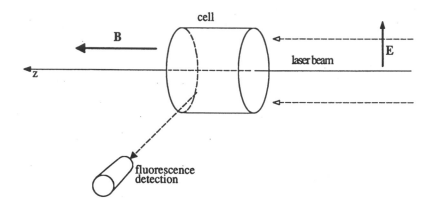

Fig. 2 *Experimental geometry for the observation of the* N L H E. *A laser beam is propagating parallel to the direction of the static magnetic field, which defines the* z *axis. The beam is linearly polarized along the* x *direction, so that both* σ *transition with* $\Delta M = \pm 1$ *can be simultaneously induced. The signal can be monitored, for instance, by measuring the intensity of the fluorescence light emitted in any direction.*

The quantization axis is chosen along the propagation direction of the beam z. We shall assume a monochromatic linearly polarized laser beam propagating along the z axis, with the electric vector parallel to the z axis, as shown in fig. 2. The effect can be monitored by measuring, for instance, the intensity of the fluorescence radiation emitted in any direction. Actually we are dealing with a four-level system, but if the light beam is linearly σ polarized, say in the x direction, the $|\mu = 0 >\ \rightarrow |m = 0 >$ transition cannot be induced, so that the level $|m = 0 >$ will not be populated (we are assuming that the energy difference between $|m = 0 >$ and $|\mu = 0 >$ is large compared to kT). Thus, practically, our vapor behaves as a V-type three-level system. If we apply an external field parallel to the beam direction, the sublevels of the excited state will be shifted with respect to each other. We shall denote by ω_+ the frequency of the $|\mu = 0 >\ \rightarrow |m = +1 >$ transition, and by ω_- the frequency of the $|\mu = 0 >\ \rightarrow |m = -1 >$ transition. The time evolution of the density matrix is described by an equation formally identical to (5)

$$(17) \qquad \frac{d\rho}{dt} = -\frac{i}{\hbar}[\mathcal{H}_0 + \mathcal{V}, \rho] + \Gamma\rho.$$

The matrices are written explicity

$$\rho = \begin{bmatrix} \rho_{++} & \rho_{+0} & \rho_{+-} \\ \rho_{0+} & \rho_{00} & \rho_{0-} \\ \rho_{-+} & \rho_{-0} & \rho_{--} \end{bmatrix} ; \quad \mathcal{H}_0 = \begin{bmatrix} \hbar\omega_+ & 0 & 0 \\ 0 & 0 & 0 \\ 0 & 0 & \hbar\omega_- \end{bmatrix} ;$$

(18)
$$\mathcal{V} = \begin{bmatrix} 0 & \hbar W e^{-i\omega t} & 0 \\ \hbar W e^{i\omega t} & 0 & \hbar W e^{i\omega t} \\ 0 & \hbar W e^{-i\omega t} & 0 \end{bmatrix} ;$$

$$\Gamma\rho = \begin{bmatrix} -2\gamma\rho_{++} & -\gamma\rho_{+0} & -2\gamma\rho_{+-} \\ -\gamma\rho_{0+} & 2\gamma(\rho_{++} + \rho_{--}) & -\gamma\rho_{0-} \\ -2\gamma\rho_{-+} & -\gamma\rho_{-0} & -2\gamma\rho_{--} \end{bmatrix} ,$$

We have denoted by $|0>$ the ground state, and by $|->$ and $|+>$ the $|m = -1>$ and $|m = +1>$ sublevels of the excited state, respectively. The term W in the interaction Hamiltonian is formally identical to (8)

(19)
$$W = -\frac{\sqrt{I}}{\hbar\sqrt{2c\varepsilon_0}} < +|\hat{\mathbf{e}} \cdot \mathbf{P}|0 > = -\frac{\sqrt{I}}{\hbar\sqrt{2c\varepsilon_0}} < -|\hat{\mathbf{e}} \cdot \mathbf{P}|0 >$$

Again, it is possible to chose the relative phases of $|0>$, $|+>$ and $|->$ so that W is real. Its square modulus is

(20)
$$|W|^2 = W^2 = \frac{I}{2\hbar^2 c\varepsilon_0} |< +|\hat{\mathbf{e}} \cdot \mathbf{P}|0 >|^2 = \frac{I}{2\hbar^2 c\varepsilon_0} |< -|\hat{\mathbf{e}} \cdot \mathbf{P}|0 >|^2.$$

A general procedure for evaluating the steady state solution of (17) at arbitrary light intensities is the development of the matrix elements of ρ into Fourier series in $exp(\pm in\omega t)$. In fact, since both σ^+ and σ^- transitions can be induced by the light beam, multiple quantum transition between $|0>$ and $|+>$ (as well as between $|0>$ and $|->$), involving odd numbers of photons, are allowed by the angular momentum conservation. This is why the subdiagonal elements of ρ will oscillate at odd multiples of ω, while only even n appear in the diagonal elements, ρ_{+-} and ρ_{-+}. An exact solution is quite involved. However, since multiple quantum transitions induced by a light beam in the (optical) frequency region of the main resonance are negligible even at high intensities, it will be reasonable to assume that each steady state matrix elemnt oscillates at a single frequency. We shall thus look for approximate solutions of (17) of the form

(21)
$$\rho_{++} = \sigma_{++}, \quad \rho_{--} = \sigma_{--}, \quad \rho_{00} = 1 - \sigma_{++} - \sigma_{--},$$

$$\rho_{+0} = \sigma_{+0}e^{-i\omega t}, \quad \rho_{0-} = \sigma_{0-}e^{+i\omega t}, \quad \rho_{+-} = \sigma_{+-},$$

$$\rho_{0+} = \sigma_{+0}{}^*, \quad \rho_{-0} = \sigma_{0-}{}^*, \quad \rho_{-+} = \sigma_{+-}{}^*,$$

where σ is, again, a constant Hermitian matrix. Inclusion of the components of the matrix elements oscillating at higher frequencies would lead to a Bloch-Siegert effect, which is negligible at optical frequencies. The steady state gain coefficient is obtained, as for the two-level case, from the condition that the number of photons absorbed at equilibrium from the light beam equals the number of spontaneously emitted photons, i.e.

$$(22) \qquad -\frac{\alpha I}{\hbar\omega} = 2\gamma N(\rho_{++} + \rho_{--}),$$

where N is again the total number of atoms per unit volume. According to (22), we are interested in the linear combination $(\rho_{++} + \rho_{--}) = (\sigma_{++} + \sigma_{--})$. This has the form

$$(23) \qquad \rho_{++} + \rho_{--} = 2W^2 \frac{A\Delta\omega^2 + B}{A\Delta\omega^4 + C\Delta\omega^2 + D}$$

where

$\omega_0 = (\omega_+ + \omega_-)/2$;

$\Delta\omega = \omega - \omega_0$ is the offset of the radiation frequency relative to the center of the doublet;

$\delta\omega = \omega_+ - \omega_0 = \omega_0 - \omega_-$ is half of the splitting between the two upper levels, and depends on the intensity of the applied static electric or magnetic field;

$A = \delta\omega^2 + \gamma^2$;

$B = (W^2 + \gamma^2 + \delta\omega^2)^2$;

$C = W^4 + 6W^2\delta\omega^2 + 6W^2\gamma^2 - 2\delta\omega^4 + 2\gamma^4$;

$D = 4W^6 + 5W^4\delta\omega^2 + 9W^4\gamma^2 + 2W^2\delta\omega^4 + 8W^2\gamma^2\delta\omega^2 + 6W^2\gamma^4 + (\delta\omega^2 + \gamma^2)^3$;

Equation (23) is the general steady state solution of our three-level problem, in absence of Doppler broadening. According to (22), the gain coefficient is obtained by multiplying (23) by $-2\gamma N\hbar\omega/I$. The complexity of this expression does not allow a direct physical insight. Some phusical insight, however, can be gained by considering the three following special cases:

i) $\delta\omega = 0$, i.e. $\omega_+ = \omega_-$, the two upper levels are completely degenerate

Since the frequency of the light beam is off-resonance by the same amount with respect to both the collapsed $|0> \rightarrow |+>$ and $|0> \rightarrow |->$ transitions we have $\rho_{++} = \rho_{--}$, and $|\rho_{+0}| = |\rho_{0-}|$, i.e. $|\sigma_{+0}| = |\sigma_{0-}|$. These conditions imply $\sigma_{0-} = -\sigma_{+0}$. Equation (23) reduces to

$$\rho_{++} + \rho_{--} = \frac{2W^2}{\Delta\omega^2 + \gamma^2 + 4W^2},$$

which is identical to (10), with the substitution $W^2 \rightarrow 2W^2$. The gain curve is given again by equation (14), with γ_S given by (13), where

(24)
$$I_S = \frac{\hbar^2 c \varepsilon_0 \gamma^2}{2| < +|\hat{\mathbf{e}} \cdot \mathbf{P}|0 > |^2}$$

At this limit, in fact, the levels $|m = +1 >$ and $|m = -1 >$ are degenerate and, if we choose x as the quantization axis, the new $|m' = 0 >$ level is an eigenstate of the Hamiltonian, so that our system behaves as a two-level system. Since the light beam is polarized parallel to the new quantization axis, we have

$$| < m' = 0|\hat{\mathbf{e}} \cdot \mathbf{P}|\mu = 0 > |^2 = 2| < m = +1|\hat{\mathbf{e}} \cdot \mathbf{P}|\mu = 0 > |^2$$
$$= 2| < m = -1|\hat{\mathbf{e}} \cdot \mathbf{P}|\mu = 0 > |^2.$$

This explains also the substitution of W^2 by $2W^2$ in the expression for the populations.

ii) $|\delta\omega|^2 >> \gamma^2$, $|\delta\omega|^2 >> W^2$.

In these conditions there is no coherence between the two upper levels. We can disregard, both in the numerator and in the denominator, the terms whose order of magnitude is lower than σw^4. If we now confine ourselves to $|\Delta\omega| \approx |\delta\omega|$. equation (23) becomes

(25)
$$\rho_{++} + \rho_{--} \approx \frac{4W^2\delta\omega^2}{(\Delta\omega^2 - \delta\omega^2)^2 + 4\delta\omega^2(\gamma^2 + 2W^2)} \approx$$
$$\approx \frac{W^2}{(\Delta\omega - \delta\omega)^2 + \gamma^2 + 2W^2} + \frac{W^2}{(\Delta\omega + \delta\omega)^2 + \gamma^2 + 2W^2}$$

The gain profile is thus approximated by the sum of two practically non-overlapping Lorentzians, centered at ω_+ and ω_- respectively. The expression of each Lorentzian is again (14), with

(26)
$$I_S = \frac{\hbar c \varepsilon_0 \gamma^2}{| < +|\hat{\mathbf{e}} \cdot \mathbf{P}|0 > |^2}$$

The factor 2 between the saturation intensities (26) and (24) is due to the fact that, for large$|\delta\omega|$, only one of the atomic $|0 > \rightarrow |+ >$ and $0 > \rightarrow |- >$ transitions can be excited at a time. Thus, only the accordingly polarized half of the beam intensity excites the transition, and twice the beam intensity as in case i) is required for achieving the same saturation effects. Since equation (25) is practically zero when $|\Delta\omega|$ is not close to $|\delta\omega|$, we can consider this expression as valid for any $\Delta\omega$.

iii) $\Delta\omega = 0$, any value of $\delta\omega$

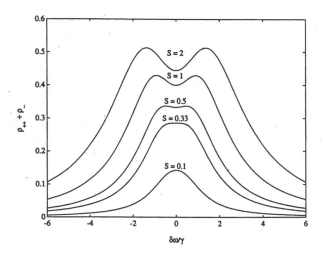

Fig. 3 *The total population of the excited state of the three level system is reported as a function of δω, i.e. one half of the splitting between the two sublevels of the excited states, for various values of the saturation parameter $S = (W/\gamma)^2$. The linearly polarized exciting light beam is assumed to be tuned to the center of the doublet. A bell shaped curve centered at δω = 0 is observed for S < 1/3. At higher radiation intensities a relative minimum is observed at δω = 0, while two maxima are observed at $\delta\omega = \pm\gamma[2\sqrt{S(1+S)} - (1+S)]1/2$.*

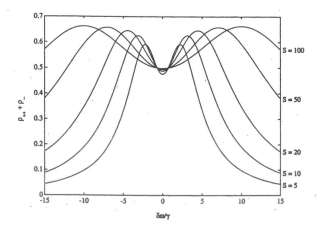

Fig. 4 *Population of the excited state of the three level system at high saturation parameters. The theoretical maxima of 0.5 for a two level system and of 2/3 for a three level system are practically reached at δω = 0 and $\delta\omega \approx \pm\gamma\sqrt{S}$, respectively. If δω is fixed at an intermediate value, a population decrease of the excited level is observed as the intensity of the excited beam increases.*

In these conditions ρ_{++} and ρ_{--} must be equal, since, as in case i), the light beam is off-resonance by the same amount with respect to both the $|0> \to |+>$ and the $|0> \to |->$ transition. Again, we must have $\sigma_{0-} = -\sigma_{+0}$. Equation (23) reduces to

$$\rho_{++} + \rho_{--} = \frac{2W^2}{\delta\omega^2 + \gamma^2 + 4W^2 \frac{W^2+\gamma^2}{W^2+\gamma^2+\delta\omega^2}}$$

and the gain coefficient becomes

$$(27) \qquad \alpha = -\frac{2\gamma N\omega_0| < +|\hat{e} \cdot \mathbf{P}|0 > |^2}{c\varepsilon_0 \hbar} \cdot \frac{1}{\delta\omega^2 + \gamma^2 + 4\gamma^2 S \frac{\gamma^2 S+\gamma^2}{\gamma^2 S+\gamma^2+\delta\omega^2}}$$

where again S is the saturation parameter I/I_S, and

$$I_S = \frac{2\hbar^2 c\varepsilon_0 \gamma^2}{| < +|\hat{e} \cdot \mathbf{P}|0 > |^2},$$

so that $S = W^2/\gamma^2$ in this case.

Case iii) is the experimentally most interesting case. The computed profile of (27) as a function of $\delta\omega$ is shown in figs. 3 and 4 for different values of S. Here and in subsequent figures the sum of the populations of the sublevels of the upper state, rather than the gain coefficient, is shown as a function of the experimental parameters. Experimentally, $\delta\omega$ can be varied by sweeping the external magnetic or electric field around zero. It is seen from fig. 3 that the gain profile is a symmetrical bell shaped curve at low field intensities. At higher intensities it remains symmetrical about $\delta\omega = 0$ but it displays a minimum at $\delta\omega = 0$ and two maxima at

$$(28) \qquad \delta\omega_{max}(S) = \pm\gamma \, [2\sqrt{S(1+S)} - (1+S)]^{1/2}.$$

We see from the above expression that the two maxima are observed only if $S > 1/3$. At the limit of high saturation parameters $(S >> 1)$ eq. (28) becomes $\delta\omega_{max}(S) \approx \pm\gamma\sqrt{S}$, this proportionality of $\delta\omega_{max}$ to the square root of S at high S values is shown in fig. 4. Qualitatively, the behavior of figs. 3 and 4 can be interpreted as due to the competition of the following two phenomena: on the one hand, the increasing external field is causing a transition from a two-level to a three-level system, which, alone, would lead to an increase in the gain coefficients. On the other hand, the external field is also shifting the two transitions out of resonance, and this latter phenomenon is dominant at $|\delta\omega| > |\delta\omega_{max}(S)|$. An interesting saturation phenomenon can be observed in fig. 4. Here, for $S > 20$, the population of the upper state, $\rho_{++} + \rho_{--}$, has practically reached its theoretical maxima of 0.5 at $\delta\omega = 0$ and of $2/3$ at $\delta\omega = \pm\delta\omega_{max}(S)$. The absorption coefficient $\alpha(\delta\omega_{max})$ is thus larger by a

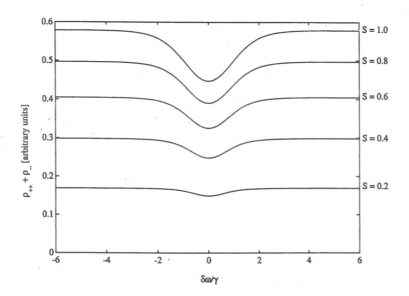

Fig. 5 Population of the excited state of the three level system at different values of the saturation parameter S, when the absorption line is Doppler broadened. A Doppler width ten times larger than the natural width has been assumed.

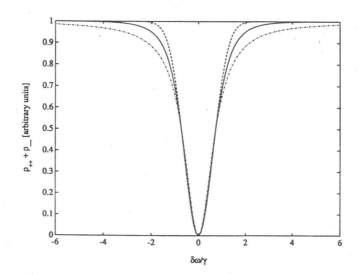

Fig. 6 Calculated NLHE signal for a saturation parameter S = 0.1 (solid line). A Lorentzian (dashed line) and a Gaussian (dash-dotted line) curve having the same height and width of the signal are also reported.

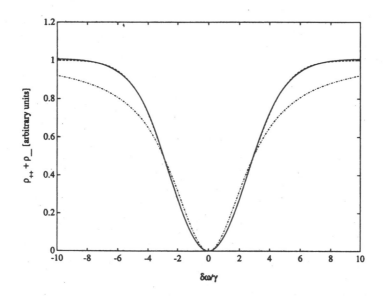

Fig. 7 Calculated NLHE signal for a saturation parameter $S = 10$ (solid line). A Lorentzian (dashed line) and a Gaussian (dash-dotted line) curve having the same height and width of the signal are also reported. The signal is much closer to a Gaussian shape than the signal of fig. 6.

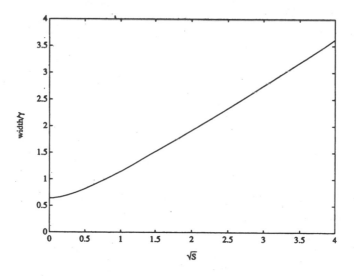

Fig. 8 Width of the NLHE signal as a function of the square root of the saturation parameter, in the presence of a large Doppler broadening. At high light intensities the width of the signal is practically determined by the power broadening.

factor 4/3 than the absorption coefficient at zero field. For any fixed $\delta\omega$, with $0 < |\delta\omega| < |\delta\omega_{max}(20)|$, we see that the population of the upper state decreases with increasing radiation intensity. The explanation is that increasing radiation intensity broadens the levels; at fixed level spacing this causes a transition from a three level system back towards a two level system.

4. - The Three-Level Case: Doppler Broadened Lines and the Rate Equations

Within the same approximations used for the two-level case, the gain coefficient for a Doppler-broadened three-level system can be evaluated as

$$(29) \qquad \alpha = -\frac{2\gamma N\omega_0| < +|\hat{e} \cdot P|0 > |^2}{c\varepsilon_0\hbar} \cdot \frac{\sqrt{\pi log2}}{\Delta\omega_D} \cdot \frac{A\sqrt{D} + B\sqrt{A}}{\sqrt{ACD + 2(AD)^{3/2}}},$$

where A, B, C and D have the same meaning as in equation (23). The computer evaluated behavior of $\rho_{++} + \rho_{--}$, in the presence of a large Doppler broadening, as a function of $\delta\omega$, i.e. of the applied external field, is shown in fig. 5 several values of the saturation parameter $S = W^2/\gamma^2$. The shapes of the Hanle signals of fig. 5 are shown in more convenient scales in fig. 6 for $S = 0.1$ and fig. 7 for $S = 10$. In both figures the vertical axis is in arbitrary units in order to normalize the height of the signal. The Lorentzian (dashed line) and Gaussian (dash-dotted line) curves of same half height width as the signal are shown in both figures. It is interesting to note that the signal shape approaches a Gaussian curve as the saturation parameter is iscreased. The computed dependence of the signal width on $\sqrt{S}$ is shown in fig. 8, this dependence becomes linear at high saturation parameters, where $\Delta\omega \approx \gamma\sqrt{S} = W$. As for the classical Hanle effect, the width of the signal is practically determined by the homogeneous contribution to the line width only, and independent of the Doppler broadening.

Equation 29 itself is quite complex, but it has two simple, interesting limiting cases,

i) $\delta\omega = 0$, i.e. zero field:

$$(30) \qquad \alpha = -\frac{2N| < +|\hat{e} \cdot P|0 > |^2\omega}{\varepsilon_0 c\hbar\Delta\omega_D} \cdot \frac{\sqrt{\pi log2}}{\sqrt{1 + 2I/I_S}},$$

ii) $|\delta\omega|^2 >> \gamma^2$, $|\delta\omega|^2 >> W^2$:

$$(31) \qquad \alpha = -\frac{2N| < +|\hat{e} \cdot P|0 > |^2\omega}{\varepsilon_0 c\hbar\Delta\omega_D} \cdot \frac{\sqrt{\pi log2}}{\sqrt{1 + I/I_S}},$$

which correspond to the minima and to the asymptotic parts of fig. 5. In both (30) and (31) I_S is given by (26). The simplicity of these expression is due to the lack of coherence at both limits. In the case of molecules, where high J values are usually involved both in the upper and lower state of the transition,

$\delta\omega$ can be large (and thus the coherence can be destroyed) also when no relevant line shift is observed. This is due to the simultaneous shift of the m sublevels in the upper and lower state.

At the limiting case of very high light intensities the ratio between α for the non-degenerate three-level system and α for the two-level system is $\sqrt{2}$, which is higher than the factor $4/3$ which we found for the homogeneously broadened case. The reason is that, while only one hole is burned into the velocity distribution at zero external field ($\delta\omega = 0$), two different holes are burned at high field. Thus a further important enhancing phenomenon is added to the population distribution between two or three levels.

Equations (30) and (31) can be directly obtained by solving a system of rate equations involving only populations, and no coherences. Thus, in the case of Doppler broadened lines with a Doppler width much larger than the natural line width, a more simple approach to the nonlinear Hanle effect is available [32]. This approach, however, can provide only the height of the signal, which is determined by the difference between the behavior at zero field and the behavior at the high field limit. The complete evaluation of the steady-state density matrix is still needed if the solution at intermediate fields is required. This is needed, for instance, for a precise evaluation of the signal width.

5. - The General Case

Up to now, the only cases which we have considered are the two-level system and the V-type three-level system. The extension of the method developed for homogeneously broadened lines to higher J values, and to more complicated relaxation patterns is, in principle, straightforward. The method can also be easily modified in order to describe the $NLHE$ on a laser transition. Here one must include terms describing the pumping mechanism into the matrix $\Gamma\rho$, which appears in equation 17. However, one should remember that the unknowns of the resulting system of linear equations are all the populations and half of the coherences of the density matrix, whose dimensions are $(2J' + 1 + 2J'' + 1) \times (2J' + 1 + 2J'' + 1)$. In the preceding expression J' is the angular momentum of the upper level, and J'' the angular momentum of the lower level of the transition. It is thus apparent how fast the solution becomes more and more complicated with increasing J values.

It is also interesting to note that, whereas the $|J' = 1 > \leftarrow |J'' = 0 >$ transition can be treated as a V-type three-level system, the $|J' = 0 > \leftarrow |J'' = 1 >$ transition cannot be treated as a Λ-type three level system. In fact, if we assume the experimental geometry of fig. 2, and absence of a redistribution mechanism within the ground state, all atoms are pumped into the $|\mu = 0 >$ sublevel, which cannot absorb σ radiation. As a result, no absorption is observed. Absorption can be restored by population redistribution within the ground state (which could be due, for instance, to collisional relaxation or to RF induced transitions). In any case, the presence of the $|\mu = 0 >$ sublevel of the ground state cannot be neglected, as we did for the $|m = 0>$ state of the upper level in the V-type case. The number of simultaneous equation is thus increased by the presence of a further population and possible further coherences. Moreover, the necessary presence of a population redistribution mechanism in the ground state further complicates the solution.

In spite of the difficulties mentioned above for obtaining an exact solution, which is required for evaluating the width and shape of the signal, it is usually possible to draw some general conclusions regarding the height of the signal.

We have seen that, for a homogeneously broadened line, saturation tends to equally populate all the sublevels involved in the transition. Changing the external field from zero to high values changes the most convenient quantization axis, and in a sense, the effective numbers of sublevels involved in the transition. For instance, we have seen that the V-type system behaves as a two-level system at zero field, and as three-level system at high fields; in any case, at complete saturation, all the sublevels tend to be equally populated. This remains true as long as no important pumping effects, as the ones encountered for the $|J' = 0 > \leftarrow |J'' = 1 >$ case, are present. Thus, since the number of sublevels of the upper and of the lower state differs, at must, by two (actually, the difference is either 0 or 2), we expect the $NLHE$ signal to be negligible, if at all present, when $J' = J''$, and to become weaker and weaker with increasing J, when $J' \neq J''$.

On the other hand, for a Doppler broadened line a further absorption enhancement effect is due to the fact the Zeeman shift brings different velocity groups into resonance with the laser radiation, as we saw in the preceding section. Thus, we expect the absorption enhancement due to the $NLHE$ to remain important also for transition involving high J values.

6. - The Nonlinear Hanle Effect in Laser Emission

The $NLHE$ is particularly important in laser emission, because an increase in the saturation intensity can lead to a population increase in the upper state of the laser transition, and therefore to an increase in the emitted power. Since the relevant parameter is the ratio of the population of the upper state at high fields to the population at zero field, the rate equation approach described in ref. 32 is very often perfectly adequate for evaluating the power increase of the laser. For evaluating how much the output power of a laser is affected by the $NLHE$, one must take into account also all the other parameters relevant to the operation of an active oscillator. The $NLHE$ plays a role because the output power W of a laser depends not only on the gain coefficient α of the active medium but also on its saturation intensity I_S. For a Doppler broadened transition and single mode laser operation W is proportional to

(32) $$W \propto \alpha_0^2 I_S$$

where α_0 is the unsaturated gain coefficient. We have shown in the preceding sections that I_S, which is inversely proportional to the matrix element of the dipole moment for the transition, is increased by the removal of degeneracy. In particular, for a Doppler broadened transition and for any J', J'' values, the average I_S is increased by a factor 2. W depends also on the offset of the frequency of the cavity mode ν_c with respect to the center of the transition frequency ν_0. Another critical parameter for the $NLHE$ is the ratio $\gamma/\Delta\nu_c$ between the homogeneous line width and the line width of the cavity mode ($\Delta\nu_c = \nu_c/Q_c$, where Q_c is the cavity figure of merit). Further, W is affected by the actual design of the cavity (Brewster windows, waveguide ...) which can force the EM field to oscillate with a specific polarization. Finally, one must remember that a laser is an active oscillator with a threshold, and the effective power enhancement

(33) $$PE = (\alpha_0^2 I_S)_{nd}/(\alpha_0^2 I_S)_d,$$

where the subscripts nd and d label the non-degenerate and the degenerate case, can be perturbed if the system is not properly optimized and the oscillation conditions are close to threshold. The most important parameter, however, is the ratio $\gamma/\Delta\nu_c$, because the power enhancement is negligible when this ratio is larger than one. Let us assume an axial magnetic field and linearly polarized light (Brewster windows). If degeneracy is removed, all the atoms are in resonance with the laser field only for one half of the allowed transition (either only σ^+ or only σ^-). Thus, $(I_S)_{nd} = (I_S)_d$, and $(\alpha_+^0 + \alpha_-^0)_{nd} = (\alpha_\pi^0)_d$ (α_+^0 and α_-^0 represent the gain coefficients of differnt atoms with opposite Doppler shifts). In conclusion

$$PE = [(\alpha_+^0 + \alpha_-^0)I_S]_{nd}/(\alpha_\pi^0 I_S)_d = 1.$$

Of course, the actual situation in a multimode laser is much more complex as the model sketched above, and the results which we have just obtained are only a guideline. It is clear, however, that a significant power enhancement is expected only when $\gamma/\Delta\nu_c < 1$.

Power enhancement due to $NLHE$ was observed in HeNe lasers almost immediately after their discovery [33]. An example of PE observed on the 638 nm line by applying a static magnetic field is shown in fig. 9. While the enhancement effect is clearly observable for short cavities with cavity modes broader than the natural line width, this is not the case for very long cavities, with very narrow modes. In this case, in fact, the Zeeman components are shifted out of the mode before the enhancement effect may occur. Thus, the $NLHE$ power enhancement has no practical application for the HeNe and similar gas lasers which require high finesse cavities [34].

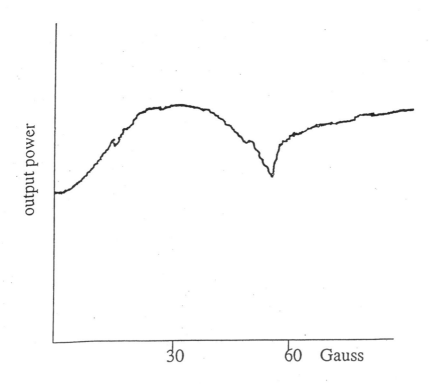

Fig. 9 *NLHE on a He-Ne laser. A power increase is first observed. Then, a power decrease corresponds to the Zeeman components being shifted out of the cavity mode. At higher fields, a new power increase corresponds to a jump to two-mode operation.*

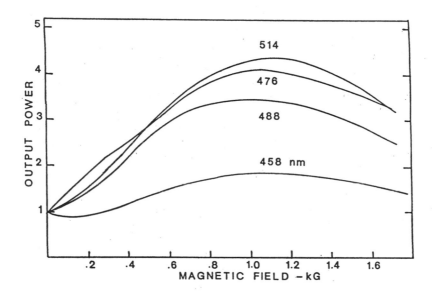

Fig. 10 *Magnetic field dependence of the output power of several lines emitted by the Ar ion laser.*

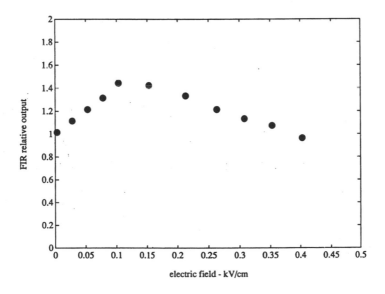

Fig. 11 *NLHE on the 96.5μm transition of CH_3OH. The polarization of this laser line is orthogonal to the polarization of its pump transition.*

On the contrary, the application of a static magnetic field of about 0.1 T, in order to enhance the emission power, is currently used in commercial argon and krypton lasers. A power increase by a factor up to four is observed. At first this power enhancement was tentatively attributed to an increase in the plasma electron density, but the experimental confirmation was contradictory [35]. Therefore, it was proposed that this enhancement could be due to a $NLHE$ [36]. A discriminating point is that, while an electron density increase would enhance all the visible laser lines in a comparable way, the enhancement is expected to be dependent on the J quantum numbers involved in the transition for a $NLHE$. In particular, no enhancement is expected for laser lines with a $J = 1/2$ upper level. This is the case for the $Ar^{++}458nm$ and for the $Kr^{++}676nm$ laser lines, which, actually, display no enhancement effect. The experimental results on the dependence of the output power of various argon ion laser lines on the applied magnetic field are shown in fig. 10 [37]. In the case of the Ar^+ and Kr^+ ion lasers the homogeneous line width is quite large ($\approx 400MHz$), and larger than the usual mode spacing ($50 \sim 150MHz$), thus the laser operation can be treated by the broad line approximation. Each atom simultaneously contributes to the amplification of several laser modes. According to [23, 24] $(\alpha_0)_{nd} \approx \sqrt{2}(\alpha_0)_d$ at large laser powers, thus we obtain a PE of the order of 4 form equation (33), in good agreement with the experimental observation.

The power enhancement due to the $NLHE$ is of practical importance also for the optically pumped far-infrared lasers. In this case, contrary to the visible lasers, not only $\gamma/\Delta\nu_c < 1$, but also $\Delta\nu_D/\Delta\nu_c < 1$. Moreover, in optically pumped lasers it is possible to have a $NLHE$ both on the pump and on the lasing transition, while in discharge excited lasers the $NLHE$, if present, affects only the lasing transition. In molecular lasers, the level splitting is usually due to Stark, rather than Zeeman effect.

The observation of the $NLHE$ in laser emission is particularly easy in optically pumped FIR lasers, since the absence of electrical discharges allows an easy application of a static electric field. CH_3OH laser lines affected by a linear Stark shift display power increases up to a factor four [22]. This power increase is observable only if the pump polarization and the Stark field are perpendicular to each other, in agreement with our previous discussion of the $NLHE$. The experimental case of the $96.5\mu m$ laser line of methanol is shown in fig. 11.

REFERENCES

[1] W. HANLE: Z. *Phys.* **30**, 93 (1924).

[2] G. MORUZZI: *Hanle Effect and Level Crossing Spectroscopy: an Introduction* in *Hanle Effect and level Crossing*, G. Moruzzi and F. Strumia editors, Plenum Publishing Company, in press.

[3] A. JAVAN: *Bull. Am. Phys. Soc.* **9**, 489 (1964) R.H. Cordover, A. Szöke, A. Javan: ib. page. 490.

[4] M. S. FELD - J. W. PARKS - H. R. SCHLOSSBERG - A. JAVAN: *Spectroscopy with Gas Lasers*, in *Physics of Quantum Electronics*, P.L. Kelley, B. Lax and P.E. Tannenwald eds., McGraw-Hill 1966, pages 567-580.

[5] M. S. FELD - A. SANCHEZ - A. JAVAN - B. J. FELDMAN: *Proc. of the Aussoix Conf. Colloques Intern. du CNRS* n. 217 page 87-104 (1974).

[6] M. S. FELD: in *Fundamental and Applied Laser Physics*, M. Feld, A. Javan and N.A. Kermitt eds., J. Wiley 1973, page 369-420.

[7] R. L. FORK - L. E. HARGROVE - M. A. POLLACK: *Phys. Rev. Lett.* **12**, 705 (1964).

[8] H. R. SCHLOSSBERG - A. JAVAN: *Phys. Rev. Lett.* **17**, 1242 (1966).

[9] T. F. JOHNSTON JR. - G. J. WOLGA: *Phys. Lett.* **27** A, 639 (1968).

[10] M. TSUKAKOSHI - K. SHIMODA: *J. Phys. Soc. Jap.* **26**, 758 (1969).

[11] W. FLYNN - M. S. FELD - B. J. FELDMAN: *Bull. Am. Phys. Soc.* **12**, 669 (1967).

[12] S. LEVINE - P. A. BONCZYK - A. JAVAN: *Phys. Rev. Lett.* **22**, 267-270 (1969).

[13] G. HERMANN - A. SCHARMANN: *Z. Physik* **254**, 46-56 (1972).

[14] C. LUNTZ - R. G. BREWER - K. L. FOSTER - J. D. SWALEN: *Phys. Rev. Lett.* **23**, 951-954 (1969).

[15] C. LUNTZ - R. G. BREWER: *J. Chem. Phys.* **53**, 3380-3381 (1970).

[16] R. G. BREWER: in *Fundamental and Applied Laser Physics*, M.S. Feld, A. Javan and N.A. Kermitt eds., J. Wiley 1973, page 421-436.

[17] H. R. SCHLOSSBERG - A. JAVAN: *Phys. Rev.* **150**, 267 (1966).

[18] M. S. FELD - A. JAVAN: *Phys. Rev.* **177**, 540 (1969).

[19] B. J. FELDMAN - M. S. FELD *Phys. Rev.* **A5**, 899 (1971).

[20] K. SHIMODA: *Jap. J. Appl. Phys.* **11**, 564 (1972).

[21] K. ERNST - M. INGUSCIO: *Rivista del Nuovo Cimento* **11**, n. 2, 1-66 (1988).

[23] F. STRUMIA - M. INGUSCIO - A. MORETTI: in *Laser Spectroscopy V*, A.R. McKellar, T. Oka, B.P. Stoicheff eds., page 255-259, Springer, Berlin, Heidelberg (1981).

[24] M. INGUSCIO - A. MORETTI - F. STRUMIA: *Appl. Phys. B* **28**, 89-90 (1982).

[25] M. INGUSCIO - A. MORETTI - F. STRUMIA: *Opt. Commun.* **30**, 355 (1979).

[26] N. BEVERINI - M. INGUSCIO: *Lett. Nuovo Cimento* **29**, 10 (1980).

[27] P. HANNAFORD - G. W. SERIES: *J. Phys. B* **14**, L661-666 (1981).

[28] B. BARBIERI - N. BEVERINI - G. BIONDUCCI - M. GALLI - M. INGUSCIO - F. STRUMIA: *Proc. 3rd Nat. Conf. Quantum Electron. and Plasma*, V. De Giorgio Ed.; Como (1982) pp. 337-340.

[29] P. HANNAFORD - G. W. SERIES: *Laser Spectroscopy V*, A.R. McKellar, T. Oka, B.P. Stoicheff eds., page 94, Springer Verlag, Berlin, Heidelberg, New York 1981.

[30] N. BEVERINI - K. ERNST - M. INGUSCIO - F. STRUMIA: *Appl. Phys. B* **37**, 17- 29 (1985).

[31] N. SOKABE - Y. TAMURA - K. MATSUSHIMA - A. MURAI: *Int. J. IR and MM Wave* **8**, 1145-1163 (1987).

[32] N. BEVERINI - G. MORUZZI - F. STRUMIA: *Applications of the Non-linear Hanle Effect* in *Laser Physics in Hanle Effect and Level Crossing Spectroscopy*, G. Moruzzi and F. Strumia editors, Plenum Publishing Company, in press.

[33] R.G. BREVER - J. KAINZ - J. SULLIVAN: *Appl. Opt.* **2**, 861 (1963).

[34] W. SASAKI - H. UEDA - T. OHTA: *IEEE JQE* **QE19**, 1259 (1983).

[35] C. C. DAVIS - T.A. KING: in *Advances in Quantum Electronics*, vol. 3, D.W.Goodwin ed., Academic Press, New York 1975, page 33-166.

[36] F. STRUMIA: *J. Phys. (Paris)* **44**, C7 (1983).

[37] W. BRIDGES - A. S. HALSTEAD: *Gaseous Ion Laser Research* Fin. Tech. Rep. Hughes Res. Lab. (1967) (unpublished) NTIS access # AD-814897.

Mixing and Entropy Increase
in Quantum Systems

H. NARNHOFER - A. PFLUG

W. THIRRING (*)

Abstract

This paper attempts to explain the key feature of deterministic chaotic classical systems and how they can be translated to quantum systems. To do so we develop the appropriate algebraic language for the non-specialist.

1. - Introduction

Classical ergodic theory has provided us with beautiful models, the K-systems, which show how random elements can result from a deterministic time evolution [1-5]. However, the deeper roots of statistical mechanics are in the atomistic domain which is governed by quantum mechanism [6]. Thus to make contact with physics one has to widen the conceptual framework so that it also includes the noncommutative algebras [7] which represent the quantum mechanical observables. In this paper we shall try to explain how this can be done and how one can classify the chaotic properties of quantum systems. Since finite quantum systems are quasiperiodic they do not show any convergence to equilibrium but are at best ergodic, we have to deal in quantum theory with a thermodynamic limit. There we shall find the analogue to the classical K-systems and the same kind of random features [8-10].

1.1 - *Mixing of a Viscous Fluid*

To illustrate the essential features of a dynamical system with deterministic and reversible time evolution showing nevertheless an increase of disorder in the limit of sufficiently long times, we consider first the classical case of stirring repeatedly an (extremely viscous) fluid in a closed container. As internal friction

(*)   Institut für Theoretische Physik Universität Wien

stops immediately any inertial motion in this situation, the (stepwise) time evolution of the fluid particles can be described by the iteration of a given invertible map of the configuration space (which is the interior of the container) onto itself. For an incompressible fluid, the map should preserve the local volume element. To study the effect of mixing, we must distinguish different regions of the fluid by adding a (small, but finite) quantity of "indicator", e.g. a tiny drop of dye. The actual state of the fluid can be described by a (normalized) concentration function $\rho(\lambda)$ of the dye on the configuration space Λ. Λ is a compact subset of R^n with nonvoid open interior; the spatial dimension $n \in N$ should remain arbitrary. "Normalized" means $\int_\Lambda \rho(\lambda) d\mu(\lambda) = 1$ where $d\mu(\lambda)$ is the "ordinary" volume element in R^n. The fraction of dye molecules in an open region Γ (with respect to their total number in Λ) is then given by the "expectation" value $\int_\Lambda \chi_\Gamma(\lambda)\rho(\lambda) d\mu(\lambda) = \int_\Gamma \rho(\lambda) d\mu(\lambda)$ of the observable $\chi_\Gamma(\lambda)$ which is simply the characteristic function of Γ. This means $\chi_\Gamma(\lambda) = 1$ if $\lambda \in \Gamma, \chi_\Gamma(\lambda) = 0$ if $\lambda \notin \Gamma$. The (physically plausible) restriction to states given by concentration functions $\rho(\lambda)$ means that we are not able to detect whether a set of volume zero (for example the points = molecules along a boundary) has been removed from or added to the fluid. Such a set has always vanishing volume and cannot contain a nonzero fraction of the total number of "dye" molecules which stem from an "open" drop with strictly positive measure. The observable $\chi_\Gamma(\lambda)$ is a yes-no experiment, because it can only be true, i.e. $= 1$, or false, i.e. $= 0$. As 0 and 1 are the only numbers with $z^2 = z$, we can express the yes-no-character of χ_Γ by the algebraic "projector" property $\chi_\Gamma^2 = \chi_\Gamma$ in the set of *complex* functions. In the noncommutative quantum case the reality of an observable does not follow from the relation $P^2 = P$, only the additional assumption $P = P^*$ (corresponding to orthogonal projections) will yield yes-no-experiments.

The set of observables $\{\chi_\Gamma | \Gamma$ coincides with an open subset of Γ up to a set of measure zero$\}$ has the algebraic property that the product of two projector functions $(\chi_{\Gamma_1} \cdot \chi_{\Gamma_2})(\lambda) := \chi_{\Gamma_1}(\lambda) \cdot \chi_{\Gamma_2}(\lambda)$ is the projector $\chi_{\Gamma_1 \cap \Gamma_2}(\lambda)$ onto the intersection of the two open sets. $\chi_{\Gamma_1 \cup \Gamma_2}$ can also be expressed algebraically, namely as $\chi_{\Gamma_1} + \chi_{\Gamma_2} - \chi_{\Gamma_1} \cdot \chi_{\Gamma_2}$. This algebraic structure is in accordance with the logic of *classical* measurements because $\Gamma_1 \cap \Gamma_2$ means a logical "and" whereas $\Gamma_1 \cup \Gamma_2$ corresponds to the logical "or" in the set of elementary yes-no observables.

If $\mathcal{P}$ is a collection of essentially open subsets of Λ containing also the intersection of all pairs of its elements, the set

$$\mathcal{A}_0(\mathcal{P}) = \left\{ \sum_{k=1}^{\ell} c_k \chi_{\Gamma_k} | \ell \in N, c_k \in C, \Gamma_k \in \mathcal{P} \right\}$$

of observable functions is a *-algebra because it allows complex conjugation, finite linear combinations and the formation of products. If $\Lambda \in \mathcal{P}$, then $\mathcal{A}_0(\mathcal{P})$ has a unit element $1 \equiv \chi_\Lambda$. For a more detailed discussion of observables, states and their time evolution we choose an explicit example, namely.

1.2 - *The Baker Transformation: Kneading the Surface of a Doughnut*

A (unbaked) dough is an extremely viscous fluid which often does not need a container to keep its external shape. We therefore consider the kneading of a thin, square sheet of dough; to get rid of boundary effects, we will identify opposite points on the circumference. It is not possible to embed the resulting toroidal dough manifold (which is topologically equivalent to the surface of a doughnut, but flat in its internal geometry) in R^3 without deforming the volume element $d\mu$. Although this embedding would be possible in some R^n (with sufficiently large n), we choose, as configuration space Λ, the compact quotient space $T^2 = R^2/Z^2 = \{(x,y) \in [0,1) \times [0,1)\}$ of the euclidean plane. The baker transformation

$$B : T^2 \rightarrow T^2, (x,y) \rightarrow \begin{cases} (2x, \frac{y}{2}) & \forall 0 \leq x < \frac{1}{2}, \ 0 \leq y < 1 \\ (2x-1, \frac{y+1}{2}) & \forall \frac{1}{2} \leq x < 1, \ 0 \leq y < 1 \end{cases}$$

corresponds to the stretching of the dough in x-direction by a factor 2 (due to incompressibility, it will simultaneously shrink in the y-direction by the same factor) and cutting it in two congruent rectangular pieces of length one. The right piece is finally put above the left one and opposite points are again identified, so that the shape and topology of the configuration space remains invariant. The inverse transformation

$$B^{-1} : T^2 \rightarrow T^2, (x,y) \rightarrow \begin{cases} (\frac{x}{2}, 2y) & \forall 0 \leq x < 1, \ 0 \leq y < \frac{1}{2} \\ (\frac{x+1}{2}, 2y-1) & \forall 0 \leq x < 1, \ \frac{1}{2} \leq y < 1 \end{cases}$$

is given by the "orthogonal" procedure of stretching in the y-direction and cutting parallelly to the x-axis. To get an idea how a state ω_ρ (defined by a "normalized" concentration function $\rho : \Lambda \rightarrow R^+, \int_\Lambda \rho(\lambda)d\mu(\lambda) = 1$ of e.g. salt on the surface) will be "mixed through" in the course of time, we start with a $\rho(\lambda) = const \cdot \chi_\Gamma(\lambda)$ concentrated in a "small" localized open region Γ with diameter $<< 1$. Under the repeated action of the "kneading", i.e. by the n-th iterate B^n of the baker transform B, Γ expands (exponentially) in x-direction and is simultaneously contracted (exponentially) along the y-axis. Finally it will form a system of narrow rings in x-direction around the tours with (exponentially) increasing number and (exponentially) small separation, which fills the whole configuration space in a quasihomogeneous manner. In spite of this (exponentially) fast spreading of the "salted region" $B^n(\Gamma)$, over the surface of the doughnut, the baker can concentrate all salt again by starting to knead in the orthogonal direction (given by B^{-1}). If he stops exactly after n "inverse" steps, the salt is again concentrated in $B^{-n}(B^n(\Gamma)) = \Gamma$ and can be easily removed from this small localized region. But if he doesn't take care and forgets to stop in time, the salted region Γ becomes wrapped around the torus in the y-direction by the inverse kneading process! Thus mixing occurs in time-reversible situations asymptotically for both directions $n \rightarrow \pm\infty$ in time!

The example of the spreading "salted" set Γ shows how a time evolution α^* can be defined for all states ω_ρ specified by a density function $\rho : \alpha^* \omega_\rho = \omega_{\alpha^* \rho}$ with $\alpha^* \rho(\lambda) := \rho(B^{-1}(\lambda))$. If $\rho = const \cdot \chi_\Gamma$ the $\alpha^* \rho(\lambda) = const \cdot \chi_\Gamma(B^{-1}(\Lambda)) = const \cdot \chi_{B(\Gamma)}(\lambda)$. For reasons of simplicity and convention we use the same symbol α^* for the evolution of the state ω_ρ (which is a functional on the set of observable functions) and the corresponding density functions $\rho : \Lambda \to R^+$. Sometimes it is more convenient to go from the "Schrödinger picture" defined above (where the time evolution affects only the states ω_ρ of the dough) to the equivalent "Heisenberg picture" with time-independent states ω_ρ and a temporal development $\chi_\Gamma \to \alpha\chi_\Gamma$ in the set of observables. The physical equivalence of the two pictures requires that in both cases we get the same predictions for the outcome of measurements, i.e. the same expectation values of all observables. This property is guaranteed by the definition $\alpha\chi_\Gamma(\lambda) = \chi_\Gamma(B(\lambda)) = \chi_{B^{-1}(\Gamma)}(\lambda)$, because the invariance of the volume element $d\mu(\lambda) = d\mu(B(\lambda))$ allows a change of the integration variables:

$$\alpha^* \omega_\rho(\chi_\Gamma) = \int \chi_\Gamma(\lambda)\rho(B^{-1}(\lambda))d\mu(\lambda) = \int \chi_\Gamma(B(\lambda))\rho(\lambda)d\mu(\lambda) = \omega_\rho(\alpha\chi_\Gamma).$$

The "Schrödinger" time evolution α^* in the set of states is therefore the dual transformation of the "Heisenberg" time development α on the set of observables, justifying the previous notation.

To study the asymptotic behaviour of observables and states when discrete time goes to infinity, we have to specify the *-algebra $A_0(P)$ characterized by a class P of sets which generate "linearly" (i.e. by unions) all regions on which the measurements can "project". As the baker transformation B should create a time automorphism of $A_0(P)$, B^{-1} and B must leave the class P invariant because $\alpha\chi_\Gamma = \chi_\Gamma \circ B = \chi_{B^{-1}(\Gamma)}$ for all $\Gamma \in P$, the analogous relation holding for α^{-1}. Therefore we choose $P = S$ where S is given by all finite intersections of "stripe sets" $S_{i,k} \subseteq T^2$ with $i \in Z$ and $k \in \{0,1\}$ defined as

$$S_{0,0} = \{(x,y) \in T^2 | 0 \leq x < \frac{1}{2}\}$$

$$S_{0,1} = T^2 \backslash S_{0,0}$$

$$S_{i,k} = B^i(S_{0,k}) \; \forall i \in Z \backslash \{0\}, k \in \{0,1\}.$$

As B is measure preserving, the volume $\int \chi_{S_{i,k}}(\lambda)d\mu(\lambda)$ of every stripe set equals $1/2$. Given an index set I with $n(I)$ (pairwise *different*) elements of integers, the volume of $\Gamma_I := \bigcap_{i \in I} S_{i,k_i}$ is reduced by a factor $1/2$ if Γ_I is intersected with a stripe set $S_{j,k}, j \notin I$. Therefore the volume of Γ_I itself equals $2^{-n(I)}$.

By construction, S is invariant under both B and B^{-1} and contains all squares from "chessboard lattices" on the torus with lattice constant $2^{-n}, n \in N$. Finite unions of such squares will allow to approximate all "decent" subsets of the torus (e.g. all open ones) in the sense that the "volume" (which is the area in our case) of the difference set tends to zero. This means that the corresponding observables (i.e. the characteristic functions on the given sets), will converge "in the mean", i.e. for all averages given by states ω_ρ where ρ is an integrable density. Enlarging $A_0(S)$ by all these "weak" limit points, we get the function space $L^\infty(T^2)$ consisting of those functions which are "essentially" bounded (i.e. bounded on every set $\Lambda' \subseteq T^2$ for which the "volume" $T^2 \backslash \Lambda'$ is zero). $L^\infty(T^2)$ is a so-called C^*-algebra, i.e. a $*$-algebra of complex functions with a norm (which, in our case, is given by the "essential supremum" where sets of volume zero can be neglected). This norm fulfills $\|a^*a\| = \|a\|^2$ for all elements $a \in L^\infty(T^2)$ and is complete in this norm (i.e. all Cauchy sequences have a limit point). The closure of $A_0(S)$ in this norm, $A(S)$, is a *strict* subalgebra of $L^\infty(T^2)$, and contains only projections on sets which are *finite* unions of elements $\in S \cdot \chi_\Delta \in L^\infty(T^2)$, the projection on the triangle $\Delta = \{(x,y) \in T^2 | 0 \leq x \leq y\}$ is therefore *not* contained in $A(S)$, although χ_Δ is the supremum of a monotonically increasing sequence of projections all contained in $A(S)$. The C^*-algebra $L^\infty(T^2)$ has the additional property that, in contrast to $A(S)$, all monotonically increasing sequences of projections admit a supremum in $L^\infty(T^2)$. Such a C^*-algebra is called a von Neumann algebra and it can always be represented as an algebra of operators on a Hilbert space closed in the weak topology which is given by expectation values with respect to the "state vectors" in this Hilbert space. In the sequel, we will omit the "generating" class S in the notation for $A(S)$ and write simply A for the minimal C^*-algebra of observables on the surface of the doughnut.

To study the limiting behaviour of states ω_ρ and observables χ_Γ when (discrete) time tends to infinity, we can assume that the density ρ is constant on the squares of a sufficiently fine chessboard lattice with lattice constant 2^{-n}. By linearity, it is enough to consider a (normalized) ρ concentrated on a small lattice square $\Gamma, \rho = 2^{2n} \cdot \chi_\Gamma$. After m time steps, $m \geq n$, the tranformed density $(\alpha^{*m})\rho = 2^{2n} \cdot \chi_{B^m(\Gamma)}$ is only a function of y taking the values 0 and 2^{2n}, and has a period of $2^{-p}, p = m - n$. The integral of $\alpha^m \rho$ over a single horizontal stripe of vertical width 2^{-p} in a stripe set $S_{p,k}, p > 0$, yields therefore 2^{-p} and the expectation value $\int (\alpha^*)^m \rho(\lambda)\chi_{\Gamma'}(\lambda)d\mu(\lambda)$ becomes equal to $\int \chi_{\Gamma'}(\lambda)d\mu(\lambda)$ if Γ' is a set from another chessboard lattice with lattice constant $2^{-p'}$ and $p' \leq p$. As $A_0(S)$ is linearly generated by all $\chi_{\Gamma'}$ of this type we conclude that $(\alpha^*)^m \omega_\rho$ approaches weakly the equilibrium state $\omega^0 = \omega_{\rho_0}$ with $\rho^0 = \chi_{T^2} = \mathbf{1}$ so that

$$\lim_{n \to \pm\infty} (\alpha^*)^n \omega_\rho(a) = \omega^0(a)$$

for all $a \in A$ and all ω_ρ with $\rho \in L^1(T^2)$. The generalization to $n \to -\infty$ is obvious. This property is equivalent to the weak convergence $\alpha^n a \rightharpoonup \omega^0(a) \cdot \mathbf{1}$, defined as

$$\lim \omega_\rho(\alpha^n a) = \omega^0(a) \cdot \omega_\rho(\mathbf{1}) \quad \forall \rho.$$

For $n \to \infty$ this is exactly the behaviour we will expect from a many particle system approaching thermal equilibrium.

Our example shows that even a completely reversible system, where both time directions are equivalent, can have this property.

The weak convergence of $\alpha^n a$ can be generalized by linearity to b that are not necessarily ≥ 0 and normalized, so that

$$\lim_{n \to \pm\infty} \omega^0(\alpha^n a \cdot b) = \omega^0(a)\omega^0(b) \quad \forall a, b \in \mathcal{A}.$$

1.3 - *From Doughnuts to Magnets: the Baker Tranformation as a Translation on a One-Dimensional Ising-Lattice*

We have chosen the C^*-algebra of observables $\mathcal{A}$ for the kneading process of the doughnut as the closure (with respect to almost uniform convergence on T^2) of a *-algebra $\mathcal{A}_0(S)$. $\mathcal{A}_0(S)$ consists of (finite) sums over (finite) products of "stripe" observables

$$a_i(\lambda) = a_i^{(0)} \cdot \chi_{S_{i,0}} + a_1^{(1)} \cdot \chi_{S_{i,1}}$$

specified by $i \in Z$ and a pair $(a_i^{(0)}, a_i^{(1)})$ of complex numbers. The characteristic functions χ_Γ have the (algebraic) "projector" property $\chi_\Gamma^2 = \chi_\Gamma$ and, in addition, fulfil the relation $\chi_{\Gamma_1} \cdot \chi_{\Gamma_2} = 0$ if Γ_1 and Γ_2 are "essentially" disjoint. Therefore

$$(a_i \cdot b_i)(\lambda) = a_i^{(0)} \cdot b_i^{(0)} \chi_{S_{i,0}} + a_i^{(1)} \cdot b_i^{(1)} \chi_{S_{i,1}}$$

and the algebraic properties of the function $a_i(\lambda)$ are represented in a natural way by the diagonal matrix

$$a_i = \begin{pmatrix} a_i^{(1)} & 0 \\ 0 & a_i^{(0)} \end{pmatrix}.$$

If $a_i^{(0)} = a_i^{(0)} = 1$, a_i is the unit matrix and $a_i(\lambda)$ equals χ_{T^2}, the unity in $\mathcal{A}_0(S)$. By construction, S is invariant under B and B^{-1} so that we can define a *-automorphism α of $\mathcal{A}_0(S)$ by

$$\alpha \chi_{S_{i,k}} := \chi_{S_{i,k}} \circ B = \chi_{B^{-1}(S_{i,k})} = \chi_{S_{i-1,k}}$$

and an (algebraic) extension on the whole of $\mathcal{A}_0(S)$.

$\mathcal{A}_0(S)$ is therefore (algebraically) isomorphic to the incomplete tensor product spanned by finite linear combinations of $\otimes_{i \in Z} a_i$, where $(a_i)_{i \in Z}$ is a sequence of complex *diagonal* 2×2-matrices with only a finite number of elements different from the identity

$$1_i = \begin{pmatrix} 1 & 0 \\ 0 & 1 \end{pmatrix}.$$

The C^*-algebra $\mathcal{A}$, which is the completion of $\mathcal{A}_0(S)$ in the norm given by uniform convergence of functions almost everywhere on T^2 (excluding sets of volume zero) can therefore be written as an infinite tensor product $A = \bigotimes_{i \in \mathbb{Z}} D_i^2$ where $D_i^2 = D^2$ is the C^*-algebra of diagonal comples 2×2 matrices,

$$D^2 = \frac{1}{2} \left((a^{(0)} + b^{(0)}) \cdot 1_2 + (a^{(0)} - b^{(0)}) \cdot \sigma_z \right).$$

(The countably infinite tensor product of C^*-algebras is defined as the norm closure of the linear space generated by products $\bigotimes_{i \in \mathbb{Z}} a_i$ where only a finite number of factors is different from 1_i.) If S_x (or S_y) is the collection of the "vertical" stripes $S_{i,k}, i \leq 0$ (or the "horizontal" stripes $S_{i,k}, i > 0$) we denote the norm closures of $A_0(S_x)$ (or $A_0(S_y)$) by A_x (or A_y). The elements of A_x resp. A_y are functions on T^2 which depend only on the coordinate x resp. y being constant in the orthogonal direction. It is obvious that

$$A_x = \bigotimes_{i \in \mathbb{Z}, i \leq 0} D_i^2 \quad \text{and} \quad A_y = \bigotimes_{i \in \mathbb{Z}, i > 0} D_i^2,$$

$A = \bigotimes_{i \in \mathbb{Z}} D_i^2$ being thus equal to $A_x \otimes A_y$. The time automorphism α given by the action of the baker tranformation can be written as

$$\alpha \left(\bigotimes_{i \in \mathbb{Z}} a_i \right) = \bigotimes_{i \in \mathbb{Z}} b_i, \, b_i = a_{i+1},$$

on the set of products with only a finite number of sectors different from 1_i. By algebraic extension and formation of limits, α can be extended to an automorphism of $A = \bigotimes_{i \in \mathbb{Z}} D_i^2$. It corresponds to a left shift on the one-dimensional lattice $\mathbb{Z}$ of commutative Ising-type observables which are algebraically isomorphic to the functions on the surface of the doughnut. Obviously, $A_x \otimes 1$ is a subalgebra invariant under α whereas $1 \otimes A_y$ does *not* have this property!

How can a state ω_ρ given by a density function $\rho \in L^1(T^2)$ be expressed in the discrete Ising picture? Every (normalized) density function can be approximated by a convex linear combination of (normalized) densities constant on a finite intersection of "stripes". We will therefore restrict the discussion to dentities proportional to a projection $p^I = \bigotimes_{i \in \mathbb{Z}} p_i^I$, where p_i^I is a *one*-dimensional diagonal projector only for i from a (finite) index set I with $n(I)$ elements, all other factors p_i^I being equal to 1_i. The integral over such a projector function p^I on the torus T^2 equals $2^{-n(I)}$: For the projector on an elementary stripe set this integral is $1/2$ and every multiplication with a projector on another elementary stripe set (with different index i) reduces the volume of the intersection by a factor $1/2$. $2^{n(I)} p^I$ is therefore a correctly normalized density function. As integration is a linear functional over $A_0(S)$ and multiplicative on the product p^I of stripe projections, it will be multiplicative on every product vector $\bigotimes_{i \in \mathbb{Z}} a_i$ in $A_0(S)$. For such product functions the integral equals $\prod_{i \in \mathbb{Z}} (\frac{1}{2} \text{Tr} \, a_i)$, where the factor $1/2$ stems from the integral over an elementary stripe projector and Tr means the trace of the (diagonal) 2×2 matrix (i.e., the sum over the diagonal elements). The integral of an observable over the torus T^2 (which corresponds to the expectation value in the state ω_{ρ^0} given by the (normalized) equilibrium density $\rho^0(\lambda) \equiv 1$) can therefore be expressed, in the Ising picture of the doughnut, by a product state

$$\omega^0 = \bigotimes_{i \in \mathbf{Z}} \omega_i^0, \ \omega_i^0 = \begin{pmatrix} 1/2 & 0 \\ 0 & 1/2 \end{pmatrix} \forall i \in \mathbf{Z} \text{ over } \mathcal{A} = \bigotimes_{i \in \mathbf{Z}} D_i^2.$$

From the action on product vectors $\otimes_{i \in \mathbf{Z}} a_i$, namely,

$$\omega^0 \Big(\bigotimes_{i \in \mathbf{Z}} a_i \Big) = \prod_{i \in \mathbf{Z}} \mathrm{Tr}(\omega_i^0 a_i)$$

(where $\omega_i^0 a_i$ means the matrix product) ω^0 can be extended linearly to the whole of $\mathcal{A}$. The normalized density function $p^I = 2^{n(I)} \cdot \bigotimes_{i \in \mathbf{Z}} p_i^I$ (with p_i^I being a one-dimensional diagonal projector for $i \in I$ where I is a finite index set with $n(I)$ elements, all other p_i^I being equal to $\mathbf{1}_i$) corresponds therefore to the product state

$$\omega^I = \bigotimes_{i \in \mathbf{Z}} 2^{n(I)} (\omega_i^0 p_i^I),$$

where $\omega_i^0 p_i^I$ means the matrix product. Absorbing the $n(I)$ factors 2 in the matrix $\omega_i^0 \forall i \in I, \omega^I$ can be written as

$$\omega^I = \bigotimes_{i \in \mathbf{Z}} \omega_i \text{ with } \omega_i = p_i^I \forall i \in I \text{ and } \omega_i = \omega_i^0 = \begin{pmatrix} 1/2 & 0 \\ 0 & 1/2 \end{pmatrix} \forall i \in \mathbf{Z} \backslash I.$$

What is the physical essence behind the mathematics telling us that there is an algebraic and measure-theoretic isomorphism between kneading the surface of a doughnut and shifting "classical" spins along an infinite lattice in one space dimension? The construction is "operational", because it is formulated entirely in the language of physical observables, time evolutions and states given by spatial densities. The underlying "surface points" of the doughnut were not used, because no physical operation will allow us to "measure" what is happening in a set of points (i.e. molecules) of the dough with vanishing volume. But the algebraic and measure-theoretic aspects of observables and states are not the whole story of physics: pairs of points, although of volume zero, determine distances and therefore define the local topology on the surface of the doughnut. The isomorphism with the ising lattice "inverts" these topological properties because it identifies projections on very fine (vertical or horizontal) stripe sets with observables far to the left or right in the one-dimensional Ising lattice: The "mixing" transition to *smaller* stripes correspond to the translation to *larger* distances. The convergence of observables and states to equilibrium values means in the *doughnut picture* that large fluctuations on *small scales* will be averaged out by observations described by integrable densities. In the lattice picture this convergence means that in the course of time, the roles of the local observations and the behaviour at infinity are somehow interchanged: Initially the structure of an observable or a state given by a (local) perturbation of the equilibrium is *concentrated* in *localized* regions. Outside these domains they become asymptotically trivial. If the time development corresponds to a shift, then the "structured" part of observables and states will diffuse out to a certain side of infinity for sufficiently large (positive or negative) times, whereas the trivial parts will flow in from the other direction of infinity. It is clear that, in quantum theory, time evolution will not yield arbitrarily fine stripe sets in phase space (because the uncertainty relation will not allow volumes smaller than $\hbar$) but in the "rescaled" topology of the lattice, a generalization to a noncommutative quantum system is possible.

1.4 - *The One-Dimensional "Heisenberg" Magnet and the GNS Construction*

The "Ising representation" of the classical mixing system on the surface of the doughnut can easily be generalized to a quantum mechanical system of particles with spin $1/2$ at fixed positions on a one-dimensional lattice $\mathbf{Z}$. The commutative algebra $\mathcal{A}$ of observable is replaced in this case by the infinite tensor product

$$\mathcal{B} = \bigotimes_{i \in \mathbf{Z}} M_i^2$$

where $M_i^2 = M^2$ is the nonabelian C^*-algebra of complex 2×2 matrices. A general element of M_i can be written as $\sum_{j=0}^{3} a_i^{(j)} \sigma_i^{(j)}$ where $a_i^{(j)}$ are complex numbers, $\sigma_i^{(0)}$ is the unit matrix $\mathbf{1}_i$ and $\vec{\sigma}_i$ are the Pauli matrices. $\mathcal{B}$ is not commutative and its orthogonal projections, i.e. the elements with $P = P^* = P^2$, will in general not give a projection when multiplied, because $P_1 P_2 P_1 P_2$ (with $P_i = P_i^* = P_i^2, i = 1, 2$) has only the projector property if one can interchange the order of the inner factors. Noncommuting projection oberservables correspond to "incompatible" yes-no-experiments, for example, measurements of the spin projections along different directions in space. If P_1 and P_2 fulfil the "inclusion relation" $P_1 P_2 = P_1$ analogous to the classical case they commute because $P_1 P_2 = P_1 = P_1^* = P_2^* P_1^* = P_2 P_1$. Inclusion thus defines a *partial* order $P_1 \subseteq P_2$ in the class of projections. The projections in a general C^*-algebra can always be represented as a collection of closed subspaces of a Hilbert space. For two projections P_1 and P_2 the intersection of the corresponding subspaces defines a "maximal" projector P (with respect to the order given by inclusion) contained both in P_1 and P_2. P can be represented algebraically by $\lim_{n \to \infty} (P_1 P_2 P_1)^n = \lim_{n \to \infty} (P_2 P_1 P_2)^n$. This limit will in general not exist in norm but only as a weak one in an appropriate representation induced by the state.

On $\mathcal{B}$ we define a time automorphism α in a way completely analogous to the baker transformation (acting as a left shift on the corresponding Ising lattice) by setting

$$\alpha \Big(\bigotimes_{i \in \mathbf{Z}} a_i \Big) = \bigotimes_{i \in \mathbf{Z}} b_i, \qquad b_i = a_{i+1},$$

on the set of product vectors in $\mathcal{B}$ which span $\mathcal{B}$ linearly. To generalize the notion of classical *densities* on the torus (giving zero weight to sets of volume zero) we consider, for the moment, only states over β which are limits of "local" perturbations of the tracial equilibrium state

$$\omega^0 = \bigotimes_{i \in \mathbf{Z}} \omega_i^0, \quad \omega_i^0 = \begin{pmatrix} 1/2 & 0 \\ 0 & 1/2 \end{pmatrix}.$$

ω^0 corresponds, for the classical Ising lattice, to the uniform distribution on the torus and is, in any case a state invariant under the time automorphism α given by the lattice transformation. Whereas in the doughnut example ω^0 is the only invariant state, we can generalize the translation invariant states on the Ising lattice allowing products of the type

$$\omega^{(\beta)} = \bigotimes_{i \in \mathbf{Z}} \omega_i^{(\beta)}, \quad \omega_i^{(\beta)} = \begin{pmatrix} \frac{1}{2}e^{-\beta} & \\ & 1 - \frac{1}{2}e^{-\beta} \end{pmatrix}$$

corresponding to different temperatures $T = 1/\beta$. On the surface of the doughnut, $\omega^{(\beta)}$ will *not* be given by a density function ρ with respect to the "normal" volume element $d\mu(\lambda)$.

In the classical doughnut system we have noticed that the algebra $\mathcal{A}$ (which was the closure of $\mathcal{A}_0(S)$ where S was generated by the "stripe sets") was a very nice object to define the equivalence to the Ising lattice, but not rich enough to contain all "natural" observables (like the characteristic function χ_Δ on the triangle $\{(x,y) \in T^2 | 0 \leq x \leq y\}$. The "natural" and "universal" algebra of observables was the von Neumann algebra $L^\infty(T^2)$ consisting of all "weak" limit points with respect to states ω_ρ specified by density function ρ.

We will formulate now this amplification of a C^*-algebra C to a von Neumann algebra C_ω induced by a class of states (which are local "perturbations" of an equilibrium state ω invariant under the time automorphism α) in a very general way which is not dependent on the commutativity of C. We have seen that already in the classical Ising lattice there is at least a one-parameter family of "equilibrium" states $\omega^\beta, \beta \in \mathbf{R}, 0 \leq \beta \leq \infty$, which are invariant under the time evolution α. We choose therefore an arbitrary α-invariant state ω which we assume to be *faithful* (i.e $\omega(a^*a) = 0$ implies $a \equiv 0$). With the help of ω, a *nondegenerate* scalar product can be defined on C by $\langle a|b \rangle := \omega(a^*b) \forall a, b \in C$. Completing C in the norm given by the "length" $\sqrt{\omega(a^*a)}$ of a vector (this norm is "smaller" than the C^*-norm) we get an Hilbert space $\mathcal{H}_\omega$, in which C, being closed in the C^*-norm, is a *dense* subspace (with respect to the vector norm given by ω). In general, $\mathcal{H}_\omega$ will be strictly larger than C. For the doughnut example, $\mathcal{H}_\omega$ is given by $L^2(T^2)$ if $\omega = \omega^0$, the only α-invariant state over $\mathcal{A}$.

We can represent the elements of C as bounded operators on $\mathcal{H}_\omega$ by defining $\prod_\omega(a)|b \rangle = |a \cdot b \rangle \forall a, b \in C$ on a dense set; by uniform boundedness, $\prod_\omega(a)$ can be extended to the whole of $\mathcal{H}_\omega$. The vector state given by the (already normalized) vector $|1 \rangle \in \mathcal{H}_\omega$ coincides with ω itself, because $\langle 1| \prod_\omega(a)1 \rangle = \langle 1|a \rangle = \omega(1 \cdot a)$. Often we will write $|1 \rangle = |\Omega \rangle$ to emphasize its relation to ω. All other normalized elements of $\mathcal{H}_\omega$ yield vector states; on the dense set C they are given by

$$\langle b|\prod_\omega(a)|b\rangle \frac{1}{\langle b|b\rangle} = \frac{\langle b|ab\rangle}{\langle b|b\rangle} = \frac{\omega(b^*ab)}{\omega(b^*b)}\ \forall a \in C.$$

If ω is a tracial state (i.e. $\omega(ab) = \omega(ba)$, which is e.g. always the case if C is commutative)

$$\frac{\omega(b^*ab)}{\omega(b^*b)} = \frac{\omega(bb^*a)}{\omega(bb^*)}.$$

In the classical doughnut example, $C = A$, $\mathcal{H}_\omega = L^2(T^2)$ if $\omega = \omega^0$, and $(bb^*)(\lambda)$ is a nonnegative element of $L^1(T^2)$, i.e. a normalized density with respect to the equilibrium state ω^0.

We define now the von Neumann algebra C_ω as the weak closure of $\prod_\omega(A)$. It coincides with the strong closure and can be characterized by the double commutant $\prod_\omega(A)''$. (The commutant C' of C^*-algebra C of operators on a Hilbert space consists of all bounded operators commuting with all elements of C; it is always weakly closed and therefore a von Neumann algebra.) In the doughnut case $\prod_\omega(A)''$ is simply the von Neumann algebra of all essentially bounded functions $L^\infty(T^2)$ acting as multiplication operators on $L^2(T^2)$.

We can define an operator U_α on $\mathcal{H}_\omega$ by setting $U_\alpha|a\rangle$ equal to $|\alpha a\rangle\ \forall a$ from the dense set $C \subseteq \mathcal{H}_\omega$. By uniform boundedness, U_α can be extended to the whole of $\mathcal{H}_\omega$ and represents there a unitary operator:

$$\langle U_\alpha a|U_\alpha b\rangle = \omega((\alpha a)^*\alpha b) = \omega(\alpha(a^*b)) = \omega(a^*b) = \langle a|b\rangle$$

because α is a $*$-automorphism and ω is α-invariant. U_α has always at least one eigenvector, namely $|1\rangle$. $\prod_\omega(\alpha a)$ can be represented as $U_\alpha\prod_\omega(a)U_\alpha^*$ because

$$\langle b|U_\alpha\prod_\omega(a)U_\alpha^*c\rangle = \langle \alpha^{-1}b|a \cdot \alpha^{-1}(c)\rangle = \omega(\alpha^{-1}(b) \cdot a\alpha^{-1}(c))$$

$$= \omega(b \cdot \alpha(a) \cdot c) = \langle b|\prod_\omega(\alpha a)c\rangle.$$

In the doughnut case $(U_\alpha a)(\lambda) = (\alpha a)(\lambda) = a(B\lambda)$ so that $\alpha^*\rho = |(U_\alpha^*a)(\lambda)|^2 = \rho \circ B^{-1}$ gives the correct dual transformation of the densities. The temporal evolution of the state $a \rightarrow \omega(b^*ab)$ in the Schrödinger picture is given by

$$\alpha^*\omega(b^*ab) = \omega(b^*\alpha(a)b) = \omega(\alpha^{-1}(b^*)a\alpha^{-1}(b))$$

because ω was assumed to be α-invariant. If it is, in addition, tracial, $\alpha^*\omega(bb^*a)$ is given by $\omega(\alpha^{-1}(bb^*)a)$, the density operator bb^* being transformed with α^{-1}.

1.5 - *Generalization of the "Heisenberg" Magnet to the Fermi System*

As remarked in the introduction the physical example we have in mind is a Fermi system in the thermodynamic limit. This Fermi system is described by creation and annihilation operators $a^\dagger(f), a(g)$ satisfying

$$[a^\dagger(f), a(g)]_+ = \langle f|g\rangle \quad g \in L^2(R^\nu) \text{ or } g \in \ell^2(Z^\nu).$$

$L^2(R^\nu)$ or $\ell^2(Z^\nu)$ are separable Hilbert spaces. Therefore we can find an orthonormal basis $\{g_i, i \in Z\}$. For fixed i we can represent

$$a^\dagger(g_i) = \begin{pmatrix} 0 & 1 \\ 0 & 0 \end{pmatrix} \quad \text{and} \quad a(g_i) = \begin{pmatrix} 0 & 0 \\ 1 & 0 \end{pmatrix}.$$

Evidently they satisfy the anticommutation relations. For a pair (i,j) with $i < j$ we can find a representation

$$a^\dagger(g_i) = \begin{pmatrix} 0 & 1 \\ 0 & 0 \end{pmatrix} \otimes \begin{pmatrix} 1 & 0 \\ 0 & -1 \end{pmatrix}.$$

$$a(g_i) = \begin{pmatrix} 0 & 0 \\ 1 & 0 \end{pmatrix} \otimes \begin{pmatrix} 1 & 0 \\ 0 & -1 \end{pmatrix}$$

$$a^\dagger(g_j) = \begin{pmatrix} 1 & 0 \\ 0 & 1 \end{pmatrix} \otimes \begin{pmatrix} 0 & 1 \\ 0 & 0 \end{pmatrix}$$

$$a(g_j) = \begin{pmatrix} 1 & 0 \\ 0 & 1 \end{pmatrix} \otimes \begin{pmatrix} 0 & 0 \\ 1 & 0 \end{pmatrix}$$

Generalization of this procedure yields

$$a^\dagger(g_i) = \bigotimes_{k=-\infty}^{i-1} 1_k \otimes \begin{pmatrix} 0 & 1 \\ 0 & 0 \end{pmatrix} \bigotimes_{k=i+1}^{\infty} \sigma_k^z.$$

It can easily be checked that by linearity all anticommutation relations are satisfied. Taking the norm limit of all finite products we get a C^*-algebra as we did for the Heisenberg magnet. The important fact is that independent of the special choice of the orthogonal basis we end up with the same C^*-algebra. On this C^*-algebra we can define the tracial state in complete analogy to the Heisenberg magnet by

$$\omega = \bigotimes_{k\in Z} \begin{pmatrix} 1/2 & 0 \\ 0 & 1/2 \end{pmatrix}_k$$

so that

$$\omega(a^\dagger(f)a(g)) = \frac{1}{2}\langle f|g\rangle.$$

(Take $g_0 = \frac{1}{\|g\|} g_0$, $g_1 = \frac{f - \langle g_0|f\rangle g_0}{\|f - \langle g_0|f\rangle g_0\|}$.) If we have a Fermi lattice system in mind, so that $g \in \ell^2(Z^\nu)$ the natural choice for $g_i = \delta_{x_i}$. Then the left shift

$\alpha a(g_i) = a(g_{i-1})$ is quite analogous to the shift for the Heisenberg magnet. But we will see in the sequel how the above contribution allows us to consider more general time evolutions, and study their mixing behaviour.

1.6 - *The Temporal Evolution of States Restricted to Subalgebras*

To conclude this chapter, we want to consider the problem of caorse-grained observations defined by restricting the state ω to a C^*-subalgebra. As a (classical) example, we allow only observables on the surface of the doughnut which are functions of x alone and do not depend on y. In the Ising lattice picture, these observables form the algebra $A_x \otimes 1$ which is a C^*-subalgebra of the full algebra $A = A_x \otimes A_y$. We can define a conditional expectation $P_x : A \to A_x$ by cutting away the right half of the product. In the original doughnut picture, P_x would correspond to a map $L^\infty(T^2) \to L^\infty(T^1)$ defined by $(P_x f)(x) = \int_0^1 f(x, y) dy$. Although the automorphism α given by the left shift on $A_x \otimes A_y$ leaves iA_x invariant (where $iA_x = A \otimes 1_y$ is the embedding of A_x in A), it does not given an automorphism of iA_x because α is not surjective on iA_x (and α^{-1} not injective).

If ω_ρ is a state over $A = A_x \otimes A_y$ we define its restriction ω_{ρ_x} to A_x, where ρ_x is a density in $L^1(T^1)$, by setting

$$P_x^* \omega_{\rho_x} = \omega_\rho \circ iP_x,$$

i.e.

$$\int \rho_x(x) a(x, y) dx dy = \int \rho(x, y) \left(\int_0^1 a(x, y) dy \right) dx dy$$

$$= \int a(x, y) \left(\int_0^1 \rho(x, y) dy \right) dx dy.$$

$\rho_x(x)$ is therefore given by $\int_0^1 \rho(x, y) dy$.

To define a time evolution $\gamma(\alpha^*)$ on the set of restricted states which is generated by the time development α^* (in the Schrödinger picture) of the states over the full algebra $A = A_x \otimes A_y$ we set

$$P_x^* \gamma(\alpha^*) \omega_{\rho_x} = (\alpha^* \omega_\rho) \circ iP_x$$

and obtain

$$\int \gamma(\alpha^*) \rho_x(x) a(x, y) dx dy = \int \rho(B^{-1}(x, y)) \cdot \left(\int_0^1 a(x, y) dy \right) dx dy =$$

$$= \int \left(\int_0^1 \rho(B^{-1}(x, y)) dy \right) a(x, y) dx dy$$

which gives

$$\gamma(\alpha^*)\rho_x = \frac{1}{2}\left(\rho_x\left(\frac{x}{2}\right) + \rho_x\left(\frac{1+x}{2}\right)\right).$$

Thus $\gamma(\alpha^*)\rho_x$ is a functional of ρ_x itself. Replacing α by $\alpha^{-1}, \gamma((\alpha^{-1})^*)\rho_x$ is not uniquely specified by ρ_x alone.

2. - Mixing Properties

DEFINITION 2.1. A time invariant state over an algebra A with time automorphism $a \rightarrow \alpha_t(a) = a_t \forall a \in A$ is called mixing if

$$\lim_{t \rightarrow \pm\infty} \omega(a_t b) = \omega(a)\omega(b) \quad \forall a, b \in A.$$

In § 1 we have already introduced the concept of the GNS representation of A, in which the state can be written as $\omega(a) = \langle \Omega | a | \Omega \rangle$, and have also discussed that over the corresponding Hilbert space we can find a unitary operator U_t that implements the time automorphism such that $U_t a | \Omega \rangle = a_t | \Omega \rangle, U_t | \Omega \rangle = | \Omega \rangle$ and $a_t = U_t a U_t^*$. We will further assume that the state is faithful, i.e. that for $a \geq 0$ $\omega(a) = 0 \Rightarrow a = 0$. In the finite dimensional case, where ω correspond to a density matrix ρ, this is equivalent to the statement that ρ is invertible. Therefore

$$\omega(ab) = \text{Tr } \rho ab = \text{Tr } \rho(\rho^{-1}b\rho)a = \text{Tr } \rho\sigma_i(b)a = \omega(\sigma_i(b)a)$$

with $\sigma_i(b) = \rho^{-1}b\rho$ is a well defined operator, the analytic continuation of the modular automorphism group $\sigma_t(b) = \rho^{it}b\rho^{-it}$. For the infinite algebra for faithful states $\sigma_i(b)$ still is well defined, though it may become an unbounded operator, but that does not cause real trouble. The important observation that operators can be brought from the right to the left inside of the expectations values without producing uncontrolled singularities remains valid [6,7].

Using this explicit representation we will show that mixing implies weak convergence of all states to the equilibrium state and even more of all observables to their thermal expectation value:

PROPOSITION 2.2. If ω is a faithful normal time invariant state over a von Neumann algebra A the following is equivalent:

(i) ω is mixing,

(ii) $\lim_{t \rightarrow \pm\infty} \nu(a_t) = \omega(a) \forall a \in A$ and ν a normal state over A,

(iii) w-$\lim_{t \rightarrow \pm\infty} a_t = \omega(a)1 \forall a \in A$,

(iv) w-$\lim_{t \rightarrow \pm\infty} U_t = |\Omega\rangle\langle\Omega|$.

PROOF:

(i) $\Rightarrow$ (ii) $\exists\, b \in \mathcal{A}$ with $\nu(a) = \omega(ab), \omega(b) = \nu(1) = 1$.

$$\lim \nu(a_t) = \lim \omega(a_t b) = \omega(a)\omega(b) = \omega(a).$$

(ii) $\Rightarrow$ (iii)

$$\lim\langle\Omega|ba_t c|\Omega\rangle = \lim\langle\Omega|\sigma_i(c)ba_t|\Omega\rangle$$
$$= \langle\Omega|\sigma_i(c)b|\Omega\rangle\langle\Omega|a_t|\Omega\rangle = \langle\Omega|bc|\Omega\rangle\langle\Omega|a|\Omega\rangle.$$

(iii) $\Rightarrow$ (iv)

$$\lim\langle\Omega|bU_t c|\Omega\rangle = \lim\langle\Omega|bc_t|\Omega\rangle$$
$$= \langle\Omega|b|\Omega\rangle\langle\Omega|c|\Omega\rangle \qquad \forall b, c, \in \mathcal{A}$$
$$= \langle b^\dagger\Omega|\Omega\rangle\langle\Omega c\Omega\rangle \qquad c\Omega, b^\dagger\Omega \text{ dense in } \mathcal{H}.$$

(iv) $\Rightarrow$ (i)

$$\lim_{t\to\pm\infty} \omega(ab_t) = \langle\Omega|aU_t b|\Omega\rangle = \langle\Omega|a|\Omega\rangle\langle\Omega|b|\Omega\rangle.$$

REMARKS 2.3.

1. Obviously (2.2) implies that the equilibrium state (remember for the von Neumann algebra) is unique.

2. Strong convergence of the observables is impossible because $a_t \to \omega(a)$ would imply $a_t^2 \to \omega(a^2) = \omega(a)^2$. But $\omega(a^2) - \omega(a)^2 = \omega((a - \omega(a))^2) = 0$ implies $a = \omega(a)$.

3. The convergence of observables to c-numbers means that for large times the system becomes weakly asymptotically abelian, i.e.

$$\text{w-}\lim[a_t, b] = 0.$$

4. Since weak and strong topologies coincide when restricted to unitaries or projections no unitary operator (or projector) $\neq 1$ from $\mathcal{A}$ can converge to a unitary operator (or projector).

5. If one prefers the Schrödinger picture to the Heisenberg picture, one may use in (ii) $\nu_t(a) \equiv \nu(a_t)$.

We illustrate the above Proposition by examples.

EXAMPLES 2.4.

1. We want to check (2.2.ii) for the baker tranformation. Here

$$\nu(a_t) = \int \rho(x, y) f(B^t x, B^t y)\, dx dy$$

and we have

$$\lim_{t\to\infty} \nu(a_t) = \int \rho(x,y)\,dx\,dy \int f(x,y)\,dx\,dy$$

because, as discussed in § 1,

$$\text{w-}\lim_{t\to\pm\infty} f(B^t x, B^t y) = 1 \cdot \int f(x,y)\,dx\,dy.$$

2. We consider the spin system $\mathcal{A} = \otimes_{\ell=-\infty}^{+\infty} M_\ell^2$ together with the tracial state. Any other state can be written

$$\nu(\alpha^t a) = \omega(b a_t)$$

for some $b \in \mathcal{A}$. We may assume that

$$b \in \overset{-\infty}{\underset{\ell=-k-1}{\bigotimes}} 1 \overset{k}{\underset{\ell=-k}{\bigotimes}} M_\ell^2 \overset{\infty}{\underset{\ell=k+1}{\bigotimes}} 1,$$

and the same for a. Then

$$\alpha^{2k+1} a \in \overset{-\infty}{\underset{\ell=k}{\bigotimes}} 1 \overset{2k+2}{\underset{\ell=k+1}{\bigotimes}} M_\ell^2 \overset{\infty}{\underset{\ell=2k+3}{\bigotimes}} 1$$

and therefore

$$\nu(\alpha^{2k+1} a) = \omega(b)\omega(a).$$

3. A more general example is provided by the quasifree fermions that are built by creation and annihilation operators $a^\dagger(f), a(g)$ satisfying $[a^\dagger(f), a(g)]_+ = \langle f|g\rangle$. Let the time evolution be given by

$$\alpha_t a(f) = a(e^{iHt} f)$$

where in momentum space we have the Fourier transform $\tilde{f}$

$$(e^{i\tilde{H}t} f)(p) = e^{i\varepsilon(p)t} \tilde{f}(p) \qquad \text{with } \varepsilon'(p) \neq 0 \forall p.$$

We consider quasifree states ω, invariant under α_t, i.e. those states that are determined by the two point function

$$\omega(a^\dagger(f)a(g)) = \int \overline{\tilde{f}}(p)\tilde{g}(p)(\tilde{\rho})(p)\,dp$$

with $0 \leq \tilde{\rho}(p) \leq 1$. When we mean either a or $a^\dagger$, we write $a^\#$. A general expectation value $\omega(\prod a^\#(g_i))$ can be written as linear combination of terms of the form $\prod \omega(a^\dagger(g_i)a(g_j))$. Then typically

$$\nu(\alpha_t a^\dagger(f_1)a(f_2)) =$$

$$\sum \omega(\prod a^\#(g_i)a^\dagger(e^{iHt}f_1)a(e^{iHt}f_2)) =$$

$$\sum \prod \omega(a^\#(g_i)a^\dagger(e^{iHt}f_1))\omega(a^\dagger(g_j)a(e^{iHt}f_2))+$$

$$\sum \omega(\prod a^\#(g_i))\omega(a^\dagger(e^{iHt}f_1)a(e^{iHt}f_2)),$$

where we used the fact that ω is quasifree. The contributions of the first part are of the form

$$\int \bar{\tilde{g}}(p)\tilde{f}(p)e^{i\epsilon(p)t}\rho(p)dp$$

and go to zero according to Riemann-Lebesgue's lemma. Only the contributions containing $\omega(a^\dagger(e^{iHt}f_1)a(e^{iHt}f_2)) = \omega(\alpha_t a^\dagger(f_1)a(f_2))$ are time invariant and remain, so that

$$\lim \nu(\alpha_t(a^\dagger(f_1)a(f_2))) = \omega(a^\dagger(f_1)a(f_2)).$$

4. Consider in the above example a time evolution e^{iHt} that partly contains a point spectrum. Then the corresponding contribution

$$\omega(a^\dagger(g_i)\alpha_t a(f_i)) = e^{i\lambda_i t}\omega(a^\dagger(g_i)a(f_i))$$

is periodic in time and therefore not convergent. Of course, the invariant mean still would given zero, except if $\lambda_i = 0$.

5. We can also consider bosons. Here $[a^\dagger(f), a(g)]_- = \langle f|g \rangle$ and as a consequence $a^\dagger(f)$ and $a(g)$ are unbounded operators. If we pass to the Weyl operators $W(f) = \exp[i(a^\dagger(f)+a(f))]$ the system is again well defined. Quasifree states are now given by

$$\omega(a(f)a^\dagger(g)) = \int \bar{\tilde{g}}(p)\tilde{f}(p)\rho(p)dp$$

with $\rho(p) \geq 1$. The analysis remains unchanged, only that we have to be more careful with Reimann-Lebesgue, and singularities in $\rho(p)$ may destroy the argument.

This in fact happens for Bose condensation $\rho(p) = \frac{1}{1-e^{-\beta\omega(p)}} \to \infty$ for $p \to 0$, and the Bose condensate remains invariant in time.

Since the normal states over the von Neumann algebra A are only local perturbations of the equilibrium ω it does not seem surprising that ω wins out once the perturbation has gone away. However, we can make a mixing statement which involves only the observables provided we have a suitable invariant state [11-13].

PROPOSITION 2.5. If a von Neumann algebra A has a faithful mixing state ω then $\forall a, b \in A$ exists $T \in R^+$ such that $a_t b \neq 0 \forall |t| > T$.

REMARK 2.6. Intuitively (2.5) means that the support of a gets spread out in time such that it eventually overlaps with the support of every other observable.

PROOF OF 2.5. $a_t b = 0 \Leftrightarrow b^* a_t^* a_t b = 0$. Now

$$\lim_{t \to \pm\infty} \omega(b^* a_t^* a_t b) = \omega(b^* b) \omega(a^* a).$$

Since ω is faithful, both factors are > 0 and therefore there must be T such that $a_t b \neq 0 \ \forall |t| > T$. Again we illustrate the Proposition with

EXAMPLE 2.7.

1. We examine the baker transformation. We write $a = \sum a_n P_n, b = \sum b_n P_n, P_n$ some projections corresponding to rectangles ans assume that only a finite number of projections is necessary, i.e. a and b are fixed by their values in a finite number of rectangles. Then

$$a^k ab = \sum a_\ell b_n \alpha^k (P_\ell) P_n.$$

For k sufficiently large $\alpha^k (P_\ell) P_n = Q_{\ell n} \neq 0$ and these projection operators are linearly independent.

2. For the shift we assume

$$a, b \in \bigotimes_{\ell=-\infty}^{-k-1} 1 \bigotimes_{\ell=-k}^{k} M_\ell^2 \bigotimes_{\ell=k+1}^{\infty} 1.$$

Then $\alpha^{2k+1} a$ and b belong to different parts of the tensor product and therefore $\|\alpha^{2k+1} a \cdot b\| = \|a\| \cdots \|b\|$. For the Fermi lattice system the multiplication with σ^z does not charge the argument.

We look now for some uniformity in mixing. Strict uniformity is in contradiction with the group property of the time evolution:

Assume $\forall a \in A, \varepsilon > 0, \exists \ T$ such that

$$|\omega(ab_t) - \omega(a)\omega(b)| \leq \varepsilon \|b\| \ \forall \ b \in A, \ t \text{ with } |t| > T.$$

But take $b = \alpha_{-t} a$. Then $\omega(a^2) - \omega(a)^2 \leq \varepsilon \|a\| \forall \ \varepsilon$ or $a = \omega(a) \ \mathbf{1}$. We have to restrict the allowed class of operators b. Clearly the clustering is uniform over finite subalgebras, but for some systems it is uniform even over suitable infinite subalgebras. In this case some of the weak convergence properties can be strengthened to strong convergence.

DEFINITION 2.8. A dynamical system (A, α_t) is called a K-system if there exists some subalgebra $A_0 \subset A$ such that

(i) $\alpha_t A_0 \supset A_0 \forall\, t \geq 0$,

(ii) $\bigvee_{t=0}^{\infty} \alpha_t(A_0) = A$,

(iii) $\bigwedge_{t=0}^{-\infty} \alpha_t(A_0) = \lambda\, \mathbf{1}$, "triviality of the tail".

Here $A \vee B$ means the algebra generated by A and B and $A \wedge B$ their intersection. We can assume $t \in Z$ or $t \in R$.

EXAMPLE 2.9.

1. In the introduction we have discussed in detail the baker transformation. A possible choice for the algebra A_0 is $A_y = \{f(y)\}$. Then $\alpha^{-1} f(y) = f(2y)$ and therefore $\bigwedge_{n=0}^{-\infty} \alpha^n A_0 = \{f(y); f(y) \text{ periodic in } y \text{ with period } 2^{-n} \text{ for all } n \geq 0\} = \{const\}$.
Therefore (2.8.iii) is satisfied. On the other hand,

$$\alpha f(y) = f\left(\frac{y}{2}\right) \theta\left(\frac{1}{2} - x\right) + f\left(\frac{1}{2} + \frac{y}{2}\right) \theta\left(x - \frac{1}{2}\right).$$

Continuing to α^n we notice that the strips in x corresponding to the products $\alpha f(y), \cdots \alpha^n f(y)$ become finer and finer and eventually any measurable function can weakly be approximated by functions built from $\alpha^k f(y), k \leq n$, which proves (ii). Notice that A_0 can also be chosen differently. Start with some function $f(x, y)$ and define $\overline{A}_0 = \{\bigvee_{n=0}^{-\infty} \alpha^n f(x, y)\}$. Evidently $\overline{A}_0 \subset \alpha \overline{A}_0$. But a general theorem tells us that $\bigwedge_{n=0}^{\infty} \alpha^{-n} \overline{A}_0 = \lambda\, \mathbf{1}$. The proof is based on the observation that $f(x, y)$ is nearly in some $\alpha^n A_0$, so $\overline{A}_0$ is almost $\subset A_n$, and in the abelian case this approximation suffices to show that $\bigwedge_{n=0}^{-\infty} \alpha^n \overline{A}_0$ is arbitrarily close to $\bigwedge_n A_0$, so it is trivial. Furthermore, one finds that unless $f = const, \alpha^n f, n \in Z$, generates all measurable functions. So the special choice of A_0 is not really critical.

2. The generalization to the infinite tensor product of spins with the shift as automorphism is rather trivial, once we have realized the equivalence of the baker transformation with the Bernoulli shift. We choose

$$A_0 = \bigotimes_{\ell=-\infty}^{0} M_\ell^2 \bigotimes_{\ell=1}^{\infty} 1_\ell, \cdots \quad \alpha^m A_0 = \bigotimes_{\ell=-\infty}^{m} M_\ell^2 \bigotimes_{\ell=m+1}^{\infty} 1 \supset A_0.$$

Evidently $\bigvee_{n=0}^{\infty} \alpha^n A_0 = A$. The triviality of the tail also follows easily if we define the algebra with the appropriate care, namely consisting of those operators that can be strongly approximated by

$$a_k \in \bigotimes_{\ell=-\infty}^{-k-1} 1 \bigotimes M_{-k}^2 \cdots \bigotimes M_k^2 \bigotimes_{\ell=k+1}^{\infty} 1 \cdots.$$

3. The shift on the Fermi lattice system

The problem is very similar to example 2. The only difference comes from the fact that odd elements (products of an odd number of creation and annihilation operators) of A have to be approximated by

$$a_k \in \overset{-k-1}{\underset{\ell=-\infty}{\bigotimes}} \sigma_\ell^z \bigotimes M_{-k}^2 \cdots \bigotimes M_k^2 \overset{\infty}{\underset{\ell=k+1}{\bigotimes}} 1 \cdots.$$

Again we deal with a K-system.

4. The continuous Fermi system with quasifree evolutions

Here we consider the algebra of creation and annihilation operators with an evolution $\alpha_t a(f) = a(U_t f)$. If U is the shift, i.e. if $\tilde{U}_t f(p) = e^{ipt} \tilde{f}(p)$, we have the continuous version of example 3. In this case $t \in R$ does not really make a difference, we take $A_0 = V_{f(x)=0, x \geq 0} a(f)$ and again we have a K-system. So we might try to transform the Fermions with some quasifree evolution by an isomorphism into the Fermi system with the shift. But first this need not be possible, and second it seems worthwhile to study the system without this detour, because then we learn something about the delicacy of the tail.

Inspired by the remark how to find $\overline{A}_0$ for the baker transformation we pick some f and define

$$A_0 = \overset{-\infty}{\underset{k=0}{\bigvee}} a(U^k f) = \underset{g \in \mathcal{H}_0}{\bigvee} a(g),$$

where $\mathcal{H}_0$ is the Hilbert space spanned by $U^k f, k = 0, -1 \cdots$. Similarly $\mathcal{H}_n$ is the Hilbert space spanned by $U^k f, k = n, n - 1, \cdots$. Evidently $\mathcal{H}_1 \supseteq \mathcal{H}_0$ and so $A_1 = \alpha A_0 \supseteq A_0$. But we have to check that $\mathcal{H}_1 \neq \mathcal{H}_0$ and $\bigwedge_{n=0}^{-\infty} \mathcal{H}_n = \{0\}$. Let us for a moment consider the continuous group of the shift and start with $\tilde{f}(p) = e^{-p^2}$. Then

$$\mathcal{H}_0^\perp = \{g; F_g(x) = \int \overline{\tilde{g}}(p) e^{ipx} e^{-p^2} dp = 0 \ \forall x \leq 0\}.$$

$F_g(x)$ is entire in x and since $= 0$ for $x \leq 0$, so it is $\forall x \in R$. Its Fourier transform too is $\equiv 0$ anf therefore $g \equiv 0$. So $\mathcal{H}_x^\perp = \{0\}$ and $\mathcal{H}_x = \mathcal{H} \forall x$ and we do not have any increase or decrease properties of A_0. Thus in contradistinction to example 1, A_0 cannot be constructed with an arbitrary f, at least not for a continuous group though for the shift an appropriate A_0 exists. Let us therefore turn to a discret sequence. Then

$$\overset{-\infty}{\underset{n=0}{\bigwedge}} \mathcal{H}_n = \{g; \underset{f \in \mathcal{H}_n}{\sup} \langle g|f \rangle = 1 \forall n\}.$$

This intersection is empty iff

$$\lim_{n \to \infty} \underset{h = \sum_{k=0}^{-\infty} c_k U^k f, \|h\|=1}{\sup} |\langle g|U^{-n} h \rangle| = 0 \ \ \forall g \in \mathcal{H}.$$

We shall show in the Appendix that in fact now we succeed provided the spectrum of U is absolutely continuous and we choose f appropriately, where the restrictions on f become stronger the smaller we choose the unit of the time interval. Thus if the Hamiltonian has some absolutely continuous spectrum we can at least construct a sub-K-system.

5. In all considerations on the underlying one particle space the statistics did not really enter. Therefore we may also consider bosons with quasifree evolution, for example, phonons with absolutely continuous spectrum. We get a K-system, provided we stay as in (2.4.5) away from bose condensation. Otherwise the condensate will contribute to the tail.

We shall now consider how mixing improved for a K-system.

THEOREM 2.10. If ω is an invariant state over a K-system, then it is K-mixing, which means $\forall\, \varepsilon > 0, a \in A \exists T$ such that

$$|\omega(a_t b) - \omega(a)\omega(b)| \le \varepsilon\|b\| \ \forall\, b \in A_0, t > T.$$

REMARKS 2.11.

1. Because of the group property of α_t a similar statement holds $\forall\, b \in A_{t'} = \alpha_{t'}(A_0)$ for any $t' \in R$. Therefore K-mixing implies mixing.

2. For A being the quasilocal algebra in quantum field theory and α the shift this uniform clustering has been deduced by Powers [14] from the quasilocal structure.
Araki gave a seemingly strengththened version with $\|b\|$ replaced by $\omega(b^\dagger b)^{1/2}$.

3. A K-system is mixed by the time evolution in such a way that the only time invariant elements are multiples of the identity. One can even show that there are no other time invariant finite subalgebras. On the other hand, there may be infinite time invariant subalgebras. In particular, the tensor product of two K-systems is again a K- system. If (A, α, A_0) and (B, β, B_0) satisfy (2.10) so does $(A \otimes B, \alpha \otimes \beta, A_0 \otimes B_0)$. This splittability opens the possibility that a quasifree time evolution gives a K-system though it can be considered as the product of independent parts.

For the proof of (2.10) we need [15, 16, 8]

LEMMA 2.12. In the GNS representation of (A, α_t) with ω define a family of projections by

$$P_t \mathcal{H}_\omega = A_t|\Omega\rangle,$$

then the K-properties (2.6) are equivalent to

(i) $P_t = \alpha_t(P_0) \ge P_0 \forall\, t \ge 0$,

(ii)   st-$\lim_{t\to\infty} P_t = 1$,

(iii)   st-$\lim_{t\to-\infty} P_t = |\Omega\rangle\langle\Omega|$.

PROOF: (i) is obvious and for (ii) or (iii) remember that if the family P_t converges weakly to a projector Q it does so strongly since

$$\langle\psi|(P_t - Q)(P_t - Q)|\psi\rangle = \langle\psi|P_t(1 - Q) + Q(Q - P_t)|\psi\rangle.$$

(ii) follows since $\bigvee_t A_t = A$ implies that for every vector $|\psi\rangle \in \mathcal{H}_\omega$ and $\varepsilon > 0 \; \exists \; a_t \in A_t$ for some t with $\||\psi\rangle - a_t|\Omega\rangle\| < \varepsilon$. For (iii) consider a weak limit point of $P_t a|\Omega\rangle$. It must be in $\bigwedge_{t=0}^{-\infty} A_t|\Omega\rangle$ which (iii) states to be $|\Omega\rangle$. Thus

$$P_t a|\Omega\rangle \rightharpoonup |\Omega\rangle\langle\Omega|a|\omega\rangle \quad \forall a \in A$$

and weak convergence on a dense set suffices for our claim.

REMARKS 2.13.

1. The P_t cannot belong to A, projections from A cannot converge to projections.
2. P_0 converts weak convergence of vectors from $\mathcal{H}_\omega$ to strong convergence for $t \to -\infty$ and $1 - P_0$ does so for $t \to \infty$. (ii) and (iii) are equivalent to

$$\text{st-}\lim_{t \to \infty}(1 - P_0)U_t a|\Omega\rangle = 0 \; \forall a \in A, \quad \text{st-}\lim_{t \to -\infty}P_0 U_t a|\Omega\rangle = |\Omega\rangle\omega(a) \; \forall a \in A.$$

Since $P_0|\Omega\rangle = |\Omega\rangle$ this does not contradict w-$\lim_{t\to\pm\infty}U_t = |\Omega\rangle\langle\Omega|$.

PROOF OF 2.10. $\forall \; b \in A_0$ we have

$$\omega(a_t b) = \langle\Omega|a U_t P_0 b|\Omega\rangle = \langle\Omega|a P_{-t} U_t b|\Omega\rangle.$$

Now (2.9.iii) means $\forall \; \varepsilon \; \exists \; T$ such that

$$\|\langle\Omega|a P_{-t} - \omega(a)|\Omega\rangle\| \leq \varepsilon \; \forall \; t > T \quad \text{and} \quad \||U_t b|\Omega\rangle\| = \omega(b^* b)^{1/2} \leq \|b\|.$$

Thus

$$|\omega(a_t b) - \omega(a)\omega(b)| < \varepsilon\|b\| \; \forall \; b \in A_0, t > T.$$

REMARKS 2.14.

1. (2.7) implies that a K-system can have only one invariant state.
2. Every normal state ν over A can be written $\nu(b) = \nu(ab)$ with some $a \in A$ and the time evolution of the state is $\nu_t(b) = \nu(b_t) = \omega(ab_t) = \omega(a_{-t}b)$. Thus (2.7) is equivalent to the statement that every state when restricted to A_0 converges strongly to ω for $t \to -\infty$. (For $t \to \infty$ it does so only weakly.)

3. - The Entropy Increase

J. von Neumann proposed for the entropy of a state given by a density matrix ρ

$$(3.1) \qquad S(\rho) = - \operatorname{Tr} \rho \ln \rho,$$

and this famous expression can be singled out by a few plausible axioms. It does not change under a unitary time evolution $\rho \to U_t \rho U_t^* = \rho_t$ and it will become infinite by passing to a thermodynamic limit. One might argue that the total entropy of an infinite system is a useless notion and only the entropy density or the information contained in a subsystem has physical significance. The entropy $S(\omega_{|B})$ of the state ω restricted to a subalgebra B can change with time but it might increase or decrease. This corresponds to the fact that a subsystem may exchange energy with the rest. As one learns in thermodynamics in this situation the entropy need not increase. Instead the second law requires the free energy to decrease [17, 18]. The difference of the free energies of a state ρ and an equilibrium state ω divided by the temperature is given by the relative entropy

$$(3.2) \qquad S(\omega|\rho) = \operatorname{Tr} \rho(\ln \rho - \ln \omega)$$

since for $\omega = e^{-\beta(H - F_\omega)}$ it becomes $\beta \operatorname{Tr} \rho H - S(\rho) - \beta F_\omega$. Thus the second law requires

$$(3.3) \qquad S(\omega|\rho_t) \le S(\omega|\rho_{t'}) \ \forall \ t \ge t'.$$

Araki has generalized the definition (3.2) for states over von Neumann algebra such that (3.3) can also be discussed in the relevant context of infinite systems where there is no density matrix. In general, Araki's definition requires the relative modular operator [7, 19] but in the case of a lattice system it can be expressed in simple terms. If Λ is a finite volume in space and $A = A_\Lambda \otimes A_{\Lambda^c}$ then $\omega_{|A_\Lambda}$ and $\varphi_{|A_\Lambda}$ are given by density matrices and (3.2) can be applied. If Λ_n is an increasing sequence of volumes with $\bigvee_n \Lambda_n = $ all of space then the A_Λ lead in (3.2) to a monotonic sequence and

$$S(\omega|\rho) = \lim_{n \to \infty} S(\omega_{|A_{\Lambda_n}}|\rho_{|A_{\Lambda_n}}).$$

However, even in finite dimensions $S(\omega|\rho)$ may be infinite (take $\rho = \frac{1}{2}\begin{pmatrix} 1 & 0 \\ 0 & 1 \end{pmatrix}, \omega = \begin{pmatrix} 1 & 0 \\ 0 & 0 \end{pmatrix}$). To avoid these difficulties we shall assume $\exists \ \lambda_{1,2} \in R$ with $\lambda_1 \rho \le \omega \le \lambda_2 \rho$. Then we get some nice behaviour of $S(\omega|\rho)$ which we quote without proof [7, 19].

PROPERTY OF THE RELATIVE ENTROPY 3.4. Restricted to states with $\lambda_1 \rho \le \omega \le \lambda_2 \rho$

(i) $S(\omega \circ \gamma | \rho \circ \gamma) \leq S(\omega | \rho)$ for every Schwarz positive map $\mathcal{B} \rightarrow \mathcal{A}$. (i.e. $\gamma(a^*a) \geq \gamma^*(a)\gamma(a)$. $\omega \circ \gamma, \rho \circ \gamma$ are then states over $\mathcal{B}$. Typical examples of completely positive maps are automorphisms, embeddings and conditional expectations.)

(ii) $S(\omega | \rho)$ is weakly lower semicontinuous.

(iii) $S(\omega | \rho)$ is strongly continuous.

(iv) $S(\omega | \rho) \geq \frac{1}{2} \| \omega - \rho \|^2$.

For the question of the second law (3.4) has the following consequences:

COROLLARIES 3.5.

(i) If γ has a right inverse $\gamma \circ \gamma_1 = 1$ which is Schwarz positive, then $S(\omega \circ \gamma | \rho \circ \gamma) = S(\omega | \rho)$ since then $S(\omega | \rho) = S(\omega \circ \gamma \circ \gamma_1 | \rho \circ \gamma \circ \gamma_1) \leq S(\omega \circ \gamma | \rho \circ \gamma)$. In particular, if $\mathcal{B} = \mathcal{A}$ and γ an automorphism, then $S(\omega | \rho)$ does not change. Thus for the total system there is no change in free energy. But also a conditional expectation does not change S since the inclusion is the right inverse.

(ii) The semicontinuity means if $\omega_t \rightharpoonup \omega_\infty$ and $\rho_t \rightharpoonup \rho_\infty$, then $S(\omega_\infty | \rho_\infty) \leq \liminf S(\omega_t | \rho_t)$. We know that the existence of a faithful mixing state ω implies that all states converge weakly to ω. But since $S(\omega | \omega) = 0$ this weak semicontinuity tells nothing since $S(\omega | \rho) \geq 0$ for all states. This is in accordance with (i) since $S(\omega_t | \rho_t)$ does not change with time and does not converge to zero.

(iii) The strong continuity tells us that $S(\omega | \rho_t)_{\mathcal{A}_0}$ converges to zero for $t \rightarrow -\infty$. On the other hand, $S(\omega | \rho_t)_{|\mathcal{A}_0} = S(\omega_t | \rho_t)_{|\mathcal{A}_0} = S(\omega | \rho)_{|\mathcal{A}_t}$ and $\mathcal{A}_t \supset \mathcal{A}_{t'} \forall\, t \geq t'$ gives an inclusion $i_{t,t'} : \mathcal{A}_t \leftarrow \mathcal{A}_{t'}$ such that

$$S(\omega | \rho)_{\mathcal{A}_{t'}} = S(\omega_{|\mathcal{A}_t} \circ i_{t,t'} | \rho_{|\mathcal{A}_t} \circ i_{t,t'}) \leq S(\omega_{|\mathcal{A}_t} | \rho_{|\mathcal{A}_t}) \equiv S(\omega | \rho)_{|\mathcal{A}_t}.$$

Thus $S(\omega | \rho_t)_{|\mathcal{A}_0}$ increases monotonically with t from zero to $S(\omega | \rho)_{|\mathcal{A}_\infty = \mathcal{A}}$.

(iv) Property (iv) gives the converse to (iii) namely there is no hope that the free energy of ρ converges to the one of ω unless ρ converges strongly to ω.

(v) For finite algebras $\mathcal{A}_f$ all topologies coincide and therefore

$$\lim_{t \rightarrow \pm\infty} S(\omega | \rho_t)_{|\mathcal{A}_f} = 0.$$

Thus there is no monotonicity in this case and there is also no possibility that $\rho_t|_{\mathcal{A}_f}$ evolves according to a semigroup of Schwarz positive maps since this would imply monotonicity.

Summarizing we can say the following. For finite subalgebras the free energy approaches its equilibrium value for both $t \rightarrow +\infty$ and $t \rightarrow -\infty$. For some infinite subalgebras it behaves monotonically, going on one side to its minimal, on the other to its maximal value.

REMARKS 3.6.

1. By finite algebra we mean that it has as linear space a finite basis. As set it has infinitely many elements. An example would be the Fermi algebra generated by the a_f with a finite number of f's. Roughly speaking this corresponds classically to a compact region in phase space which has a finite volume when measured in units of $\hbar$.

2. Even the casual reader will notice at this point that we proved exactly the opposite of what we wanted, namely that the free energy of A_0 increases rather than decreases. However, at this point everything is reversible and we have not yet specified whether the physical time corresponds to α_t or α_{-t}. In fact, classically for every K-system (A, α_t) also (A, α_{-t}) is a K-system. This is obvious because one characterization of a K-system is the positivity of the dynamical entropy $h(\alpha, B)$ of every nontrivial subalgebra B. h will be discussed subsequently but we shall only mention here $h(\alpha^{-1}, B) = h(\alpha, \beta)$ which proves our claim. In the pure quantum case where A has no center one can argue as follows: Denote by B_n the relative commutant of $A_n, B = \{a \in A : [a, A_n] = 0\}$ then

(i) $A_n \supset A_m \Leftrightarrow B_n \subset B_m \forall\, n \geq m,$

(ii) $\bigvee_n A_n = A \Rightarrow \bigwedge_n B_n = z\, 1,$

(iii) $\bigwedge_n A_n = 1 \Rightarrow \bigvee_n B_n := B \subset A.$

Thus there is at least a sub-K-system with contracting rather than expanding subalgebras. The free energy of B_0 is actually converging to its equilibrium value for $t \to +\infty$.

EXAMPLE 3.7. We consider over the spin-algebra $A = \otimes_{i=-\infty}^{-\infty} M_i^{(2)}$ a product state $\rho = \otimes_{i=-\infty}^{\infty} \rho_i$ where the ρ_i are given by the following density matrices:

$$\rho_i = \begin{pmatrix} 1/2 & \\ & 1/2 \end{pmatrix} \forall |i| > m, \qquad \rho_i = \begin{pmatrix} r_i & \\ & 1 - r_i \end{pmatrix} \forall |i| \leq m.$$

The trace state ω is of the form

$$\omega = \bigotimes_{i=-\infty}^{\infty} \omega_i, \quad \omega_i = \begin{pmatrix} 1/2 & \\ & 1/2 \end{pmatrix} \forall\, i \in Z.$$

When restricted to the subalgebra $A_{k,j} = \bigotimes_{i=k}^{j} M_i^{(2)}$ we have

$$F := S\left(\omega_{|A_{k,j}} \,|\, \rho_{|A_{k,j}}\right) = \sum_{i=-m\bigvee k}^{m\bigwedge j} \left(\ln 2 + r_i \ln r_i + (1-r_i)\ln(1-r_i)\right) =: \sum_i \gamma_i,$$

if $a \bigvee b$ (resp. $a \bigwedge b$) designate the bigger (resp. the smaller) of two integers. If the time evolution α_t is the shift $\alpha_t A_{k,j} = A_{k+t,j+t}$ we get

$$F_t := S\left(\omega_{|\alpha_t A_{k,j}} \,|\, \rho_{|\alpha_t A_{k,j}}\right) = \sum_{i=-m\bigvee k+t}^{i=m\bigwedge j+t} \gamma_i.$$

Since $\gamma_i \geq 0$ one notices

(i)   If k and j are finite, $\lim_{t\to\pm\infty} F_t = 0$.

(ii)   If $k = -\infty, j < \infty$, then $\lim_{t\to-\infty} F_t = 0, \lim_{t\to+\infty} F_t = F$.

(iii) If $j = \infty, k > -\infty$, then $\lim_{t\to-\infty} F_t = F, \lim_{t\to+\infty} F_t = 0$.

Thus for the expanding algebra (with $k = -\infty$) we get the unphysical time dependence and for the contracting algebra ($j = \infty$) the physical one for the free energy. Actually in the present case F is $S(\omega) - S(\rho)$ since the tracial state corresponds to $\beta H = 0$. The reason for this behaviour is the following: What decreases the entropy below its equilibrium value are the correlations in the state ρ, ω being the most chaotic state. Only these observables which can feel the correlations have an entropy below the maximum. Thus the entropy reaches its maximal value whenever the algebra has left the region with correlations. This happens for the contracting or finite algebra if $t \to \infty$. On the other hand, the expanding algebra eventually sees all correlations and thus gets an entropy below the equilibrium entropy. Similarly for the more physical time evolution of the continuous system (2.9.4) we get an entropy increase for the a_f's with f's corresponding to outgoing waves. Ingoing waves show an entropy decrease since they will reach the regions where the correlations are located. Generally one can say that in a deterministic time evolution correlations cannot disappear and therefore the entropy should not increase. However they become less accessible and unless one is specially equipped one loses them out of sight. This means that for a finite subalgebra the entropy will approach its equilibrium value.

So far the entropy increase was a transient phenomenon and it stops once the state has returned to equilibrium. One can also define a perpetual entropy increase in a stationary state by the following consideration. The decrease of $-S\left(\omega_{|A_{-\infty,n}} + \rho_{|A_{-\infty,n}}\right)$ with increasing n means only that the entropy of $\rho_{|A_{-\infty,n}}$ does not increase as much as the entropy $\omega_{|A_{-\infty,n}}$. This change of the entropies sounds strange because $S\left(\omega_{A_{-\infty,n+1}}\right) = S\left(\omega \circ \alpha_{|A_{-\infty,n}}\right)$ and S was invariant under an automorphism α. However, since the entropy of an infinite system is infinite, these expression do not mean much and we have to define the entropy increase by the limit $k \rightarrow -\infty$ of $S(\omega_{|A_{k,n+1}}) - S(\omega_{|A_{k,n}})$. More generally, one might attempt to give the following definition of this so-called dynamical entropy of a subalgebra B

$$h_\omega(\alpha, B) = \lim_{k \to -\infty} \left(S(\omega_{|\bigvee_{j=k}^{n+1} \alpha^j B}) - S(\omega_{|\bigvee_{j=k}^{n} \alpha^j B}) \right).$$

It would lead us too far to explain when this expression is well defined and how one can improve it if is not [20, 21]. In any case we have a mathematically well-defined quantity $h_\omega(\alpha, \beta)$ which measures the information gained about B by another observation after one time unit. Albeit the deterministic time evolution the system has lost part of its memory about B and if $\lim_{n \to \infty} h_\omega(\alpha^n, B) = S(\omega_{|\beta}$ then this means that after a long time the memory loss was 100% and one gains maximal information by a new observation. It turns out that for abelian algebras A the K-properties are equivalent to

$$\lim_{n \to \infty} h_\omega(\alpha^n, B) = S(\omega_{|B}) \qquad B \subset A.$$

One can also define in quantum theory entropic K-systems by the requirement that the memory loss of every subalgebra be 100% [22]. Generally this condition is not equivalent to the assumption that we are dealing with an algebraic K-system: There are noncommutative algebraic K-systems which are not entropic K-systems but they contain entropic K-subsystems [20].

4. - Conclusions

Our findings so far can be summarized by the following logical implications between the various properties:

mixing   ⟺all states converge weakly to equilibrium

⇑

K-mixing   ⟺all states restricted to suitable subalgebras converge strongly
to equilibrium

⇕

The entropy of suitable subalgebras converges for all states
monotonically to its equilibrium value

⇕

The memory loss is 100% for all finite subalgebras

In the quantum case the last equivalence requires some assitional assumptions.

So far we have translated the concept of mixing and of a K-system from the classical theory to the quantum theory and we have shown that this in fact is possible: We encounter these structures provided we pass to the thermodynamic limit of an infinite system. The K-properties appear in different guises. For the baker tranformation mixing seems a deep property whereas in the form of the Bernoulli shift one might consider it as the trivial effect that a local perturbation diffuses to infinity and eventually cannot be observed any more. However, one should keep in mind that it is not necessarily a particle which escapes to infinity. In a system with quasiparticles like phonons we have also a K-automorphism as time evolution but what disappears with time are the correlations in the notion of the particles in the underlying cristal. This situation corresponds better to our picture of a motion becoming chaotic.

Apart from these examples mathematics allows us to construct other K-systems that are not equivalent to the shift, i.e. constructed by tensor products. Already in the classical situation not every K-system can be translated into a Bernoulli shift, but in the appropriate interpretation it is at least close to a Bernoulli shift. In the quantum case it seems that the noncommutativity makes the structure much richer. Unfortunately the construction of these other examples for K-systems uses techniques that do not show any relations to an explicit construction of a time evolution that results from local Hamiltonians in the thermodynamic limit. So it is an open problem whether these systems also describe physical situations. But so far it seems that nature frequently uses possibilities that mathematics offers to her.

We turn to the complain that the really job of statistical mechanics is to explain why only so few thermodynamically stable states exist. For instance, examining the quasifree evolution of the phonons, all space translationally invariant quasifree states, i.e. those states that are determined by the two point functions $\omega(a^\dagger(p)a(q)) = \omega(p)\delta(p - q)$ allow to define an algebraic K-system, these states are suitable in the sense that when locally disturbed they eventually return to ω even if $\omega(p)$ is not a thermal distribution. Since we hardly can imagine to have better mixing properties than those of a K-system we might have to look for other characteristics of disorder, either by considering a different alebra, containing also macroscopic observables, or allowing a randomly disturbed time evolution. This, in fact, was done in the concept of dynamical stability or the theory of passivity, where time dependent perturbations of the time evolution are taken into account and those states are the preferred equilibrium states for which the effect of these perturbations disappear in the adiabatic limit. Remarkably enough, this condition already singles out the equilibrium states, provided we assume some additional properties of the time evolution that is asymptotic abelianess and good clustering, exactly those properties that are guaranteed, if we are dealing with a K-automorphism.

Appendix

THEOREM: If the unitary group $R \ni t \to U(t) : \mathcal{H} \to \mathcal{H}$ has nonvoid absolutely continuous spectrum then there exists a $f \in \mathcal{H}$ and $0 < \tau_f \in R$ such that

$$\lim_{k \to \infty} \left(\sup_{\tilde{f} \in \mathcal{H}(f,\tau), \|\tilde{f}\|=1} |\langle g, U(k \cdot \tau)\tilde{f}\rangle| \right) = 0 \quad \forall\, g \in \mathcal{H} \text{ and all } \tau > \tau_f,$$

where the subspace $\mathcal{H}(f,\tau)$ is the linear span of $\{U(n \cdot \tau)f | n \in N \cup \{0\}\}$.

PROOF: In the spectral representation $U(t)$ is represented by e^{itx} and if its spectrum is absolutely continuous, $\mathcal{H}$ contains a $L^2(\theta, dx)$-subspace with $\theta \subset R$ some open interval. Consider f (resp. g) $\in L^2(\theta, dx), \|f\| = \|g\| = 1$, with compact support and twice (resp. once) differentiable. By partial integration we get no boundary terms and thus

$$(A.1) \qquad k\tau |\langle g|U(-k\tau)f\rangle = |\int_\theta dx\, g^*(x) f(x) \frac{\partial}{\partial x} e^{ik\tau}| \le \|f'\| + \|g'\| =: \gamma_1$$

and

$$(A.2) \quad (k\tau)^2 |\langle f|U(-k\tau)f\rangle = |\int_\theta dx |f(x)|^2 \frac{\partial^2}{\partial x^2} e^{ik\tau}| \le 2(\|f''\| + \|f'\|^2) =: \gamma_2.$$

A vector $\tilde{f}$ from $\mathcal{H}$ can be written $\tilde{f} = \sum_{n=0}^\infty c_n U(n\tau)f$ and though the $U(n\tau)f$ are not orthogonal we see from (A.2) that for sufficiently large $\tau \sum_n |c_n|^2 < \infty$: Since

$$b := \sum_{k \ne 0} |\langle f|U(k\tau)f\rangle| \le \frac{\gamma_2}{\tau_2} \sum_{k \ne 0} \frac{1}{k^2}$$

we can make $b < 1$ and

$$b \sum_{n=0}^\infty |c_n|^2 \ge \sum_{m=0}^\infty \sum_{0 \le n \ne m}^\infty c_m^* \langle f|U(\tau(n-m)|f\rangle c_n\rangle \ge \sum_{n=0}^\infty |c_n|^2 - 1.$$

Here we used first that with Cauchy-Schwarz

$$|\sum_{m,n} c_m^* a_{m-n} c_n| = \sum_{m,n} |c_m|^2 |a_{m-n}| = \sum_m |c_m|^2 \sum_n |a_n|$$

and then $\|\tilde{f}\| = 1$. This proves

$$\sum_{n=0}^\infty |c_n|^2 \le \frac{1}{1-b}.$$

Now

$$|\langle g|U(k\tau)f\rangle| \le (\sum_n |c_n|^2)^{1/2} (\sum_n |\langle g|U((n+k)\tau)f\rangle|^2)^{1/2}$$

$$\le \frac{\gamma_1}{\sqrt{1-b}} (\sum_{n=0}^\infty \frac{1}{(n+k)^2})^{1/2} = O(\frac{1}{\sqrt{k}}).$$

Thus, if $\gamma_1 < \infty$, then

$$\lim_{k \to \infty} \sup_{\tilde{f} \in \mathcal{H}(f,\tau), \|\tilde{f}\|=1} \langle g | U(k\tau) \tilde{f} \rangle = 0.$$

Now the g's with $\gamma_1 < \infty$ are dense in the $L^2(\theta, dx)$ part of $\mathcal{H}$ and the rest of $\mathcal{H}$ is orthogonal to $\mathcal{H}(f,t)$.

REFERENCES

[1]   V.I. ARNOLD - A. AVEZ, Ergodic Properties of Classical Mechanics, Benjamin, New York (1968).

[2]   I.P. CORNFELD - S.V. FOMIN - YA G. SINAI, Ergodic Theory, Springer, Berlin, Heidelberg, New York (1982).

[3]   P. WALTERS, An Introduction to Ergodic Theory, Springer, New York, Heidelberg, Berlin (1982).

[4]   A. LASOTA - M.C. MACKEY, Probabilistic Properties of Deterministic Systems, Cambridge (1985).

[5]   A.N. KOLMOGOROV, Dokl. Acad. Sci. USSR. **119**, 861 (1958).

[6]   W. THIRRING, Quantum Mechanics of Large Systems, Springer, New York, Wien (1983).

[7]   O. BRATTELI - D.W. ROBINSON, Operators Algebras and Quantum Statistical Mechanics I, II, Springer, new York, Heidelberg, Berlin (1979).

[8]   G.G. EMCH, Commun. Math. Phys. **49**, 191 (1976).

[9]   B. KÜMMERER - W. SCHRÖDER, Commun. Math. Phys. **90**, 251 (1983).

[10] H. NARNHOFER - W. THIRRING, Algebraic K-Systems, Vienna Preprint UWThPh-1989-17.

[11] R. LONGO - C. PELIGRAD, Journ. of Funct. Anal. **58** (1984).

[12] O. BRATTELI - G.A. ELLIOTT - D.W. ROBINSON, J. Math. Soc. Jap. **37**, 115 (1985).

[13] H. NARNHOFER - W. THIRRING - H. WIKLICKY, Journ. Stat. Phys. **52**, 1097 (1988).

[14] R.T. POWERS, Ann. of Math. **86**, 138 (1967).

[15] B. MISRA - I. PRIGOGINE - M. COURBAGE, Physica **98A**, 1 (1979).

[16] S. GOLDSTEIN - B. MISRA - M. COURBAGE, Journ. Stat. Phys. **25**, 111 (1981).

[17] H. SPOHN, Journ. Math. Phys. **19**, 1227 (1978).

[18] H. SPOHN - J. LEBOWITZ, Adv. Chem. Phys. **38**, 109 (1978).

[19] H. NARNHOFER - W. THIRRING, Fizika **17**, 257 (1985).

[20] H. NARNHOFER, Vienna Preprint UWThPh- 1989-5, to be publ. in Quantum Probability 5, L. Accardi and W. von Waldenfels ed.

[21] A. CONNES - H. NARNHOFER - W. THIRRING, Commun. Math. Phys. **112**, 691 (1987)

[22] H. NARNHOFER - W. THIRRING, UWThPh- 1988-40, to be publ. in Commun. Math. Phys.

Color Symmetry and Color Confinement

KAZUHIKO NISHIJIMA (*)

Foreword

It is a great pleasure for me to dedicate this article to Luigi Radicati on the occasion of his seventieth anniversary.

I recall our old friendship and in particular his warm hospitality at the Scuola Normale Superiore with gratitude and nostalgia.

The main part of the present article was completed during my sojourn in Pisa in an academic atmosphere surrounding him. It is concerned with the question of color confinement.

1. - Introduction

The quark model of hadrons has been so successful that one can no longer think of any other substitute for it. Yet, isolated quarks have never been observed to date [1]. In order to establish this model, therefore, one has either to detect isolated fractionally-charged quarks with an improved detector or to offer an interpretation of why they are confined and escape from detection. In this article we shall make the latter choice and give an interpretation of quark confinement, thereby extending the concept of quark confinement to that of color confinement. We shall argue that only color singlet particles are subject to observation when a certain limitation on the number of quark flavors is met.

The hypothesis of color confinement is certainly consistent with all the known experimental facts, and it is one of the most basic problems in particle physics to account for the mechanism of color confinement within the framework of non-abelian gauge theories. Early attempts at color confinement are based mainly on classical gauge theories eventually exploiting topological quantization and often an appeal has been made to stress the similarity between the confining vacuum and the type II superconductor [2]. In such an approach hadronic strings are identified with the Landau-Ginzburg-Abrikosov vortices of quantized

(*)   Research Institute for Fundamental Physics Kyoto University, Kyoto 606, Japan.

magnetic flux in the superconducting vacuum with magnetic monopoles at both ends [3]. Thus the Yang-Mills vacuum could be a coherent superposition of magnetic monopoles instead of Cooper pairs, and the confining phase resembles that of a type II superconductor with electric and magnetic fields interchanged.

Another powerful approach to this problem is the lattice gauge theory, and the confinement criterion has been formulated in terms of the area law for Wilson's loop correlation function [4]. The area law leads to the linear potential between a heavy quark and a heavy antiquark in the quenched approximation to neglet dynamical quark loops. The linear potential is no longer valid, however, when a quark-antiquark pair can be created from the vacuum, since it is energetically more favorable to split the string between the heavy quark pair thereby attaching a light quark and a light antiquark to the two split ends than to stretch the string indefinitely. This is physically expected, since an uncritical extension of the linear potential to macroscopic distances leads to too strong a van der Waals force between hadrons as suggested by Matsuyama and Miyazawa [5].

So far we are concerned with either semi-classical theories or quantum-mechanical but discrete systems, and we shall proceed to quantum-mechanical continuum theories. In discussing color confinement the most basic problem is its interpretation. When a proper interpretation is given the problem is half solved, so that we shall first look for a possible interpretation within the framework of known field theories. In QED quantization of the electromagnetic field in a covariant gauge leads to transverse, longitudinal and scalar photons, but the latter two are never subject to observation. We shall interpret this fact as an example of confinement, and an attempt is made to generalize this idea to QCD. The basic ingredients of confinement of longitudinal and scalar photons in QED consist of indefinite metric resulting in the cancellation of their contributions among themselves and the subsidiary (Lorentz) condition to eliminate indefinite metric from observables. The confinement of longitudinal and scalar photons is kinematical in nature since they can be eliminated by making use of a certain invariance argument without reference to detailed dynamics of the theory. In QCD, however, confinement is no longer kinematical, but we have to refer to further dynamical properties of the system as we shall see in what follows.

In order to study the metric structure of QCD as well as an appropriate subsidiary condition to eliminate indefinite metric from the theory, Becchi-Rouet-Stora (BRS) algebra is briefly reviewed in the next section. In Sec. 3 the representations of BRS algebra are studied to elaborate the metric structure of non-abelian gauge theories. In Sec. 4 the definition of confinement is given in accordance with the representations of BRS algebra. In Sec. 5 a dynamical criterion to realize color confinement is introduced, and in Sec. 6 its similarity with superconductivity is stressed. In Sec. 7 a sufficient condition to satisfy the dynamical criterion is formulated in terms of the number of quark flavors. Finally, in Sec. 8 a possible test of the idea presented in this article is briefly discussed.

2. - *BRS* Algebra

The Lagrangian density of a system consisting of non-abelian gauge fields and fundamental fermion fields consists of three terms:

$$(2.1) \qquad \mathcal{L} = \mathcal{L}_{\text{inv}} + \mathcal{L}_{GF} + \mathcal{L}_{FP}.$$

The first term represents the gauge-invariant part,

$$(2.2) \qquad \mathcal{L}_{\text{inv}} = -\frac{1}{4} F_{\mu\nu} \cdot F_{\mu\nu} - \overline{\psi}(\gamma_\mu \cdot D_\mu + m)\psi,$$

where we have adopted the Pauli metric and color and flavor indices have been suppressed. The gauge-fixing term is chosen as

$$(2.3) \qquad \mathcal{L}_{GF} = \partial_\mu B \cdot A_\mu + \frac{\alpha}{2} B \cdot B,$$

and the Faddeev-Popov (*FP*) ghost term [6] necessary to maintain the unitary character of the theory is given by

$$(2.4) \qquad \mathcal{L}_{FP} = i\partial_\mu \overline{c} \cdot D_\mu c,$$

where c and $\overline{c}$ are anticommuting hermitian fields.

For *fundamental fields* A_μ, ψ and $\overline{\psi}$ we can define the infinitesimal local gauge transformation by

$$(2.5) \qquad \begin{aligned} A_\mu(x) &\to A_\mu(x) + D_\mu\lambda(x), \\ \psi(x) &\to \psi(x) + ig(\lambda(x) \cdot T)\psi(x), \\ \overline{\psi}(x) &\to \overline{\psi} - ig\overline{\psi}(x)(\lambda(x) \cdot T), \end{aligned}$$

where the matrix T appears in the covariant derivative of ψ as

$$(2.6) \qquad D_\mu\psi = (\partial_\mu - igT \cdot A_\mu)\psi.$$

The *BRS* transformation [7] of the *fundamental fields* is defined by replacing the infinitesimal gauge function $\lambda(x)$ by $c(x)$,

$$(2.7) \qquad \delta A_\mu = D_\mu c, \quad \delta\psi = ig(c \cdot T)\psi, \quad \delta\overline{\psi} = -ig\overline{\psi}(c \cdot T).$$

For auxiliary fields B, c and $\overline{c}$ the infinitesimal local gauge transformation is not defined, and we have to define their *BRS* transformation so as to make the total Lagrangian density invariant under this transformation,

$$(2.8) \qquad \delta B = 0, \quad \delta\overline{c} = iB, \quad \delta c = -\frac{1}{2} g(c \times c),$$

where the antisymmetric croos-product is defined by using the structure constant of the gauge group.

The invariance of the theory under the BRS transformation enables us to introduce the corresponding Noether current and the conserved charge Q_B called the BRS charge. Then the BRS transformation of an arbitrary operator ϕ is given by

$$(2.9) \qquad \delta\phi = i[Q_B, \phi]_\mp,$$

where we choose the $-(+)$ sign when ϕ is even(odd) in the ghost fields c and $\bar{c}$. The charge Q_B is hermitian and nilpotent, namely,

$$(2.10) \qquad Q_B^\dagger = Q_B,$$

and

$$(2.11) \qquad Q_B^2 = 0.$$

These properties are consistent only in the presence of indefinite metric.

An alternative BRS transformation [8] can be introduced by replacing $\lambda(x)$ by $\bar{c}(x)$, namely,

$$(2.12) \qquad \bar{\delta}A_\mu = D_\mu\bar{c}, \quad \bar{\delta}\psi = ig(\bar{c}\cdot T)\psi, \quad \bar{\delta}\,\bar{\psi} = -ig\bar{\psi}(\bar{c}\cdot T).$$

Corresponding to (2.8), the auxiliary fields are transformed as

$$(2.13) \qquad \bar{\delta}\,\bar{B} = 0, \quad \bar{\delta}c = i\bar{B}, \quad \bar{\delta}\bar{c} = -\frac{1}{2}\,g(\bar{c}\times\bar{c}),$$

where $\bar{B}$ is defined by

$$(2.14) \qquad B + \bar{B} - ig(c\times\bar{c}) = 0.$$

The nilpotency of the BRS transformation is generalized to

$$(2.15) \qquad \delta^2 = \bar{\delta}^2 = \delta\bar{\delta} + \bar{\delta}\delta = 0,$$

and

$$(2.16) \qquad Q_B^2 = \bar{Q}_B^2 = Q_B\bar{Q}_B + \bar{Q}_B Q_B = 0,$$

where $\bar{Q}_B$ denotes the alternative BRS charge.

We are also aware of the fact the theory is invariant under the scale transformation

$$(2.17) \qquad c \to e^\lambda c, \quad \bar{c} \to e^{-\lambda}\bar{c},$$

and the corresponding Noether current and the conserved charge [6] are given, respectively, by

(2.18)
$$j_\mu^c = i(\partial_\mu \bar{c} \cdot c - \bar{c} \cdot D_\mu c),$$

and

(2.19)
$$Q_c = \int d^3x j_0^c(x)$$

This charge satisfies commutation relations of the form

(2.20)
$$i[Q_c, \phi] = N\phi,$$

where N denotes the number of c fields minus that of $\bar{c}$ fields involved in ϕ as factors. N is called the ghost number of the field ϕ and is clearly an integer. Since Q_c is hermitian and ϕ and $\phi^\dagger$ carry the same ghost number the eigenvalues of the antihermitian operator iQ_c are integers. This also necessitates introduction of indefinite metric.

Since Q_B carries the unit ghost number, we find

(2.21)
$$i[Q_c, Q_B] = Q_B.$$

The graded Lie algebra characterized by Eqs. (2.11) and (2.21) is called BRS algebra. In what follows we assume that the vacuum state is BRS invariant and carries zero ghost number, namely,

(2.22)
$$Q_B|0> = 0, \quad iQ_c|0> = 0.$$

Then we have the following identity because of Eq. (2.9):

(2.23)
$$<0|\; \delta T\;[AB\ldots]|> = 0.$$

This identity is referred to as the BRS identity.

Needless to say we can extend Eqs. (2.22) and (2.23) by including $\overline{Q}_B$ as

(2.24)
$$\overline{Q}_B|0> = 0,$$

and

(2.25)
$$<0|\overline{\delta}T[AB\ldots]|0> = 0.$$

3. - Representations of BRS Algebra

The facts that the hermitian operator Q_B is nilpotent and that the antihermitian operator iQ_c has real integral eigenvalues hint at the necessity of indefinite metric, and we shall recapitulate some basic features of indefinite metric [9] in what follows.

Let us consider a linear vector space $\mathcal{V}$ and define the inner product between a pair of vectors $|k>$ and $|\ell>$ by imposing the following conditions:

(i) $<\ell|k>$ is a complex number.

(3.1)   (ii) $<\ell|k> = <k|\ell>^*$,

(3.2)   (iii) $<\ell|\alpha k + \beta k'> = \alpha <\ell|k> + \beta <\ell|k'>$,

where α and β are complex numbers.

A vector space of positive-definite metric is characterized by an additional condition

$$(3.3) \qquad\qquad\qquad <k|k> \geq 0,$$

and $<k|k> = 0$, if and only if $|k> = 0$. A vector space of indefinite metric is introduced by lifting this condition. Then $<k|k>$ need not be positive, but it must be real.

Let us introduce a complete set of basis $\{|e_j>\}$ in terms of which an arbitrary vector $|x>$ can be expressed uniquely as

$$(3.4) \qquad\qquad\qquad |x> = \sum_j x_j |e_j> .$$

Then we introduce a linear operator T in $\mathcal{V}$ and define its representation, with reference to the basis $\{|e_j>\}$, by

$$(3.5) \qquad\qquad\qquad T|e_j> = \sum_k |e_k> t_{kj}.$$

Under a change of basis

$$(3.6) \qquad\qquad |e'_j> = \sum_k |e_k> u_{kj}, \qquad \text{with det } u \neq 0$$

the new representation of T defined with reference to the new basis $\{|e'_j>\}$ is related to the original one through

$$(3.7) \qquad\qquad\qquad t' = u^{-1}tu.$$

The metric matrix for the set of basis $\{|e_j>\}$ is defined by

(3.8)
$$\eta_{ij} = \;< e_i|e_j > \,.$$

Eq. (3.1) guarantees that η is an hermitian matrix,

(3.9)
$$\eta^\dagger = \eta.$$

Under a change of basis the metric matrix undergoes the following transformation instead of Eq. (3.7):

(3.10)
$$\eta' = u^\dagger \eta u.$$

The hermitian conjugate $T^\dagger$ of a linear operator T is defined, for an arbitrary pair of vectors $|k>$ and $|\ell>$, by

(3.11)
$$< k|T^\dagger|\ell > \, = \;< \ell|T|k >^* .$$

The matrix element of the operator T is not given by t but by

(3.12)
$$< e_i|T|e_j > \, = \sum_k \eta_{ik} t_{kj}.$$

Let $\tilde{t}$ be the representation of $T^\dagger$, then we have

(3.13)
$$\eta\tilde{t} = (\eta t)^\dagger = t^\dagger \eta,$$

and the representation h of an hermitian operator must satisfy

(3.14)
$$\eta h = h^\dagger \eta.$$

The product of a matrix t and a vector $|x>$ is defined by

(3.15)
$$t|x > = \sum_{i,j} t_{ij} x_j |e_i > \,.$$

Let us introduce two vectors $|f>$ and $|g>$ defined by

(3.16)
$$|f > = a|x >, \quad |g > = b|y >$$

then their inner product is given by

(3.17)
$$< g|f > = \sum_{i,j} y_i^* (b^\dagger \eta a)_{ij} x_j \equiv (y, b^\dagger \eta a x).$$

This formula is useful in the following development.

We now introduce representation matrices of Q_B and iQ_c by

(3.18)
$$Q_B|e_j> = \sum_k |e_k> q_{kj},$$

(3.19)
$$iQ_c|e_j> = \sum_k |e_k> n_{kj}.$$

Then *BRS* algebra is transcribed in the following form:

(3.20)
$$[n, q] = q, \quad q^2 = 0, \quad \text{also} \quad (q^\dagger)^2 = 0$$

The hermiticity of Q_B and of Q_c is expressed, with the help of Eq. (3.14) by

(3.21)
$$\eta q = q^\dagger \eta, \quad \eta n = -n^\dagger \eta.$$

A characteristic feature of the problem of finding representations in the indefinite metric space consists in the point that we have to find not only the representations of operators but also the metric matrix [10]. In what follows we assume that the vector space $\mathcal{V}$ is non-degenerate, namely, the metric matrix η is non-singular. Furthermore, instead of studying the abstract algebra we bring in some additional assumptions generally valid in gauge theories.

Postulate I. The eigenvalue spectrum of n consists of all the integers

In what follows we shall confine ourselves to representation in which n is diagonalized, so that

(3.22)
$$n^\dagger = n, \quad \eta n = -n\eta.$$

Now let us introduce a matrix Δ defined by

(3.23)
$$\Delta = q^\dagger q + q q^\dagger.$$

Then, by making use of Eq. (3.20), we can immediately verify

(3.24)
$$[\Delta, q] = [\Delta, q^\dagger] = 0.$$

Combining Eqs. (3.20) and (3.22) we find

(3.25)
$$[n, q^\dagger] = -q^\dagger,$$

and Eqs. (3.20), (3.23) and (3.25) lead to

(3.26)
$$[\Delta, n] = 0.$$

Thus Δ commutes with $q, q^\dagger$ and n, and it is a Casimir operator [11] in BRS algebra. Furthermore, we obtain from Eq. (3.22)

$$(3.27) \qquad\qquad [n, \eta^2] = 0,$$

and we can diagonalize n and η^2 simultaneously. By choosing u adequately in Eq. (3.10) it is possible to choose the basis so that we have the standard form without violating Eq. (3.21),

$$(3.28) \qquad\qquad \eta^2 = 1.$$

Then Eq. (3.21) may be extended to

$$(3.29) \qquad\qquad \eta q = q^\dagger \eta, \quad \eta q^\dagger = q \eta.$$

Thus for the standard choice (3.28) we have

$$(3.30) \qquad\qquad [\Delta, \eta] = 0.$$

Thus for the standard choice (3.28), Δ commutes not only with $q, q^\dagger$ and n but also with η. Irreducible representations of BRS algebra can be obtained by considering eigenstates of Δ. In what follows we shall give only the results.

BRS singlet states

We can define the BRS singlet subspace $\mathcal{V}_S$ defined by

$$(3.31) \qquad\qquad \mathcal{V}_S = \{|s> |\Delta|s>= 0, \ |s>\in \mathcal{V}\}$$

A member of such a subspace is called a BRS singlet state and it obviously satisfies the relations

$$(3.32) \qquad\qquad q|s>= q^\dagger|s>= 0.$$

Again we shall introduce a constraint imposed on $\mathcal{V}_S$.

Postulate II. The BRS singlet space $\mathcal{V}_S$ is of positive-definite metric

This postulate imposes a restriction on the eigenvalue of the ghost number. Namely, when $\Delta|s>= 0$, we also have

$$(3.33) \qquad\qquad n|s>= 0.$$

In this subspace we can introduce a complete set of basis $\{|s_j>\}$ satisfying the orthonormalization condition

$$(3.34) \qquad\qquad < s_i|s_j>= \delta_{ij}.$$

BRS doublet states

The collection of non-zero eigenstates of Δ forms another subspace $\mathcal{V}_D$ called the *BRS* doublet space, so that we have the orthogonal decomposition

$$(3.35) \qquad \mathcal{V} = \mathcal{V}_S \oplus \mathcal{V}_D.$$

Furthermore, $\mathcal{V}_D$ is expressed as a direct sum of two subspaces $\mathcal{V}_p$ and $\mathcal{V}_d$ called the parent space and the daughter space, respectively. They are defined by

$$(3.36) \qquad \mathcal{V}_p = q^\dagger \mathcal{V} = \{ q^\dagger |x> \mid |x> \in \mathcal{V} \},$$

$$(3.37) \qquad \mathcal{V}_d = q \mathcal{V} = \{ q|x> \mid |x> \in \mathcal{V} \},$$

respectively. We can also introduce a set of basis $\{|p^j>\}$ in $\mathcal{V}_p$ and $\{|d_j>\}$ in $\mathcal{V}_d$ in such a way that they satisfy

$$(3.38) \qquad <p^i|p^j> \; = \; <d_i|d_j> \; = 0,$$

$$(3.39) \qquad <p^i|d_j> \; = \delta_{ij}.$$

These are the basic properties of the three subspaces.

Before closing this section we shall give some additional remarks. First, two vectors belonging to different eigenvalues of Δ are orthogonal. To show this consider

$$(3.40) \qquad \Delta|x_1> \; = \lambda_1^2|x_1>, \quad \Delta|x_2> \; = \lambda_2^2|x_2>,$$

then, by making use of Eqs. (3.17) and (3.30), we find

$$\begin{aligned}
\lambda_1^2 <x_1|x_2> \; &= \; \lambda_1^2(x_1, \eta x_2) \\
&= (x_1, \Delta \eta x_2) \\
&= (x_1, \eta \Delta x_2) \\
&= \lambda_2^2 <x_1|x_2> .
\end{aligned}$$

Therefore, when $\lambda_1^2 \neq \lambda_2^2$, we have

$$(3.41) \qquad <x_1|x_2> \; = 0.$$

It is clear from this result that a *BRS* singlet state is orthogonal to a *BRS* doublet state.

From Eq. (3.32) we find for a *BRS* singlet state $|s>$,

$$(3.42) \qquad Q_B|s> \; = q|s> \; = 0,$$

and for a daughter state $|d> = q|x>$ we also find

(3.43) $$Q_B|d> = q|d> = q^2|x> = 0.$$

Thus both BRS singlet states and daughter states are annihilated by Q_B.

4. - The Physical Subspace

In the preceding section we have shown that introduction of indefinite metric is indispensable in order to realize BRS algebra. Then it is necessary to introduce a subsidiary condition like the Lorentz condition in order to eliminate indefinite metric from physically observable quantities.

Following Kugo and Ojima [6] we introduce the physical subspace $\mathcal{V}_{\text{phys}}$ by

(4.1) $$\mathcal{V}_{\text{phys}} = \{|x> \; |Q_B|x >= 0, \; |x >\in \mathcal{V}\}.$$

It is important to recognize that this subspace is defined without specifying the complete set of basis in $\mathcal{V}$. Similarly, the daughter space can be defined by

(4.2) $$\mathcal{V}_d = \{Q_B|x > | \; |x >\in \mathcal{V}\},$$

which is again independent of the choice of the set of basis. From the remark that we made at the end of the preceding section we find

(4.3) $$\mathcal{V}_{\text{phys}} = \mathcal{V}_S \oplus \mathcal{V}_d.$$

Although $\mathcal{V}_{\text{phys}}$ and $\mathcal{V}_d$ are defined uniquely, $\mathcal{V}_p$ and $\mathcal{V}_S$ cannot be defined uniquely. They depend on the choice of the set of basis as we shall show in what follows [12].

A standard set of basis is chosen subject to two conditions. The first one is the orthonormaization condition:

(4.4)
$$< s_i|s_j > = \delta_{ij}, \quad < s_i|d_j > = \; = < s_i|p^j > = 0,$$
$$< d_i|p^j > = \delta_{ij}, \quad < p^i|p^j > = \; = < d_i|d_j > = 0,$$

and the second one is given by

(4.5) $$Q_B|s_i > = 0, \quad Q_B|d_i > = 0.$$

We shall consider a change of basis which leaves both $\mathcal{V}_d$ and $\mathcal{V}_{\text{phys}}$ unchanged in the following form:

$$|d_i'> = \sum_j |d_j> v_{ji},$$

(4.6)
$$|s_i'> = \sum_j (|s_j> + \sum_k |d_k> c_{kj}) u_{ji},$$

$$|p^{i'}> = \sum_j (|p^j> + \sum_k |s_k> a_{kj} + \sum_k |d_k> b_{kj}) w_{ji}.$$

The new set of basis satisfies the orthonormalization condition (4.4) as well as the condition (4.5) provided that the following equations are satisfied:

$$u^\dagger u = u u^\dagger = 1,$$

(4.7)
$$w v^\dagger = v^\dagger w = 1,$$

$$a + c^\dagger = 0, \quad a a^\dagger + b + b^\dagger = 0.$$

The transformations represented by u, v and w do not mix daughter, parent and singlet spaces, but those represented by a, b and c do mix them.

The non-uniqueness of these subspaces leads to the basis-dependence of various projection operators,

$$P(\mathcal{V}_d) = \sum_i |d_i> < p^i|,$$

(4.8)
$$P(\mathcal{V}_p) = \sum_i |p^i> < d_i|,$$

$$P(\mathcal{V}_S) = \sum_i |s_i> < s_i|.$$

These projection operators certainly satisfy the condition

(4.9)
$$P(\mathcal{V}_a) P(\mathcal{V}_b) = \delta_{ab} P(\mathcal{V}_a), \quad (a, b = d, p, S)$$

but hermiticity is not satisfied for the first two, namely,

(4.10)
$$P^\dagger(\mathcal{V}_d) = P(\mathcal{V}_p), \quad P^\dagger(\mathcal{V}_p) = P(\mathcal{V}_d),$$

and

(4.11)
$$P^\dagger(\mathcal{V}_S) = P(\mathcal{V}_S)$$

Let us now introduce the projection operator to the physical subspace by

(4.12)
$$P(\mathcal{V}_{\text{phys}}) = P(\mathcal{V}_S) + P(\mathcal{V}_d),$$

then we can easily verify the relation

(4.13)
$$P^\dagger(\mathcal{V}_{\text{phys}})P(\mathcal{V}_{\text{phys}}) = P(\mathcal{V}_S).$$

Let $|f>$ and $|g>$ be physical states, then we have with the help of Eq. (4.13) the relation

(4.14)
$$< f|g> = < f|P(\mathcal{V}_s)|g >$$

The projection operator $P(\mathcal{V}_s)$ is basis-dependent, but the inner product on the *l.h.s.* is not. Hence, the expression on the *r.h.s.* is actually basis-independent despite of its appearance. In what follows we shall generalize this property to a wider class of operators.

Let us assume that T is a *BRS* invariant operator and that $|f>$ belongs to the physical subspace, then

(4.15)
$$T|f > \in \mathcal{V}_{\text{phys}}.$$

When two physical states $|f_1 >$ and $|f_2 >$ satisfy

(4.16)
$$|f_1 > -|f_2 > \in \mathcal{V}_d,$$

they are said to be equivalent, and we express it as

(4.17)
$$|f_1 > \equiv |f_2 > . \quad (\text{mod } \mathcal{V}_d)$$

Thus they form an equivalent class and a class of such states may be represented by $|f \gg$.

Let a physical state $|f >$ be transformed into another physical state $|f' >$ by a *BRS* invariant operator T,

(4.18)
$$T|f > = |f' > .$$

When $|f >$ is replaced by another member of the class $|f \gg$, $|f > +|d > = |f > +Q_B |p >$, we have

(4.19)
$$\begin{aligned} T|f > \quad &\rightarrow T|f > \quad +TQ_B |p > \\ &= |f' > \quad \pm Q_B T|p > \\ &\equiv |f' > . \quad (\text{mod } \mathcal{V}_d). \end{aligned}$$

Hence, we can trascribe the relation (4.18) into a class relation:

(4.20)
$$T|f \gg = |f' \gg .$$

The collection of classes of physical states is given by a quotient space,

(4.21)
$$H_{\text{phys}} = \mathcal{V}_{\text{phys}}/\mathcal{V}_d.$$

Since both subspaces $\mathcal{V}_{\text{phys}}$ and $\mathcal{V}_d$ are Poincaré-invariant, so is the quotient space H_{phys}. In this way a redundant degree of freedom corresponding to $\mathcal{V}_d$ can be eliminated. In taking the matrix element of a product of BRS invariant operators, we have

$$
\begin{aligned}
\ll f_2|T_2 T_1|f_1 \gg &= < f_2|T_2 T_1|f_1 > \\
&= < f_2|T_2 P(\mathcal{V}_S) T_1|f_1 > \\
&= \sum_i < f_2|T_2|s_i > < s_i|T_1|f_1 > \\
&= \sum_i \ll f_2|T_2|s_i \gg \ll s_i|T_1|f_1 \gg,
\end{aligned}
$$

(4.22)

where use has been made of the relation (4.14).

In gauge theories we interpret a class of physical states as representing physically observable states. Naturally a class of physical states $|f \gg$ should satisfy the condition

(4.23) $$ Q_B|f \gg = 0. $$

A change of basis may be expressed in terms of classes of physical states, and Eq. (4.6) leads to

(4.24) $$ |s'_i \gg = \sum_j |s_j \gg u_{ji}, $$

where u is a unitary matrix, and the orthonormalization condition is expressed by

(4.25) $$ \ll s_i|s_j \gg = \delta_{ij}. $$

The S matrix is BRS invariant and the unitarity condition for physical states can be expressed as

$$
\begin{aligned}
\ll f|g \gg &= \sum_i \ll f|S^\dagger|s_i \gg \ll s_i|S|g \gg \\
&= \sum_i \ll f|S|s_i \gg \ll s_i|S^\dagger|g \gg.
\end{aligned}
$$

(4.26)

Since this is a class relationship it has a basis-independent meaning. If Eq. (4.26) is true for a specific set of basis, so should it be for any set of basis.

Now we shall argue that a state $|f >$ which does not satisfy $Q_B|f > = 0$ is unobservable in the sense that it can never enter the intermediate states in the unitarity condition of the S matrix. In this case $|f >$ can be expressed for a certain choice of the set of basis as

(4.27) $$ |f > = \sum_i x_i|p^i > + \sum_i y_i|s_i > + \sum_i z_i|d_i >. $$

Then introduce a change of basis given in Eq. (4.6) with $u = v = w = 1$, and we can eliminate the singlet part provided that we choose the matrix a so as to satisfy

(4.28)
$$y_k + \sum_i a_{ki} x_i = 0.$$

We can always choose a to satisfy this equation unless all the coefficients x_i vanish indentically. However, the fact that $Q_B |f> \neq 0$ implies that there must be at least one non-vanishing x coefficient. Thus we have proved that for an appropriate choice of the set of basis $|f>$ belongs to the BRS doublet space and does not contribute to the intermediate states in Eq. (4.26). The unitarity condition of the S matrix is expressed entirely in terms of physical states, and unphysical states are not subject to observation. Thus BRS doublet states or unphysical states correspond to confined states just as the longitudinal and scalar photons in QED. The problem of confinement, therefore, reduces to that of showing that quarks, gluons and all other colored particles belong to BRS doublet states.

5. - Sufficient Condition for Color Confinement

As discussed in the preceding section the problem of color confinement reduces to that of showing that all the colored particles belong to BRS doublet states. In this section we shall present a sufficient condition for color confinement. For this purpose, however, we have to rely on the Lehmann-Symanzik-Zimmermann (LSZ) reduction formula. In view of the strong infrared singularities of QCD the validity of the reduction formula is doubtful and for that reason we shall assume its validity only for observable BRS singlet states.

The asymptotic condition provides the particle interpretation of fields, so that the BRS singlet states are assumed to consist of particles described by asymptotic fields. However, the situation is somewhat complicated with the auxiliary fields because of some identities such as

(5.1)
$$< A_\mu^a(x), B^b(y) > = -\delta_{ab} \partial_\mu D_F(x - y),$$

(5.2)
$$< (D_\mu c)^a(x), \bar{c}^b(y) > = i\delta_{ab} \partial_\mu D_F(x - y),$$

where D_F denotes the free massless propagator. These two-point functions represent the vacuum expectation values of time-ordered products of field operators.

These relations indicate that all the four operators in Eqs. (5.1) and (5.2) generate massless spin 0 particles, and we shall denote the corresponding asymptotic fields as

(5.3)
$$A_\lambda^a \to \partial_\lambda \chi^a, \quad B^a \to \beta^a, \quad \bar{c}^a \to \bar{\gamma}^a, \quad D_\lambda c^a \to \partial_\lambda \gamma^a,$$

and we have

(5.4) $< \chi^a(x), \beta^b(y) > \ = -\delta_{ab} D_F(x-y).$

(5.5) $< \gamma^a(x), \overline{\gamma}^b(y) > \ = i\delta_{ab} D_F(x-y).$

Form Eqs. (2.7) and (2.8) we obtain the relationships:

(5.6) $\delta \chi^a = \gamma^a, \quad \delta \overline{\gamma}^a = i\beta^a.$

First, we shall introduce a sufficient condition for all the colored particles to be confined. For this purpose we shall write down a set of Ward-Takahashi identities relevant to this purpose [13].

(5.7)
$$\partial_\lambda < \delta\overline{\delta} A_\lambda^a(x), \psi^\alpha(y), \overline{\psi}^\beta(z) >$$
$$= ig T_{\alpha\beta}^a [\delta^4(x-y) S_F(y-z, \beta) - \delta^4(x-z) S_F(y-z, \alpha)],$$

(5.8)
$$\partial_\lambda < \delta\overline{\delta} A_\lambda^a(x), A_\mu^b(y), A_\nu^c(z) >$$
$$= ig M_{bc}^a [\delta^4(x-y) D_{F\mu\nu}(y-z, c) - \delta^4(x-z) D_{F\mu\nu}(y-z, b)],$$

where $M_{bc}^a = -if_{abc}$ corresponds to the adjoint representation. S_F and D_F are the propagator of the quark field and of the gauge field, respectively. In what follows we shall discuss only the quark field without loss of generality.

The operator A_λ^a generates both massless spin 1 gluon and massless spin 0 χ particle, and so does $\delta\overline{\delta} A_\lambda^a$ also in general. In order to pick out the spin 0 projection we replace A_λ^a by its spin 0 projection $\overline{A}_\lambda^a$ as

(5.9) $$\overline{A}_\lambda^a(x) = -i \int d^4y \partial_\lambda D_F(x-y) \partial_\mu A_\mu^a(y).$$

Eq. (5.7) is still valid when we replace A_λ^a by $\overline{A}_\lambda^a$, but $\overline{A}_\lambda^a$ does not generate the massless spin 1 gluon. The only massless particle that $\overline{A}_\lambda^a$ generates is the spin 0 χ particle, and $\delta\overline{\delta}\,\overline{A}_\lambda^a$ would generate the spin 0 $\delta\overline{\delta}\chi^a$ particle. However, when the condition

(5.10) $\delta\overline{\delta}\chi^a = -\overline{\delta}\gamma^a = 0$

is satisfied, $\delta\overline{\delta} A_\lambda^a$ no longer generates a massless particle. At this stage we remark that the condition (5.10) is inconsistent with the spontaneous breaking of the color symmetry. Indeed, when the color symmetry is spontaneously broken $\delta\overline{\delta}\,\overline{A}_\lambda^a$ should necessarily generate a massless spin 0 Nambu-Goldstone particle in view of the conservation law

(5.11) $\partial_\lambda(\delta\overline{\delta}\,\overline{A}_\lambda^a) = 0.$

Now assume that the quark is observable and belongs to the BRS singlet space so that we can apply the LSZ reduction formula to the quark field. We shall then show that this assumption along with the condition (5.10) leads to a contradiction. In other words we will show that the quark cannot belong to the BRS singlet space.

For this purpose we shall first factorize the three-point function as

$$(5.12) \quad < \delta\bar{\delta}A_\lambda^a(x), \psi^\alpha(y), \overline{\psi}^\beta(z) > =$$
$$\int d^4y' \int d^4z' S_F(y-y') V_\lambda^a(y', z' : x)_{\alpha b} S_F(z'-z).$$

Since we have assumed an exact color symmetry S_F does not depend on the color index α or β. We then introduce the following Fourier representations:

$$(5.13) \quad S_F(x) = \frac{-i}{(2\pi)^4} \int d^4p \; e^{ip\cdot x} S_F(p),$$

$$(5.14) \quad V_\lambda^a(y, z : x)_{\alpha\beta} = \frac{1}{(2\pi)^8} \int d^4p \int d^4q e^{ip\cdot(y-x)+iq\cdot(x-z)} V_\lambda^a(p,q)_{\alpha\beta}.$$

Then Eq. (5.7) takes the following simple form:

$$(5.15) \quad S_F(p)(p-q)_\lambda V_\lambda^a(p,q) S_F(q) = ig T_{\alpha\beta}^a (S_F(p) - S_F(q)).$$

Now we shall replace V_λ by its spin 0 projection $\overline{V}_\lambda$ defined by

$$(5.16) \quad \overline{V}_\lambda(p, q) = \frac{(p-q)_\lambda(p-q)_\mu}{(p-q)^2} V_\mu(p, q),$$

corresponding to Eq. (5.9). On account of the assumption (5.10) $\overline{V}_\lambda(p, q)$ is free of the pole at $(p-q)^2 = 0$, and we have the following relation by differentiating (5.15) with respect to p_λ and then putting $q = p$:

$$(5.17) \quad S_F(p)\overline{V}_\lambda^a(p, p)_{\alpha\beta} S_F(p) = ig T_{\alpha\beta}^a \frac{\partial}{\partial p_\lambda} S_F(p).$$

Application of the LSZ reduction formula to the quark fields leads to

$$(5.18) \quad < p, \alpha|\delta\bar{\delta} \; \overline{A}_\lambda^a(x)|p, \beta > = g T_{\alpha\beta}^a \overline{u}(p)\gamma_\lambda u(p),$$

where $|p, \alpha >$ denotes a single quark state of color α, and $u(p)$ and $\overline{u}(p)$ denote the corresponding Dirac spinors. However, when the quark state belong to the BRS singlet space the l.h.s. of Eq. (5.18) should vanish identically since

$$(5.19) \quad Q_B|p, \alpha > = Q_B|p, \beta > = 0,$$

whereas the r.h.s. of Eq. (5.18) is finite. Thus we may conclude that the quark state cannot satisfy Eq. (5.19) when the condition (5.10) is met. Then by the argument given at the end of the preceding section we conclude that the quark must be confined, and similarly all the colored particles must be confined. In other words, the condition (5.10) is a sufficient condition for color confinement, and then both elementary particles, quarks and gluons, are confined. When this is the case only composite hadrons described by BRS-invariant composite fields are subject to observation.

When this is the case, the S matrix elements are independent of the choice of gauges within a certain class. Let $\mathcal{L}_I$ and $\mathcal{L}_{II}$ be two Lagrangian densities sharing the same gauge-invariant part and consequently representing the same physical theory. Now we introduce their difference,

$$(5.20) \qquad \Delta\mathcal{L} = \mathcal{L}_{II} - \mathcal{L}_I,$$

$$(5.21) \qquad \Delta S = \int d^4x\,\Delta\mathcal{L} = S_{II} - S_I.$$

Then Green's functions in these two gauges are expressed in terms of path integrals as

$$< XYZ\ldots >_I = \frac{1}{N_I} \int \mathcal{D}A_\mu \ldots XYZ \; \exp(iS_I),$$

$$(5.22)$$

$$N_I = \int \mathcal{D}A_\mu \ldots \exp(iS_I),$$

and a similar relation in the gauge II is obtained. The above Green's functions in the two gauges are related through

$$(5.23) \qquad < XYZ\ldots >_{II} = < XYZ\ldots\exp(i\Delta S) >_I \;/\; < \exp(i\Delta S) >_I,$$

which is a generalization of the Gell-Mann-Low relation expressing Green's functions in the Heisenberg representation in terms of those in the interaction representation.

Now let us assume that the difference of actions can be expressed as

$$(5.24) \qquad \Delta S = \delta \int d^4x\,M, \quad \text{or} \quad \Delta\mathcal{L} = \delta M,$$

and consider only $\dot{B}RS$-invariant composite field operators $X, Y, Z, \ldots$, then by means of the BRS identity (2.25) we can easily show

$$(5.25) \qquad < XYZ\ldots >_{II} = < XYZ\ldots >_I .$$

Hence, the S matrix elements for hadronic reactions are the same within a class of gauges satisfying Eq. (5.24) provided that color confinement is realized. It also suggests that proof of color confinement in one particular gauge guarantees its validity for a large class of gauges. It is also worth emphasizing that the gauge-fixing term in (2.3) satisfies Eq. (5.24).

6. - Simultaneous Confinement of Colored Particles

In the preceding section we have concluded that Eq. (5.10) represents a sufficient condition for color confinement. Then a question may be immediately raised as to whether it is also necessary. In this section we shall argue on the basis of an approximate calculation that it is also necessary. This conclusion bears an important implication that all the colored particles must be confined simultaneously. For instance, if quarks are confined gluons are simultaneously confined and *vice versa*.

In order to show this we shall first study the physical meaning of the condition (5.10). The condition $\bar{\delta}\gamma^a = 0$ implies that the asymptotic field γ^a is a *BRS* daughter operator so that there must be its parent p^a with respect to the alternative *BRS* transformation, namely,

$$(6.1) \qquad \bar{\delta}p^a = \gamma^a.$$

Then, substituting (6.1) for γ^a in Eq. (5.5) and using the alternative *BRS* identity, we find

$$(6.2) \qquad < \gamma^a(x), \bar{\gamma}^b(y) > = - < p^a(x), \bar{\delta}\bar{\gamma}^b(y) > = i\delta_{ab}D_F(x-y),$$

or

$$(6.3) \qquad \bar{\delta}\bar{\gamma} = (\bar{\delta}\bar{c})^{\text{asy}} = -\frac{1}{2}\,g(\bar{c}\times\bar{c})^{\text{asy}} \neq 0.$$

This result requires the existence of a massless bound state of a pair of *FP* ghosts, and the condition for color confinement reduces to a bound state problem [14]. Here we should emphasize a similarity between color confinement and superconductivity. In the *BCS* theory superconductivity is realized when the phonon attraction between a pair of electrons dominates the Coulomb repulsion so as to form a bound state, the so-called Cooper pair. On the other hand color confinement is realized, as we have seen above, when a pair of *FP* ghosts forms a bound state. At this stage it is clear that the condition (5.10) cannot be satisfied in an abelian gauge theory such as *QED*, since the *FP* ghosts are described by free fields and they cannot form bound states. Indeed, in *QED* we have

$$(6.4) \qquad \bar{\delta}\delta\chi = \bar{\delta}\gamma = i\beta \neq 0.$$

The bound state will appear as a massless pole in Green's function $< c, c, \bar{c}, \bar{c} >$ so that $c \times c$ should also have its asymptotic field. Next let us assume the existence of the asymptotic quark field in a certain approximate sense in an infrared cut-off theory such as the lattice gauge theory in a finite box in order to define the quark mass. Then, if $\delta\psi$ also has its asymptotic field degenerate with the quark in mass, the following pair of operators:

(6.5) $$\{\psi^{\text{asy}}, \delta\psi^{\text{asy}}\}$$

forms a BRS doublet, and the quark will be confined. Then the equation

(6.6) $$\delta\psi^{\text{asy}} = ig((c \cdot T)\psi)^{\text{asy}}$$

requires the existence of a bound state of a quark and a FP ghost.

The general theory developed in the preceding section indicates that the existence of the bound FP ghost pair would imply that of the quark-ghost bound state. We shall check this statement by studying these bound state problems and shall show that the existence of the solution of one of the two bound state problems automatically leads to that of the other problem [14]. This further implies that the condition (5.10) is not only sufficient but also necessary for color confinement.

The existence of the massless dighost bound state is sufficient for color confinement, but a question is raised of the kinematical feasibility of such a bound state since the binding energy is zero. In such a case the exponential fall-off of the wave function at large distances is impossible, but a power-like fall-off is still possible. In fact, it is easy to construct a non-relativistic example. In what follows we shall investigate the two bound state problems by means of the Bethe-Salpeter (BS) equations.

1) *The quark-ghost bound state*

The existence of the quark-ghost bound state leads to quark confinement, and we shall consider the Feynman amplitude

(6.7) $$F(x, y) = < 0|T[c(x)\psi(y)]| > .$$

The BS equation for one-gluon-exchange is given by

(6.8) $$F(x, y) = G\frac{\partial}{\partial x_\mu} \int d^4x' \int d^4y'$$
$$D_F(x - x')S_F(y - y')D_{F\mu\nu}(x' - y')\gamma_\nu F(x', y'),$$

where $G = g^2(M \cdot T)$ depends on the representation of the bound state. For the color triplet bound state in which we are interested, the corresponding eigenvalue of G is given by

(6.9) $$G^{(3)} = (3/2)g^2.$$

The gluon propagator is given by

(6.10) $$D_{F\mu\nu}(x) = \frac{1}{4\pi^2} \left[\frac{1}{2}(1+\alpha)\frac{\delta_{\mu\nu}}{x^2 + i\varepsilon} + (1-\alpha)\frac{x_\mu x_\nu}{(x^2 + i\varepsilon)^2} \right].$$

The differential form of Eq. (6.8) is given by

$$(6.11) \qquad \Box_x D_y F^{(3)}(x, y) = G^{(3)} \frac{\partial}{\partial x_\mu} \left[D_{F\mu\nu}(x - y) \gamma_\nu F^{(3)}(x, y) \right],$$

where D_y denotes the Dirac operator for the quark field,

$$(6.12) \qquad D_y = \gamma_\mu \frac{\partial}{\partial y_\mu} + m.$$

The bound state represents a Dirac particle of mas m, so that we shall put

$$(6.13) \qquad F^{(3)}(x, y) = f(s) e^{iP \cdot y}, u(P),$$

where $s = x^2 + i\varepsilon$ and $u(P)$ denotes the Dirac spinor satisfying

$$(6.14) \qquad (iP \cdot \gamma + m) u(P) = 0.$$

Then $f(s)$ satisfies the equation

$$(6.15) \qquad 2s^3 f'''(s) + 6s^2 f''(s) + \lambda(3 - \alpha) s f'(s) - 2\lambda \alpha f(s) = 0,$$

where

$$(6.16) \qquad \lambda = G^{(3)}/16\pi^2 = (3/32\pi^2) g^2.$$

The normalization condition reads as

$$(6.17) \qquad \int d^4 x f(s) \Box_x f(s) < \infty.$$

The solution of Eq. (6.15) is obtained by putting

$$(6.18) \qquad f(s) = C s^b,$$

then the parameter b satisfies a cubic equation,

$$(6.19) \qquad 2b(b - 1)(b + 1) + \lambda(3b - \alpha(b + 2)) = 0.$$

2) *The dighost bound state*

The condition (5.10) implies the existence of the dighost bound state. The corresponding Feynman amplitude is given by

$$(6.20) \qquad F^{ab}(x, y) = \langle 0|T[c^a(x) c^b(y)]| \rangle.$$

The *BS* equation for one-gluon-exchange is given by

(6.21)
$$F^{ab}(x,y) = -g^2 \frac{\partial^2}{\partial x_\mu \partial y_\nu} \int d^4x' \int d^4y' D_F(x-x')D_F(y-y')$$
$$\times M^A_{ac}M^A_{bd}D_{F\mu\nu}(x'-y')F^{cd}(x',y').$$

For the color octet bound state in question we may introduce

(6.22)
$$\overline{G} = -g^2(M_1 \cdot M_2) \rightarrow (3/2)g^2,$$

which is identical with (6.9). The differential form of Eq. (6.21) then reduces to

(6.23)
$$\Box_x \Box_y F^{(8)}(x,y) = \overline{G}^{(8)} \frac{\partial^2}{\partial x_\mu \partial y_\nu} [D_{F\mu\nu}(x-y)F^{(8)}(x,y)].$$

This time the bound state is massless and its four-momentum P_μ satisfies $P_\mu^2 = 0$. Now introduce

(6.24)
$$X = \frac{1}{2}(x+y), \quad r = x-y, \quad s = r^2, \quad t = P \cdot r,$$

and put

(6.25)
$$F^{(8)}(x,y) = e^{iP \cdot X}f(s,t).$$

If we assume $f(s,0)$ to be finite, then $f(s,0)$ satisfies Eq. (6.15), so that we may put

(6.26)
$$f(s,t) = f(s)h(t).$$

Then $h(t)$ satisfies

(6.27)
$$h''(t) + \frac{\beta}{t}h'(t) + \frac{1}{4}h(t) = 0,$$

where

(6.28)
$$\beta = b(b-1)(b+1)/[b(b-1) + \frac{\lambda}{4}(1-\alpha)].$$

Its solution regular at $t = 0$ is given by

(6.29)
$$h(t) = t^{-\beta/2}J_{\beta/2}(t/2).$$

Then, because of $P_\mu^2 = 0$ the normalization condition reduces to

(6.30)
$$\int d^4x f(s,t)\Box_x f(s,t) = \int d^4x f(s,0)\Box_x f(s,0) < \infty.$$

Thus we found that both $f(s)$ and $f(s,0)$ satisfy the same equation and the same normalization condition at least in the present approximation.

This result is consistent with the general statement in the preceding section that the existence of the solution of the latter problem implies that of the former. In the present approximation the converse is also true so that the existence of the solution of the two bound state problems implies that of the other.

For small values of λ the s-dependence of the solution is controlled by Eq. (6.18), and in the Landau gauge ($\alpha = 0$) an appropriate power b can be chosen from among the three roots of Eq. (6.19), namely,

(6.31)
$$b = 1 - \frac{3}{4}\lambda + O(\lambda^2)$$
$$= 1 - \frac{9}{128\pi^2} g^2 + O(g^4).$$

We shall further assume that $f(s)$ falls off sufficiently rapidly for large values of s. If we take the power-like solution literally, the normalization integral diverges for large values of s and we must improve the approximation. The simplest improvement is achieved by replacing the coupling constant g^2 by the running coupling constant $g^2(\Lambda^2 s)$ that is supposed to increase for large values of s, namely, at large distances, where Λ is a parameter of the dimension of mass. Then the normalization integral can converge for large values of s.

Thus we may conclude that Eq. (5.10) is a necessary and sufficient condition for color confinement. Hence, when gluons are confined the necessary condition (5.10) must be satisfied, and then quarks must be confined since the sufficient condition (5.10) is satisfied. In order to establish color confinement, therefore, it is sufficient to prove that gluons are confined under a certain condition. This is the main subject in the next section.

7. - Condition for Gluon Confinement

As mentioned at the end of the preceding section it is sufficient to confine gluons in order to realize color confinement. In the present section we shall derive a sufficient condition for gluon confinement in terms of the number of quark flavors. For this purpose we shall exploit the results of Oehme and Zimmermann [15] but with an interpretation of the results different from theirs. They have analyzed the structure of the gluon propagator in detail by making use of asymptotic freedom and analyticity. Furthermore, they have introduced an ingenious idea of analyzing the asymptotic behavior of the projected transverse gluon propagator to find a restriction imposed on the number of flavors. It is the purpose of the present section to reexamine their consistency condition in the light of the confinement condition and deduce the condition for gluon confinement.

First, we shall introduce the following operator $\grave{a}$ la Oehme and

Zimmermann:

(7.1) $$A_{\mu\nu}^a = \partial_\mu A_\nu^a - \partial_\nu A_\mu^a.$$

The transverse gluon propagator is defined by

(7.2) $$< A_{\mu\nu}^a(x), A_{\rho\sigma}^b(y) > = \delta_{ab} G_{\mu\nu\rho\sigma}(x - y),$$

and its Fourier transform is given by

(7.3) $$G_{\mu\nu\rho\sigma}(x) = \frac{-i}{(2\pi)^4} \int d^4 k e^{ik \cdot x} G_{\mu\nu\rho\sigma}(k).$$

It is expressed in the following form:

(7.4) $$G_{\mu\nu\rho\sigma}(k) = (k_\mu k_\rho \delta_{\nu\sigma} - k_\nu k_\rho \delta_{\mu\sigma} - k_\mu k_\sigma d_{\nu\rho} + k_\nu k_\sigma \delta_{\mu\rho}) D(k).$$

It is convenient to introduce a dimensionless combination,

(7.5) $$R(k) = k^2 D(k),$$

and we normalize it at a space-like point $k^2 = \mu^2$ to unity,

(7.6) $$R(k) = R\left(\frac{k^2}{\mu^2}, g\right) = 1, \quad \text{for} \quad k^2 = \mu^2.$$

This function R satisfies the renormalization group (RG) equation of the following form in QCD:

(7.7) $$[D + 2\gamma_V(g)]R\left(\frac{k^2}{\mu^2}, g\right) = 0.$$

We have discarded the quark-mass-dependence of R, but as far as high energy behavior is concerned this is irrelevant. In the Landau gauge, we have

(7.8) $$D = \mu \frac{\partial}{\partial\mu} + \beta(g) \frac{\partial}{\partial g},$$

where

(7.9) $\beta(g) = g^3(\beta_0 + \beta_1 g^2 + \ldots)$, with $\beta_0 = -\dfrac{1}{48\pi^2}(11N - 2N_f)$,

(7.10) $\gamma_V(g) = g^2(\gamma_0 + \gamma_1 g^2 + \ldots)$, with $\gamma_0 = -\dfrac{1}{48\pi^2}(13N - 4N_f)$.

Here the color group is assumed to be $SU(N)$, and N_f denotes the number of quark flavors. Oehme and Zimmermann have solved Eq. (7.7) under the boundary condition (7.6) assuming asymptotic freedom ($\beta_0 < 0$) and have obtained the following asymptotic form of R for $k^2 \to \infty$:

$$(7.11) \qquad R\left(\frac{k^2}{\mu^2}, g\right) \sim C_V \left(\ell n \, \frac{k^2}{\mu^2}\right)^{-\gamma_0/\beta_0}$$

where $C_V > 0$. Thus we can write down an unsubtracted dispersion relation for $D = R/k^2$ as

$$(7.12) \qquad D(k) = \int_0^\infty dm^2 \, \frac{\rho(m^2)}{k^2 + m^2 - i\varepsilon}.$$

By evaluating its discontinuity across the real axis, they have found the asymptotic form of ρ for $k^2 \to \infty$,

$$(7.13) \qquad \rho \sim - \left(\frac{\gamma_0}{\beta_0}\right) C_V (k^2)^{-1} \left(\ell n \, \frac{k^2}{\mu^2}\right)^{-1-\gamma_0/\beta_0}$$

When $\beta_0 < 0$ and $\gamma_0 < 0$, namely $N_f \leq 9$ for $N = 3$, not only D but also R satisfies an unsubtracted dispersion relation, and consequently a superconvergence relation follows:

$$(7.14) \qquad \int_0^\infty dm^2 \rho(m^2) = 0.$$

The RG method enables us to study high energy behavior of the propagator, while confinement is a problem of low energy behavior. This superconvergence relation, however, relates the low energy behavior with the high energy behavior, so that it strikes the keynote of gluon confinement as we shall see in what follows.

Let us consider the projected propagator defined by

$$(7.15) \qquad <0|T[A_{\mu\nu}^a(x)P(\mathcal{V}_S)A_{\rho\sigma}^b(y)]|0> = \delta_{ab} G_{\mu\nu\rho\sigma}^+(x - y).$$

The BRS singlet subspace $\mathcal{V}_S$ cannot be defined in a Lorents-invariant manner, nor is it unique. It depends, for instance, on the choice of a special direction for the time-axis as in QED. The unit vector in the time direction will be denoted by n, so that we have $n^2 = -1$.

Corresponding to the expression (7.4) we have

$$(7.16) \qquad \begin{aligned} G_{\mu\nu\rho\sigma}^+(k) &= (k_\mu k_\rho \delta_{\nu\sigma} - k_\nu k_\rho \delta_{\mu\sigma} - k_\mu k_\sigma \delta_{\nu\rho} + k_\nu k_\sigma \delta_{\mu\rho}) D^+(k) \\ &+ (k_\mu n_\nu - k_\nu n_\mu)(k_\rho n_\sigma - k_\sigma n_\rho) D_1^+(k). \end{aligned}$$

Both $D^+(k)$ and $D_1^+(k)$ are functions of k^2 and $\mathbf{k}^2 = k^2 + (n \cdot k)^2$. Again, we introduce a dimensionless combination R^+ by

$$(7.17) \qquad R^+(k) = k^2 D^+(k) = R^+\left(\frac{k^2}{\mu^2}, \frac{\mathbf{k}^2}{\mu^2}, g\right).$$

This R^+ also satisfies Eq. (7.7), and as its boundary condition we postulate

$$(7.18) \qquad \lim_{g \to 0} R^+(1, 0, g) = c > 0,$$

when gluons are unconfined [16]. We are essentially assuming that *unconfined gluons tend to free gluons in the weak coupling limit*. We cannot make such an assumption for confined gluons in view of the interpretation of confinement in terms of bound states as presented at the end of the preceding section. Oehme and Zimmermann, on the other hand, assumed a condition essentially equivalent to (7.18) as a condition for the consistency of the theory.

The solution of Eq. (7.7) for R^+ then yields the following asymptotic behaviour of R^+ for $\mathbf{k} = 0, k_0 \to \infty$:

$$(7.19) \qquad R^+\left(-\frac{k_0^2}{\mu^2}, 0, g\right) \sim c\, C_V \left(\ell n\, \frac{k_0^2}{\mu^2}\right)^{-\gamma_0/\beta_0}$$

The function D^+ obeys an unsubtracted dispersion relation of the following form:

$$(7.20) \qquad D^+(k) = \int_0^\infty dm^2\, \frac{\rho^+(m^2, \mathbf{k}^2)}{k^2 + m^2 - i\varepsilon}.$$

Since, however, the BRS singlet subspace is of positive-definite metric, we have

$$(7.21) \qquad \rho^+(m^2, \mathbf{k}^2) \geq 0.$$

From Eq. (7.19) and the assumption $\beta_0 < 0$ and $\gamma_0 < 0$, we can again infer that both D^+ and R^+ satisfy unsubtracted dispersion relations leading to the superconvergence relation

$$(7.22) \qquad \int_0^\infty dm^2 \rho^+(k_0^2, 0) = 0.$$

Combination of Eqs. (7.21) and (7.22) leads to

$$(7.23) \qquad \rho^+(k_0^2, 0) = 0.$$

The spectral function $\rho^+(m^2, \mathbf{k}^2)$ is decomposed into a sum of the pole and branch cut contributions as

(7.24)
$$\rho^+(m^2, \mathbf{k}^2) = a(\mathbf{k}^2)\delta(m^2) + \sigma(m^2, \mathbf{k}^2)$$
$$= a\delta(m^2) + \sigma(m^2, \mathbf{k}^2),$$

where we have exploited the fact that the pole term is Lorentz-invariant as it is the case in QED. Eq. (7.23) then reduces to

(7.25)
$$a = 0, \qquad \sigma(m^2, 0) = 0,$$

and the first equation indicates that a gluon state does not belong to the BRS singlet space. Hence, for the color group $SU(3)$ gluons are confined when $N_f \le 9$ corresponding to $\beta_0 < 0$ and $\gamma_0 < 0$.

In terms of the number of generations we may conclude that gluons and consequently all the colored particles are confined when the number of generations does not exceed four [16]. It should be stressed that Eq. (7.23) leads to $c = 0$ contradicting the assumed unconfined gluons in (7.18).

In the present article, color confinement means that the unitarity condition for the S matrix is saturated by color singlet hadronic states, and this result has a significance independent of the choice of gauge as discussed in Sec. 5. When both elementary particles, gluons and quarks, are confined, the theory would be trivial there not observable composite hadrons, since the only observable state would be the vacuum state. However, we shall show their existence on the basis of a plausible assumption. Consider Green's functions of BRS-invariant composite operators:

(7.26)
$$G = \langle ABC \ldots \rangle,$$

where $\delta A = \delta B = \delta C \ldots = 0$. We shall further assume that these operators carry some non-vanishing additive quantum numbers, so that we have

(7.27)
$$\langle A \rangle = \langle B \rangle = \langle C \rangle = \ldots = 0.$$

The only assumption that we make is given by

(7.28)
$$G \ne 0,$$

at least for some combinations of the operators. This certainly implies the existence of observable BRS singlet states other than the vacuum state because of Eq. (4.22). They must represent states consisting only of observable composite particles usually referred to as hadrons.

8. - Implications of Confinement

In the preceding section it has been shown that color confinement is realized when $\beta_0 < 0$ and $\gamma_0 < 0$. As the number of quark flavors is increased γ_0 changes its sign and apparently a phase transition takes place in the sense that the wave function renormalization constant of the gauge field in the Landau gauge changes drastically. Namely, we have

$$(8.1) \qquad Z_3^{-1} = \int dm^2 \rho(m^2) = \begin{cases} 0, & \text{for } \beta_0 < 0, \ \gamma_0 < 0, \\ \infty, & \text{for } \beta_0 < 0, \ \gamma_0 > 0, \end{cases}$$

as is clear from Eqs. (7.13) and (7.14).

The conditions $\beta_0 < 0$ and $\gamma_0 < 0$ correspond to $N_f \leq 16$ and $N_f \leq 9$, respectively, in the color $SU(3)$ theory, while they correspond to $N_f < 11$ and $N_f \leq 6$, respectively, in the color $SU(2)$ theory. When $\beta_0 < 0$ and $\gamma_0 > 0$, we do not know whether color confinement is realized or not, but as far as we know the theory seems to be consistent with unconfined gluons. Therefore, this transition must represent the one from the confinement phase to the deconfinement phase. We cannot check this phase transition by real experiments, however, since the number of quark flavors is fixed in Nature. In order to detect this phase transition we can possibly exploit the Monte Carlo simulation of the lattice gauge theory. As mentioned in the introduction, however, we cannot utilize Wilson's area law to check the phase transition in the presence of dynamical quark loops. Then what could we utilize to check it? One possibility is to study glueballs. They are stable or metastable only when gluons are confined, but they are unstable and decay immediately into massless gluons when they are not confined. Thus glueballs would cease to exist as the number of quark flavors is increased to change the sign of γ_0. This is only an example, but we can think of many other similar tests.

Perhaps it is also worth emphasizing that the integral representation (7.12) can be cast into the following form when the superconvergence relation (7.14) holds to realize color confinement:

$$(8.2) \qquad D(k) = \int\limits_0^\infty dm^2 \, \frac{\tau(m^2)}{(k^2 + m^2 - i\varepsilon)^2},$$

where

$$(8.3) \qquad \tau(m^2) = \int\limits_0^{m^2} d\mu^2 \rho(\mu^2).$$

When $\tau(m^2)$ has a peak for small values of m^2, say, at κ^2, the one gluon exchange potential can be represented by a linear potential at small distances, but it falls off rapidly at large distances [17]. This is consistent with the analysis by Matsuyama and Miyazawa [5] who showed that an uncritical extension of the linear potential to macroscopic distances leads to too strong a van der Waals force between hadrons.

In this connection it would also be interesting to check the phase transition by observing the change of the static quark-antiquark potential in the presence of dynamical quark loops by increasing the number of quark flavors.

REFERENCES

[1] L. LYON, Phys. Reports, **126** (1985) 225.

[2] H.B. NIELSEN - P. OLESEN, Nucl. Phys. **B61** (1973) 45.

[3] Y. NAMBU, Phys. Rev. **D10** (1974) 4262.

[4] K.G. WILSON, Phys. Rev. **D14** (1974) 2455.

[5] S. MATSUYAMA - H. MIYAZAWA, Prog. Theor. Phys. **61** (1979) 942.

[6] T. KUGO - I. OJIMA, Phys. Rev. Lett. **73B** (1978) 495; Prog. Theor. Phys. Suppl. No. 66 (1979).

[7] C. BECCHI - A. ROUET - R. STORA, Ann. of Phys. **98** (1976) 287.

[8] G. CURCI - R. FERRARI, Phys. Lett. **63B** (1976) 91.

I. OJIMA, Prog. Theor. Phys. **64** (1980) 625.

L. BONORA - P. PASTI - M. TONIN, Nuovo Cim. **68A** (1981) 307.

[9] N. NAKANISHI, Prog. Theor. Phys. Suppl. No. 51 (1972).

[10] K. NISHIJIMA, Nucl. Phys. **B238** (1984) 601.

[11] K. NISHIJIMA, Prog. Theor. Phys. **80** (1988) 897.

[12] K. NISHIJIMA, Prog. Theor. Phys. **80** (1988) 905.

[13] K. NISHIJIMA, Prog. Theor. Phys. **74** (1985) 889.

[14] K. NISHIJIMA - Y. OKADA, Prog. Theor. Phys. **72** (1984) 294.

[15] R. OEHME - W. ZIMMERMANN, Phys. Rev. **D21** (1980) 471, 1661.

[16] K. NISHIJIMA, Prog. Theor. Phys. **75** (1986) 1221.

[17] K. NISHIJIMA, Prog. Theor. Phys. **77** (1987) 1035.

Plasma Modes, Periodicity and Symplectic Structure

F. PEGORARO (*)

Abstract

A group theoretical formulation of the so-called "ballooning" or "extended variable" representation, commonly adopted for the study of the linear eigenmodes of short-wavelength excitations in a toroidal, axisymmetric electromagnetic plasma, is presented. With the help of this novel formulation, this representation is generalised so as to include the non-linear evolution of the excitations.

1. - Introduction

The behaviour of a high temperature plasma in a strong magnetic field depends on the global geometrical and topological features of its configuration. Its properties are anisotropic because charged particles move essentially freely along field lines, while they are effectively tied to them in the perpendicular directions. These features are reflected in the behaviour and spatial structure of the plasma collective excitations. These generally exceed the thermal fluctuation level as plasmas are in most cases far from thermodynamic equilibrium. In particular, short-wavelength collective excitations (micro-instabilities) are held responsible for the observed enhanced particle and energy diffusion across the field lines in inhomogeneous magnetically confined plasmas of the type that are of interest for controlled thermonuclear fusion experiments.

It has long since been recognised that the most dangerous excitations are those with an amplitude which is approximately constant along the equilibrium magnetic field of the configuration. The reason for this can be easily shown, e.g. in the case of magnetic plasma excitations, as a strong restoring force would be produced by the tension of the field lines which are stretched when the perturbed magnetic field varies along the equilibrium field. Whether, or to

(*) Scuola Normale Superiore, 56100 Pisa, Italy. JET Joint Undertaking, Abingdon, Oxon. OX14 3EA, UK.

what degree, this condition can be satisfied, depends on the global structure of the magnetic configuration.

In many experiments, the largely unrestrained motion of the plasma particles along the field lines has led to the adoption of closed confinement configurations. The resulting topology is characteristically that of a (full) torus with helically shaped magnetic field lines winding along toroidal surfaces (magnetic surfaces) as sketched in Fig. 1a. Generally the pitch of these helices is not constant across the toroidal surfaces (Fig. 1b), i.e. the magnetic field is sheared, and non-rational surfaces are covered ergodically by the field lines. Sheared magnetic configurations of this type are advantageous for plasma stability.

To illustrate this point we refer to a ubiquitous form of instability in magnetically confined plasmas. Charged particles, as they move along curved magnetic field lines, e.g. in the configuration depicted in Fig. 1a, experience a centrifugal acceleration which is equivalent to an effective gravity. If this acceleration points in the direction opposite to the gradient of the plasma pressure, the plasma can decrease its energy simply by exchanging regions of high and low pressure. This instability is similar to the Rayleigh-Taylor instability [1] which occurs in inhomogeneous or stratified fluids. However, the motion of a highly conductive plasma is constrained by the (approximate) conservation of the magnetic flux. A convenient way of accounting for this constraint is by imagining the plasma to be made of tubes of constant magnetic flux that maintain their identity as they move. If the magnetic field is sheared, these tubes can be thought of as forming sheaths of skewed elastic strings. This net of strings cannot be opened to allow the exchange of different plasma regions without the individual strings being stretched.

In a toroidally confined plasma the direction of the centrifugal acceleration relative to that of the pressure gradient changes from the inside to the outside of the torus as indicated in Fig. 2, so that the energy of the configuration is most efficiently reduced by displacing only that portion of the magnetic tube where these two directions are opposite. Such a deformation involves wavelengths that are short on the scale of the torus minor radius and stretches the magnetic tube leading to a restorig force.

These qualitative arguments are taken here as a justification for writing the linearised, local dispersion relation for magnetic plasma excitations (plasma modes) of the type described above in the heuristic form

(1) $$\omega^2 = k_\parallel^2(r, \theta) \; c_A^2 - g(r, \theta)/r_p.$$

Here ω is the mode frequency, $k_\parallel(r, \theta)$ is the local value of the component of the mode wave vector along the equilibrium magnetic field, r and θ are coordinates that label the magnetic surfaces and the poloidal angle respectively (see Fig. 2), c_A is the plasma Alfvén velocity (i.e. the propagation velocity of magnetic perturbations in a plasma), $-g$ is the effective gravity projected

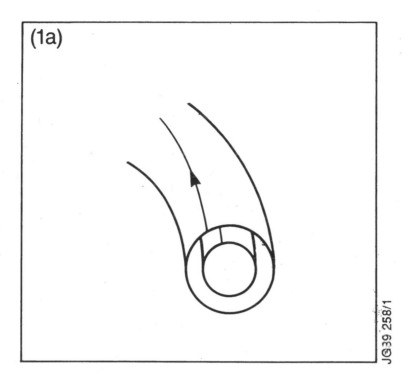

Fig. 1a - Section of a toroidal magnetic equilibrium configuration. Two magnetic surfaces and two magnetic field lines (thin lines) are displayed.

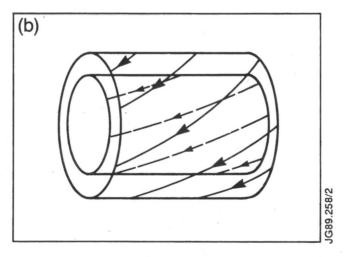

Fig. 1b - Section of two (straightened) magnetic surfaces displaying the change in the pitch of the magnetic field lines (thin lines).

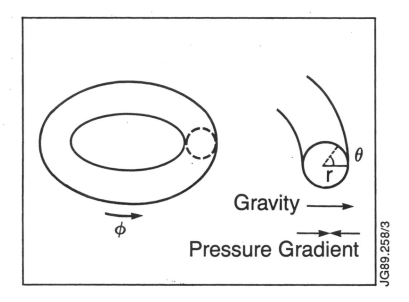

JG89.258/3

Fig. 2 - Toroidal and poloidal angles φ and θ and radial coordinate r. Relative orientation of the effective gravity and of the pressure gradient at the outside (θ = 0, right) and at the inside (θ = π, left) of the torus.

onto the direction of the pressure gradient and r_p is the characteristic scale length of the pressure gradient. The dispersion relation (1) applies to small amplitude excitations, and, if they are growing in time, is restricted to their linear phase. Being local (its r.h.s. is a function of r and $θ$), Eq. (1) cannot be used to determine the mode frequency, which can only be obtained by solving the relevant dispersion equation for the full mode spatial eigenfunction, but can be interpreted as relating $k_\parallel$ to $ω$, for example in an eikonal $(WKBJ)$ approximation scheme [2]. The term quadratic in $k_\parallel$ arises from the restoring force due to the tension of the magnetic field lines and depends on r because the direction of the magnetic field changes with r (magnetic shear). It also depends on $θ$ since the relative orientation of the centrifugal acceleration and the pressure gradient is a function of $θ$ so that energy is to be gained if the amplitude of the mode is localised on the outside $(θ = 0)$. If, for modes so localised, the therm $-g/r_p$ prevails in the r.h.s. of (1), $ω^2$ is negative and a purely growing instability occurs, which is called the "ballooning" instability [3] in the plasma physics literature in view of its spatial appearance.

2. - Small amplitude eigenfunctions

The onset of plasma excitations can be described in terms of a dispersion equation for the mode eigenfunction, obtained by linearising the appropriate set of dynamical equations aroud a prescribed equilibrium configuration. In the case of the magnetic modes introduced above, the simplest significant set of dynamical equations is provided by the so-called "ideal magnetohydrodynamic" approximation. This portrays the plasma as a perfectly conducting fluid. This description has a rather limited validity, but will suffice for the scope of this presentation which simply aims to elucidate the general mathematical properties and spatial structure of the excitations.

The plasma confinement experiments that are considered in the present analysis consist of toroidal magnetic equilibrium configurations (see Fig. 1) which are approximately axisymmetric. We may thus consider excitations characterised, during their linear phase, by a well defined toroidal (the long way aroud the torus) mode number n. We thus write

$$\text{(2)} \qquad \xi(t,\phi,\theta,r) = \hat{\xi}(\theta,r) \ \exp(in\phi - i\omega t),$$

where ϕ is the toroidal angle and we have chosen a scalar quantity ξ to represent the excitation. For instance ξ may be defined as the radial (i.e. normal to the magnetic surfaces) component of the plasma displacement vector associated with the perturbed magnetic field. On the contrary the excitation amplitude $\hat{\xi}(\theta,r)$ contains several poloidal (the short way around the torus) mode numbers, as the different poloidal harmonics are coupled by the lack of rotational symmetry in the poloidal plane. For the mode amplitude to be poloidally localised, the expansion of $\hat{\xi}(\theta,r)$ in poloidal harmonics must extend to large poloidal mode numbers m. This condition can be consistent with the requirement that the excitation be approximately constant along the equilibrium field lines only if the toroidal mode number n is also large (short-wavelength excitations). More precisely the n/m ratio, with m the central value of the poloidal mode number spectrum, must match the pitch ι of the magnetic field lines. This is defined as the ratio on each toroidal surface between the number of turns a field line must follow along the short and the long way around the torus respectively before closing onto itself. Thus we expect $\hat{\xi}(r,\theta)$ to be a non-factorisable function of r and θ and the spectrum of poloidal mode numbers to change with r so as to accomodate the variation of ι across magnetic surfaces. A residual mismatch between the pitch of the mode and that of the field lines must remain on magnetic surfaces which are not "mode rational" i.e. for which $n/\iota(r)$ is not an integer.

Since n is large, and provided the dimensionless shear parameter s

$$\text{(3)} \qquad s \equiv d \ln \iota / d \ln r$$

is not small, the distance between adjacent mode rational surfaces is much smaller that the characteristic scale length of variation of the equilibrium configuration (such as r_p in Eq. (1)). The latter is generally of the order of the minor radius of the torus. Thus, when solving the mode dispersion equation for $\hat{\xi}(r, \theta)$ we will adopt a two-scale approach [2] corresponding to an asymptotic expansion for large n. The additional radial coordinate is defined as

(4) $$S = n/\iota - m^0,$$

with $m^0 = n/\iota(r_0)$ and r_0 a reference mode rational surface. The distance between two adjacent mode rational surfaces corresponds to $\Delta S = 1$. Following this procedure we consider S and r to be independent variables and write

(5) $$\hat{\xi}(r, \theta) = X(r, S, \theta) \ \exp(-im^0\theta),$$

where, for the sake of convenience, part of the mode poloidal dependence has been factorised. The dependence of the new amplitude X on r and S must be determined separately by solving the mode dispersion equation in successive orders of approximation for large n. The dependence on S describes the response of the mode to the shear of the magnetic field lines, whereas the dependence on r is due to the radial change of the equilibrium. To leading order in the asymptotic expansion, only the S dependence comes into play.

The detailed derivation of the dispersion equation requires algebraic steps that we think convenient to bypass by referring to a model equation (see e.g. ref. [4]) which retains the features essential for the present analysis. Then, to leading order in $1/n$, we write the mode dispersion equation as

$$\left(\frac{\partial}{\partial\theta} + iS\right)\left(1 - s^2 \frac{\partial^2}{\partial S^2}\right)\left(\frac{\partial}{\partial\theta} + iS\right) X(S, \theta)$$

(6) $$+ \Gamma\left[\cos\theta + is \ \sin\theta \ \frac{\partial}{\partial S}\right] X(S, \theta)$$

$$+ \Omega^2\left[1 - s^2 \frac{\partial^2}{\partial S^2}\right] X(S, \theta) = 0.$$

The function $\Gamma = \Gamma(r)$, which is a constant on the S scale, is proportional to the radial gradient of the plasma pressure. The poloidal variation of the effective gravity is represented by the trigonometric functions of θ, while Ω is a properly normalised frequency. The operator $(\partial/\partial\theta + iS)$ plays the role of the mode wave number along the equilibrium field and the linear term in S is due to the shear of the magnetic field. If, formally, we substitute $-ik_\parallel$ for $(\partial/\partial\theta + iS)$ and disregard the S derivatives, we recover the heuristic dispersion relation (1) with $\Gamma \ \cos\theta \to g$.

3. - Periodicity and extended poloidal variable

It would appear that Eq. (6) can be reduced to an ordinary differential equation in θ by means of the transformation

(7)
$$X'(S, \theta) = X(S, \theta) \, \exp(iS\theta),$$

and by subsequently expanding $X'(S, \theta)$ into plane-wave solutions in S. However $X(S, \theta)$ is a single valued function of the poloidal angle θ. This periodicity reintroduces an S-dependence through the condition

(8)
$$X'(S, \theta + 2\pi) = X'(S, \theta) \, \exp(i2\pi S).$$

A solution to this problem was found in almost the same months by different researchers, including the present author in collaboration with Dr. T.J. Schep (see refs. [5-8]). Setting aside differences in their formulation, all the solutions that were presented relied on introducing an "extended" poloidal variable $\hat{\theta}$ ranging from $-\infty$ to $+\infty$, and an extended mode amplitude $\hat{X}(\hat{\theta})$ from which the physical amplitude $X(S, \theta)$ can be reconstructed through a summation procedure.

The approach of ref. [5], which is described in detail in ref. [9], is based on the observation that the dependence on S introduced by the exponential factor in (8) is periodic with period one. Then, solutions for $X'(S, \theta)$ can be sought that, aside for a phase factor, are periodic in S with period one.

These are obtained by writing

(9)
$$X'(S, \theta) = \int\limits_{-\pi}^{+\pi} \frac{d\alpha}{2\pi} \, X'_\alpha(S, \theta) \, \exp(iS\alpha),$$

with

(10)
$$X'_\alpha(S, \theta) \, \exp(iS\alpha) = \sum_{p=-\infty}^{+\infty} X'(S + p, \theta) \, \exp(-ip\alpha),$$

where

(11)
$$X'_\alpha(S + 1, \theta) = X'_\alpha(S, \theta)$$

and α, $-\pi < \alpha \leq \pi$, plays the role of a radial wave number. The S-periodic amplitude X'_α is subsequently expanded in a Fourier series

(12)
$$X'_\alpha(S, \theta) = \sum_{m=-\infty}^{+\infty} X_{\alpha,m}(\theta) \, \exp(-i2\pi mS).$$

When (8) is imposed, the following relationship between the Fourier coefficients $X_{\alpha,m}(\theta)$ is found

$$(13) \qquad X_{\alpha,m}(\theta) = X_{\alpha,0}(\theta + 2\pi m) \equiv \hat{X}_{\alpha}(\hat{\theta}),$$

finally leading to the "extended poloidal variable" representation

$$(14) \qquad X'_{\alpha}(S,\theta) = \sum_{m=-\infty}^{+\infty} \hat{X}_{\alpha}(\hat{\theta}) \, \exp(-i2\pi mS),$$

with $\hat{\theta} = \theta + 2\pi m$ the extended poloidal variable.

Equation (14) can be inverted in the form

$$(15) \qquad \hat{X}_{\alpha}(\hat{\theta}) = \int_{-1/2}^{1/2} dS \, X'_{\alpha}(S,\theta) \, \exp(i2\pi mS).$$

Combining (15) and (10) with (7), and (14) and (9) with (7) we have

$$(16) \qquad \hat{X}_{\alpha}(\hat{\theta}) = \int_{-\infty}^{+\infty} dS \, X(S,\theta) \, \exp[iS(\hat{\theta} - \alpha)]$$

and

$$(17) \qquad X(S,\theta) = \sum_{m=-\infty}^{+\infty} \int_{-\pi}^{\pi} \frac{d\alpha}{2\pi} \, \hat{X}_{\alpha}(\hat{\theta}) \, \exp[-iS(\hat{\theta} - \alpha)].$$

A corresponding representation can be obtained in terms of an extended S-variable by expanding the θ-periodic amplitude $X_{\alpha}(S,\theta) \equiv X'_{\alpha}(S,\theta) \, \exp(-iS\theta)$ in poloidal harmonics and subsequently using Eq. (11) to find a relationship analogous to (13) between the poloidal Fourier coefficients. The resulting expressions are

$$(18) \qquad X_{\alpha}(S,\theta) = \sum_{\ell=-\infty}^{+\infty} \overline{X}_{\alpha}(\overline{S}) \, \exp(i\ell\theta),$$

with $\overline{S} = S + \ell$ the extended S variable, and

$$(19) \qquad \overline{X}_{\alpha}(\overline{S}) = \int_{-\pi}^{+\pi} \frac{d\theta}{2\pi} \, X_{\alpha}(S,\theta) \, \exp(-i\ell\theta).$$

It is easy to verify that the representation in the extended poloidal amplitude and that in the extended S-variable are related by the Fourier transformation

$$(20) \qquad \overline{X}_\alpha(\overline{S}) = \int_{-\infty}^{+\infty} \frac{d\hat{\theta}}{2\pi} \, \hat{X}_\alpha(\hat{\theta}) \, \exp(-i\overline{S}\hat{\theta}),$$

and

$$(21) \qquad \hat{X}_\alpha(\hat{\theta}) = \int_{-\infty}^{+\infty} d\overline{S} \, \overline{X}_\alpha(\overline{S}) \, \exp(i\overline{S}\hat{\theta}).$$

Furthermore the norm is preserved as

$$(22) \quad \int_{-1/2}^{+1/2} dS \int_{-\pi}^{+\pi} \frac{d\theta}{2\pi} \, |X_\alpha(S,\theta)|^2 = \int_{-\infty}^{+\infty} d\overline{S} \, |\overline{X}_\alpha(\overline{S})|^2 = \int_{-\infty}^{+\infty} \frac{d\hat{\theta}}{2\pi} \, |\hat{X}_\alpha(\hat{\theta})|^2.$$

Representations (14) and (18) associate a function of a single variable on the interval $-\infty, +\infty$ to the functions $X'_\alpha(S,\theta)$ and $X_\alpha(S,\theta)$. When applied to the dispersion equation (6) they reduce it into an ordinary differential equation which includes the periodicity conditions.

The presence of the trigonometric terms in θ in Eq. (6) makes the use of the extended poloidal representation more suitable. Then, for each α-component, we obtain

$$\frac{\partial}{\partial\hat{\theta}} \, [1 + s^2(\hat{\theta} - \alpha)^2] \, \frac{\partial}{\partial\hat{\theta}} \, \hat{X}_\alpha(\hat{\theta}) + \Gamma[\cos\hat{\theta} + s(\hat{\theta} - \alpha) \, \sin\hat{\theta}] \, X_\alpha(\hat{\theta})$$

$$(23)$$

$$+\Omega_\alpha^2 [1 + s^2(\hat{\theta} - \alpha)^2] \, \hat{X}_\alpha(\hat{\theta}) = 0.$$

Equation (23) is to be solved for the eigenvalue Ω_α^2 with the condition that for $|\hat{\theta}| \to \infty$, $\hat{X}_\alpha(\hat{\theta})$ vanishes sufficiently rapidly* for the inverse transformation (15) to converge for all S.

The linear stability condition against these "ballooning" excitations is obtained by requiring that an initial perturbation does not grow (exponentially) i.e. by imposing that the eigenvalue Ω_α^2 is positive for all values of α. This leads to an instability threshold of the form $\Gamma = \Gamma_{th}(s)$, which expresses the balance between the combined destabilising effect of the pressure gradient and of the magnetic curvature, Γ, and the opposing effect of the magnetic shear s.

---

* Weaker conditions can be adopted when this method is extended [10] to the description of excitations characterised by the presence of a boundary layer.

4. - Extended variable representations

The extended variable representations can be conveniently reinterpreted by adopting a group theoretical formalism. First we notice that the differential operators $\partial/\partial\theta + iS$ and $\partial/\partial S$ in (6) satisfy the commutation relation of the Heisenberg algebra [11]

$$(24) \qquad \left[\frac{\partial}{\partial S}, \frac{\partial}{\partial\theta} + iS\right] = i.$$

This is the same symplectic condition [12] that is satisfied by a coordinate and its conjugate momentum in phase space, which hints directly at an interpretation for the Fourier relationships (19) and (20). We denote the three-dimensional Lie algebra generated by the operators $\partial/\partial S$, $\partial/\partial\theta + iS$ and i by L.

Transformation (7), leads to an equivalent realisation of this algebra in terms of the operators $\partial/\partial S - i\theta$, $\partial/\partial\theta$ and i. We denote the algebra generated by the latter operators by L'. Then we find

$$(25) \qquad [L, L'^*] = [L', L^*] = 0,$$

where a star denotes complex conjugation. We are thus led to consider the invariance properties of the dispersion equation (6) under the action of the group H generated by L'^*, i.e. by the operators $\partial/\partial S + i\theta$, $\partial/\partial\theta$ and $-i$. Its action on the mode amplitude $X(S, \theta)$ is defined by

$$(26a) \qquad \Lambda_1 \ X(S, \theta) = X(S + \lambda_1, \theta) \ \exp(i\lambda_1\theta)$$

$$(26b) \qquad \Lambda_2 \ X(S, \theta) = X(S, \theta + \lambda_2)$$

$$(26c) \qquad \Lambda_3 \ X(S, \theta) = X(S, \theta) \ \exp(-i\lambda_3)$$

with Λ_1, Λ_2 and Λ_3 elements of the one dimensional subgroups generated by $\partial/\partial S + i\theta$, $\partial/\partial\theta$ and $-i$ respectively, and $\lambda_1, \lambda_2, \lambda_3$ the parameters that label the transformations.

Due to its θ-dependent terms, the dispersion equation (6) is invariant only under the subgroup of H defined by the condition that $\lambda_2/2\pi$ be an integer. In addition, for the mode amplitude to remain periodic in θ, we must further reduce the invariance group to integer values of λ_1. We are thus left with the discrete subgroup G of H generated by the transformations corresponding to $\theta \to \theta + 2\pi$ and to $S \to S + 1$.* Since G is Abelian, we can expand the mode amplitude $X(S, \theta)$ into common eigenfunctions of the transformations $X(S, \theta) \to X(S, \theta + 2\pi)$ and $X(S, \theta) \to X(S + 1, \theta) \ \exp(i\theta)$. Periodicity in θ requires that the eigenvalue under the first transformation be equal to one. The eigenvalue of the second transformation can be written as

---

* Since these transformations commute we do not need the constant phase transformations of H.

$\exp(i\alpha)$, where α coincides with the "radial" mode number in Eq. (9) and labels the irreducible unitary representations** [11] of G on the space of the functions $X_\alpha(S,\theta) \exp(iS\alpha)$. The transformation (7) leads to an equivalent representation of G on the functions $X'_\alpha(S,\theta) \exp(iS\alpha)$.

The extended variable representations (14) and (18) are thus simply a consequence of the restriction of the invariance group from H to G and arise from the decomposition of irreducible representations of H into irreducible representations of G. The possibility of employing two different extended variable representations stems from the fact that the group H is not Abelian. In the case of Eq. (18) the starting point is the set of the (periodic) eigenfunctions of the transformations (26b) generated by the operator $\partial/\partial\theta$. The restriction of the invariance group couples these eigenfunctions, leading to the summation over ℓ, but, since the subgroup G is Abelian, allows for the simultaneous implementation of the discrete invariance $S \to S+1$, which introduces a relationship between the Fourier coefficients. In the case of Eq. (14) (rewritten in terms of the amplitude $X_\alpha(S,\theta)$ instead of $X'_\alpha(S,\theta)$) the starting point is the set of the eigenfunctions of the transformations (26a) generated by the operator $\partial/\partial S + i\theta$. The restricted invariance results now in the coupling between eigenfunctions with eigenvalues differing by 2π times an integer number and in the relationship (13) between their coefficients.

5. - Representation product

The extended variable representation can be generalised to the product of the amplitudes of the excitations. Let $X_\alpha(S,\theta) \exp(iS\alpha)$ and $Y_{\alpha'}(S,\theta) \exp(iS\alpha')$ belong to the α and to the α' representations of G respectively. The product $Z_{\alpha''}(S,\theta) = X_\alpha(S,\theta) Y_{\alpha'}(S,\theta)$ satisfies the periodicity conditions

$$(27) \qquad Z_{\alpha''}(S,\theta + 2\pi) = Z_{\alpha''}(S,\theta),$$

and

$$(28) \qquad Z_{\alpha''}(S+1,\theta) = Z_{\alpha''}(S,\theta) \exp(-i2\theta),$$

where $\alpha'' = \alpha + \alpha'$. Thus $Z_{\alpha''}(S,\theta) \exp(iS\alpha'') = X_\alpha(S,\theta) Y_{\alpha'}(S,\theta) \exp[iS(\alpha + \alpha')]$ belongs to the irreducible representation labelled by $\langle\alpha''\rangle = \alpha'' - 2\pi k$ (with $k = 0, \pm 1$ such that $-\pi < \langle\alpha''\rangle \le \pi$) of the group $G^{(2)}$ generated by the transformations $Z(S,\theta) \to Z(S,\theta + 2\pi)$ and $Z(S,\theta) \to Z(S+1,\theta) \exp(i2\theta)$.

The group $G^{(2)}$ can be seen as a subgroup of the group G_2. This is isomorphic to G and its action is obtained from that of G by substituting $2S$

** An ambiguity in denominations can arise here. We recall that the representation of the group action on a linear space, and the representation of a function of two variables in terms of a function of a single extended variable are different mathematical objects.

for S. The group $G^{(2)}$ is obtained from G_2 by restricting the transformations $2S \rightarrow 2S + p$ to even values of p.

The group G_2 acts on functions $K(2S, \theta)$ periodic in θ. These can be expanded into components belonging to irreducible representations of G_2 which can be expressed as functions of a single extended variable following a procedure analogous to that developed in the previous two sections. The new radial scale length is half that of the group G and corresponds to the distance between the mode rational surfaces of perturbations with toroidal number $2n$, as consistent with the multiplication of the mode amplitudes in Eqs. (2) and (5).

The generalisation of the extended variable representation to quadratic terms in the excitation amplitude is then obtained by finding the relationship between the eigenfunctions of $G^{(2)}$ and those of G_2. Since $G^{(2)}$ is a subgroup of G_2, this relationship must involve the coupling of (two) eigenfunctions of G_2 belonging to different representations. It is easily seen that

$$(29) \qquad Z_{\alpha''}(S, \theta) \, \exp(iS\alpha'') = K_\beta(2S, \theta) \, \exp(i2S\beta)$$
$$+ \, K_{\beta \pm \pi}(2S, \theta) \, \exp[i2S(\beta \pm \pi)],$$

where $K_\beta(2S, \theta + 2\pi) = K_\beta(2S, \theta)$, $K_\beta(2S + 1, \theta) = K_\beta(2S, \theta) \, \exp(-i\theta)$, $\beta = \alpha''/2 = (\alpha + \alpha')/2$, $-\pi < \beta \le \pi$ and the sign in $(\beta \pm \pi)$ must be chosen such that $-\pi < \beta \pm \pi \le +\pi$. Equation (29) is inverted in the form

$$(30) \qquad K_\beta(2S, \theta) = \frac{1}{2} \, [Z_{\alpha''}(S, \theta) + Z_{\alpha''}(S + 1/2, \theta) \, \exp(i\theta)],$$

and

$$(31) \quad K_{\beta \pm \pi}(2S, \theta) = \frac{1}{2} \, [Z_{\alpha''}(S, \theta) - Z_{\alpha''}(S + 1/2, \theta) \, \exp(i\theta)] \, \exp(\mp i2\pi S).$$

The explicit form of the generalised extended variable representations can be derived by inserting the extended variable representations of $X(S, \theta)$, of $Y(S, \theta)$ and of $K(2S, \theta)$ into Eq. (9).

In the following part of this section we list a few groups of relevant formulae:

i) *Addition rule for the radial mode number*

If $Z^i(S, \theta) \equiv X^i(S, \theta) \, Y^i(S, \theta) = X(S, \theta) \, Y(S, \theta) \, \exp(i2S\theta)$, then

$$(32) \qquad Z'(S, \theta) = \int\limits_{-\pi}^{+\pi} \frac{d\gamma}{2\pi} \, K'_\gamma(2S, \theta) \, \exp(i2S\gamma),$$

where $K'_\gamma(2S, \theta) = K_\gamma(2S, \theta) \exp(i2S\theta)$, $K'_\gamma(2S + 1, \theta) = K'_\gamma(2S, \theta)$ and

$$(33) \qquad K'_\gamma(2S, \theta) = \int\limits_{-\pi}^{+\pi} \frac{d\alpha}{2\pi} \, [X'_{\gamma+\alpha}(S, \theta) \, Y'_{\gamma-\alpha}(S, \theta)$$

$$+ X'_{\gamma+\alpha}(S + 1/2, \theta) \, Y'_{\gamma-\alpha}(S + 1/2, \theta)].$$

Here $X'_{\gamma+\alpha} = X'_{\langle\gamma+\alpha\rangle} \exp(-i2k\pi)$, $Y'_{\gamma-\alpha} = Y'_{\langle\gamma-\alpha\rangle} \exp(-i2k\pi)$, and $\langle\gamma \pm \alpha\rangle = \gamma \pm \alpha - 2k\pi$, with $k = 0, \pm 1$, such that $-\pi < \langle\gamma \pm \alpha\rangle \leq \pi$. The amplitudes $X'_{\langle\gamma+\alpha\rangle}(S, \theta)$ and $Y'_{\langle\gamma-\alpha\rangle}(S, \theta)$ are obtained from $X'(S, \theta)$ and $Y'(S, \theta)$ according to Eq. (9).

ii) *Product of extended poloidal variable representations*

If $Z'_{\alpha''}(S, \theta) = X'_\alpha(S, \theta) \, Y'_{\alpha'}(S, \theta)$, then

$$(34) \quad Z'_{\alpha''}(S, \theta) = \sum\limits_{m=-\infty}^{+\infty} [\hat{K}_\beta(\hat{\theta}) + \hat{K}_{\beta\pm\pi}(\hat{\theta}) \, \exp(\pm i2\pi S)] \, \exp(-i2\pi m2S),$$

where $\beta = \alpha''/2$ and $\hat{K}_\beta(\hat{\theta})$, with $\hat{\theta} = \theta + 2\pi m$, is the extended poloidal variable representation of $K'_\beta(2S, \theta) = K_\beta(2S, \theta) \exp(i2S\theta)$, with $K_\beta(2S, \theta)$ from Eq. (30). Equation (34) is inverted by

$$(35) \qquad \hat{K}_\beta(\hat{\theta}) = \int\limits_{-1/2}^{1/2} dS \, Z'_{\alpha''}(S, \theta) \, \exp(i2\pi m2S),$$

and

$$(36) \qquad \hat{K}_{\beta\pm\pi}(\hat{\theta}) = \int\limits_{-1/2}^{1/2} dS \, Z'_{\alpha''}(S, \theta) \, \exp[i2\pi \, (m \mp 1/2) \, 2S].$$

The relationship between the extended poloidal variable representation of $Z'_{\alpha''}(S, \theta)$ and those of $X'_\alpha(S, \theta)$ and $Y'_{\alpha'}(S, \theta)$ is expressed by the convolution product

$$(37) \qquad \hat{K}_\beta(\hat{\theta}) = \sum\limits_{p=-\infty}^{+\infty} \hat{X}_\alpha(\hat{\theta}^+) \, \hat{Y}_{\alpha'}(\hat{\theta}^-),$$

$$(38) \qquad \hat{K}_{\beta\pm\pi}(\hat{\theta}) = \sum\limits_{p=-\infty}^{+\infty} \hat{X}_\alpha(\hat{\theta}^+ \mp 2\pi) \, \hat{Y}_{\alpha'}(\hat{\theta}^-),$$

where $\hat{\theta} = \theta + 2\pi m$, $\hat{\theta}^+ = \hat{\theta} + 2\pi p = \theta + 2\pi(m+p)$, and $\hat{\theta}^- = \hat{\theta} - 2\pi p = \theta + 2\pi(m-p)$.

iii) *Product of extended radial variable representations*

If $Z_{\alpha''}(S,\theta) = X_\alpha(S,\theta)\, Y_{\alpha'}(S,\theta)$, then

$$(39) \quad Z_{\alpha''}(S,\theta) = \sum_{\ell=-\infty}^{+\infty} [\overline{K}_\beta(\overline{2S}) + (-)^\ell\, \overline{K}_{\beta\pm\pi}(\overline{2S})\, \exp(\pm i\overline{2S}\pi)]\, \exp(i\ell\theta),$$

where $\beta = \alpha''/2$ and $\overline{K}(\overline{2S})$, with $\overline{2S} = 2S + \ell$, is the extended radial variable representation of $K_\beta(2S,\theta)$ from Eq. (30). Equation (39) is inverted by

$$(40) \quad \overline{K}_\beta(\overline{2S}) = \frac{1}{2}\int_{-\pi}^{+\pi} \frac{d\theta}{2\pi}\, [Z_{\alpha''}(S,\theta) + Z_{\alpha''}(S+1/2,\theta)\,\exp(i\theta)]\,\exp(-i\ell\theta),$$

and

$$(41) \quad \overline{K}_{\beta\pm\pi}(\overline{2S}) = \frac{1}{2}\int_{-\pi}^{+\pi} \frac{d\theta}{2\pi}\, [Z_{\alpha''}(S,\theta)$$
$$- Z_{\alpha''}(S+1/2,\theta)\,\exp(i\theta)]\,\exp(\mp i2\pi S - i\ell\theta).$$

The relationship between the extended radial variable representation of $Z_{\alpha''}(S,\theta)$ and those of $X_\alpha(S,\theta)$ and $Y_{\alpha'}(S,\theta)$ is expressed by the convolution product

$$(42) \quad \overline{K}_\beta(\overline{2S}) = \frac{1}{2}\sum_{p=-\infty}^{+\infty} [\overline{X}_\alpha(\overline{S}^+)\, \overline{Y}_{\alpha'}(\overline{S}^-)$$
$$+ \overline{X}_\alpha\left(\overline{S}^+ + 1/2\right)\, \overline{Y}_{\alpha'}\left(\overline{S}^- - 1/2\right)]$$

$$\overline{K}_{\beta\pm\pi}(\overline{2S}) = \frac{1}{2}\sum_{p=-\infty}^{+\infty} [\overline{X}_\alpha(\overline{S}^+)\, \overline{Y}_{\alpha'}(\overline{S}^-)$$
$$(43) \quad - \overline{X}_\alpha\left(\overline{S}^+ + 1/2\right)\, \overline{Y}_{\alpha'}\left(\overline{S}^- - 1/2\right)]\,\exp(\mp i2\pi S)$$

where $\overline{2S} = 2S + \ell$, $\overline{S}^+ = S + \ell/2 + p$ and $\overline{S}^- = S + \ell/2 - p$.

The functions $\overline{K}_\beta(\overline{2S})$ and $\hat{K}_\beta(\hat{\theta})$, and $\overline{K}_{\beta\pm\pi}(\overline{2S})$ and $\hat{K}_{\beta\pm\pi}(\hat{\theta})$, are related by a Fourier transformation analogous to that in Eqs. (20) and (21).

Conclusions

We have shown that the extended variable representation, employed in the plasma physics literature for the description of short-wavelength excitations in a toroidal axisymmetric plasma configuration, is related to the irreducible representation of an Abelian subgroup G of the Heisenberg group H. The Lie algebra of H commutes with the differential operators in the dispersion equation of the excitations. The subgroup G corresponds to discrete transformations in the poloidal and in the radial directions. The reduced invariance from H to G arises from the poloidal modulation of the coefficients of the dispersion equation and from the requirement that the excitation amplitude be a single valued function of the poloidal angle.

The action of G can be suitably extended so as to include products of the amplitudes of excitations with equal (or multiple) toroidal mode numbers. The expansion of these products into irreducible representations allows us to generalise the formalism of the extended variable representation to the description of the non linear evolution of the excitations.

In this paper we have explicitly developed this formalism in the case of quadratic terms. It can be extended to the general case along the same lines. In a future paper these results will be applied to the solution of the non linear dynamical equations for the magnetic excitations described in the introduction.

Acknowledgement

This paper was written on the occasion of the 70th birthday of Prof. L.A. Radicati di Brozolo whose teaching first, inspiration and guidance afterwards, and criticism always, I valued greatly over the years. Indeed it was with some guilt that I realised, while preparing this paper, that a formalism I had contributed to develop some ten years ago, could have been formulated in a more transparent, and now I see, more complete way, had I used some of the mathematical ideas he introduced me to.

REFERENCES

[1] S. CHANDRASEKHAR, *Hydrodynamic and Hydromagnetic Stability*, Clarendon Press, Oxford, 1961.

[2] C.M. BENDER - S.A. ORSZAG, *Advanced Mathematical Methods for Scientists and Engineers*, McGraw-Hill Book Company, New York, 1978.

[3] B. COPPI, Phys. Rev. Lett. **39**, 939 (1977).

[4] B. COPPI - J. FILREIS - F. PEGORARO, Ann. Phys. **121**, 1 (1979).

[5] F. PEGORARO - T.J. SCHEP, presented at the International Atomic Energy Agency Advisory Group meeting on Thermal Conduction and Transport, Kiev (1977),

also in *Plasma Physics and Controlled Nuclear Fusion Research*, Innsbruch 1978 (International Atomic Energy Agency, Vienna 1978), Vol. 1, p. 507.

[6]   J.W. CONNOR - R.J. HASTIE - J.B. TAYLOR, Phys. Rev. Lett. **40**, 396 (1978) and in Proc. R. Soc. London Ser. **A365**, 1 (1979).

[7]   Y.C. LEE - J.W. VAN DAM in Proceedings of the Finite Beta Theory Workshop, edited by B. Coppi and W. Sadowski (U.S. Department of Energy, Washington, D.C. 1977) p. 93.

[8]   A.H. GLASSER, ibid. p. 55.

[9]   F. PEGORARO - T.J. SCHEP, Phys. of Fluids **24**, 478 (1981).

[10] F. PEGORARO - T.J. SCHEP, Plasma Phys. Controlled Fusion **28**, 697 (1986).

[11] R. HERMAN, *Lie Groups for Physicists*, W.A. Benjamin, Inc. New York (1966).

[12] H. GOLDSTEIN, *Classical Mechanics*, 2nd Ed., Addison-Wesley Publ., Reading, Massachusetts (1980).

Broken Symmetries

RUDOLF PEIERLS

Symmetries, and the associated conservation laws have been used in physics for a long time, but their generality and importance has been emphasised only during the last thirty years or so. Indeed when I wrote in 1955 a book summarising the whole of theoretical physics in non-technical language I could omit any mention of the term "symmetry". Since then the study of symmetry in physics has become a definite and important field, to which Luigi Radicati has made important contributions.

Broken symmetries have also occurred in many branches of physics, but again the name "broken symmetry" has become current only fairly recently, and with it the realisation of its importance and generality. This note does not intend to list all the broken symmetries now known or suspected in physics, but to use a few typical examples to illustrate their nature.

Perhaps the oldest example of a broken symmetry is the lack of isotropy in our everyday environment. Here, in fact, the breakdown, i.e. the difference between vertical and horizontal directions, is so great that our intuition does not recognise the underlying isotropy. I am not sure who first described the isotropy of physical laws. In this case the origin of the lack of isotropy is clear: The laws of physics are isotropic, but the particular solution applicable to our earthbound situation is not. It is then proper to distinguish the general laws from the particular event. I shall refer to this type of broken symmetry as a symmetry broken in the event. Gravity and other non-isotropic effects are scale-dependent and on the atomic scale, with which much of modern physics is concerned, they are a very small perturbation.

Relativity teaches us that isotropy is part of a much wider symmetry, including the Lorentz group. This is again broken since the cosmic microwave background can be isotropic in only one Lorentz frame. There is therefore a preferred frame of reference. This is again perfectly compatible with Lorentz-invariant equations, and, as in the previous example, we must blame the breakdown on the event, in this case our universe. But one feels a little uncomfortable about this: In the first case our location on the earth's surface is clearly an accidental event; there are other place on the earth, where the vertical direction is different from ours, and the laws of physics describe them all. But we know of only one universe; if there are any others we could presumably never find out about them, so the question of their existence becomes metaphysical. So, at least to the best of our knowledge, of the many solutions which differ by a Lorentz transformation, only one is realised. Does our universe just "happen" to have that particular preferred Lorentz frame?

There are many situations in physics where a symmetry is only approximately valid. For example in nuclear theory charge independence, i.e. the invariance under rotation in isospin space is an approximate symmetry. The reason is that it is a symmetry of the strong interactions, but not of the electromagnetic forces, which in the nucleus are relatively weak, although their cumulative nature increases their importance for heavy nuclei. Isospin symmetry is therefore rather strongly broken for heavy nuclei, but it can still be used to obtain useful results, as was first shown by Radicati.

This situation is called an intrinsically broken symmetry. One is reluctant to postulate its occurrence in too many fundamental laws because it makes their form more complicated than we would like. For this reason the idea of spontaneously broken symmetries is attractive. The spontaneously broken symmetries most often quoted are those of molecules without mirror symmetry, and ferromagnets. The molecules, such as sugar exist in two modifications, one being the mirror image of the other. Each of these by itself violates the conservation of parity. Since the one kind dominates the chemistry of living organisms, all life violates parity. Here the event is some historical accident, which has not yet been described fully. But we are prepared to believe that if there is life on other planets its molecules might be the mirror images of ours.

In principle one modification of a molecule could change into the other by quantum-mechanical tunnelling. The true stationary state of lowest energy would be a symmetric superposition of the two forms. Another state has a slightly higher energy, but this energy difference is very small because it involves the tunnelling probability. A molecule initially of one form would periodically change into the other form and back. However the period of this oscillation is of astronomical length since the frequency is given by the energy difference of the odd and even states, i.e. by the tunnelling rate. So for all practical purposes the tunnelling can be ignored, and the molecule retains its unsymmetric form for all reasonable times.

In the ferromagnet the interactions between the constituents are such that in equilibrium at low temperatures the whole body is magnetised in one direction. This direction is arbitrary, as follows from the isotropy of the underlying laws, but the magnet itself is not isotropic, because of this preferred direction. Here the "event" is the initial direction of magnetisation.

It is worth remembering that this "Schulbeispiel" of a broken symmetry, as it is usually described, resembles a real ferromagnet only crudely, because it takes no account of the magnetic interaction between the spins, and ignores the consequent influence of the crystal lattice. In a real ferromagnet the direction of equilibrium magnetisation is not arbitrary, but confined to a finite number of crystallographically equivalent directions. In addition a real ferromagnet exceeding a certain size will in equilibrium contain different domains with different directions of magnetisation, making the total magnetisation zero. (I have described a perfectly pure crystal without lattice imperfections, otherwise things are even more complicated). However as an analogue the "abstract" ferromagnet of the preceding paragraph is more useful that the real one, so this will be used in the further discussion.

It is again true that in quantum mechanic the state of lowest energy of the abstract ferromagnet would be a symmetric superposition of states with all possible directions of magnetisation, but the energy differences between this state and those of other symmetries are minute, and the time taken to change the direction if we start from a particular approximate direction is enormous. The direction could not change uniformly, because this would mean all spins turning simultaneously, and this would involve a small matrix element to a very large power. Turning the magnetisation of a small part at a time would, because of the resulting misalignment, requires energy, and this would constitute a substantial barrier.

Most of the recent uses of the concept of broken symmetry apply to the vacuum, assumed to have a degenerate ground state. It is therefore instructive to pursue the analogy with ferromagnetism to an abstract ferromagnet which would extend over very large distances, perhaps the whole universe. In that case the state of lowest energy would still be that in which the direction of magnetisation is the same throughout, but this state would be very hard to reach.

Suppose the ferromagnet was divided into two domains, with different orientations of magnetisation. The boundary would have a certain surface energy, but there would be no mechanism for making this disappear in one step since this would involve changing the direction of every spin in one of the two domains, and this is clearly impossible. The only possible change is a displacement of the boundary. The boundary would have a tendency to become plane since this minimises the surface area and hence the surface energy. But once it is straightened out to a plane it could not gain energy by any further motion; it could, in principle shift to another position or direction without loss or gain of energy, but since this would involve flipping an infinite number of spins simultaneously, which has zero probability, or deforming the surface, and this requires energy.

Similarly two or more parallel boundaries will stay put, unless they are very close. If the distance between two boundaries is of the order of the range of the interatomic forces, and if the orientations of the outer domains are identical, energy would be gained by eliminating the intermediate domain. Again this cannot happen by displacing the whole of one boundary; it could happen by a local displacement of a boundary. But this involves curvature, and hence an increase in surface energy, and therefore cannot happen (except by tunnelling) unless the surfaces are close enough for the gain in interaction to outweigh the price paid in surface energy.

Similarly edges between three domains will remain; edges between four domains can disappear if two of the domains have identical orientations. Corners between eight domains will remain unless two of the domains have identical orientations. In other words a domain structure will in general remain indefinitely.

If, as in the case of the real ferromagnet, only discrete orientations occur the picture is unchanged, except that, if the number of orientations is less than eight, (as in the case of a cubic crystal in which the favoured directions are in the (1,0,0) and equivalent directions) intersections between three boundaries, in which eight domains meet, will not survive.

These statements apply directly only to the ferromagnet, but it seems almost certain that they would also apply to a degenerate vacuum, provided there are no non-local interactions contained in the Hamiltonian. One must therefore be prepared to find many domain boundaries in a degenerate vacuum, unless in the formation of the degenerate phase some general law leads to the symmetry being broken in a particular direction. This would imply that the symmetry was broken in the laws, and not as a spontaneous breaking in their solution.

The domain problem is sometimes discussed in terms only of the horizon, i.e. of the question whether different parts of the universe are in communication with each other by actions travelling at most with the speed of light. Evidently this is a necessary condition for complete alignment from an initial domain structure, but the above arguments show that it is not sufficient.

Whether in the absence of complete alignment the existence of domains would produce observable effects has to be decided for each of the models in question.

A different kind of broken symmetry occurs in the Kaluza-Klein five-dimensional form of the equations of general relativity, and in the more recent theories with many more dimensions. Here the underlying symmetry is that of general relativity, by which in particular all coordinates are equivalent. However in the actual geometry of these theories all but four degrees of freedom are "compactified", i.e. become cyclic coordinates of quiter short period. So the general-relativistic invariance becomes a broken symmetry.

Is this an intrinsic or a spontaneous breakdown? If it is intrinsic, i.e. if the basic laws show major deviations from the symmetries underlying the covariant form of the equations, there seems no reason to maintain these symmetries of form, restricting equations to multidimensional tensor form. If, on the other hand, we are dealing with a spontaneously broken symmetry, the question of domains might have to be faced.

It seems that broken symmetries are a fascinating field that, even in its general aspects is far from exhausted.

I.J.R. Aitchison has helped me to claify my thoughts but should not be held responsible for them.

Semiconductor Memory Detectors

P. REHAK (*) - E. GATTI (**)

Abstract

The importance of semiconductor detectors for elementary particle physics and X-ray astronomy is briefly rewiewed. Semiconductor detectors are divided into two groups; i) classical semiconductor diode detectors and ii) semiconductor memory detectors. Principles of newer semiconductor memory detectors are described in more details. The performance of silicon drift detectors, belonging to the group of memory detectors, is reported here.

1. - Introduction

We felt very honored when asked to contribute to the Volume of selected scientific papers dedicated to Prof. Luigi Radicati. At the same time we had trouble relating the topic we intended to discuss with the central content of the Volume which is Symmetry in Nature. Very soon we realized, however, that there are two main connections between the topic of this contribution and Symmetry in Nature.

Modern semiconductor detectors, our topic, are based on the transport of electrons in crystals. Studies of symmetry in Nature started a long time ago with the study of crystal symmetry. Present physics and cosmology theories, based on abstract symmetries have the same mathematical structure as the structure of crystal symmetry.

Secondly, the detectors are being used today in particle physics experiments

(*)   Brookhaven Nat. Lab., Upton N.Y., 11973.
(**) Politecnico di Milano, 32 Piazza Leonardo da Vinci, 20133 Milano, Italy. - This research is also supported by the Italian INFN and CNR.

This manuscript has been authored under contract number DE-AC02-76CH00016 with the U.S. Department of Energy. Accordingly, the U.S. Government retains a non-exclusive, royalty-free license to publish or reproduce the published form of this contribution, or allow others to do so, for U.S. Government purposes.

and in X-ray astronomy. The results of experiments and observations obtained with these detectors are "pictures" showing new higher Symmetries in Nature.

This contribution is organized as follow: in the second section we will identify a class of measurements where the semiconductor detectors are superior to traditional detectors. Thus, there are experiments where the semiconductor detectors provide the best solution for particle detection and measurement.

In the third section we will describe classical semiconductor detectors. In the fourth section we will introduce a newer kind of semiconductor detectors called memory semiconductor detectors. We will stress the principles of operation rather than technical details.

The fifth Section will present the experimental performance of silicon drift detectors as representative of semiconductor memory detectors. The sixth section will try to show future developments in design and use of semiconductor detectors and will conclude this contribution.

2. - New Challenges in Experimental Particle Physics

There has always been a need for detectors providing precise position information in experimental particle physics. Momenta of charged particles are determined from the measurement of curvature of tracks in a magnetic field. Better position resolution improves the measurement of the curvature and makes the determination of the particle momentum more precise. To improve the momentum measurement, however, it is also possible to increase the size of the apparatus and measure the curvature along longer segments of tracks.

The discovery of new quantum flavors years ago brought new requirements for the position resolution of tracking detectors. The life time of particles containing heavy flavors of charm or beauty is so short that in order to see their decay, a position resolution better than several μm is needed. When searching for secondary vertices indicating the decay of these particles there is no substitute for position resolution. Tracking detectors must be located as close as possible to the primary vertices and have the best position resolution in absolute terms.

Semiconductor detectors are well suited as high resolution position detectors. The density of the ionization in semiconductors by a minimum ionizing particle is by four orders of magnitude higher than the densities of ionization in gases at atmospheric pressure. The high linear ionization density allows semiconductor detectors to be much thinner than the gas detectors. There is another advantage of semiconductors for tracking coming from the range of delta rays. In semiconductors the ionization is confined to a μm diameter column. The range of the delta rays in gas is much larger than in semiconductor and the ionization may be spread away from the trajectory of the fast particle. The position information is degraded in the gas detector by the very process of the particle ionization.

It is interesting to note that use of semiconductor detectors for position measurements is relatively recent. The traditional use of semiconductor detector was in the field of nuclear spectroscopy. Conversion of the released energy into signal charge is more efficient in the semiconductors than in glasses. The same released energy produces about ten times more charge in semiconductors than in a gas. The higher number of free charges together with a lower value of Fano factor in semiconductors leads to a smaller relative statistical fluctuation and a better intrinsic energy resolution. Presently semiconductor detectors are used almost exclusively for the low energy X-ray spectroscopy and as detectors in a new field of low energy X-ray astronomy.

3. - Classical Semiconductor Detectors

The cross section of a classical semiconductor detector is shown in Fig. 3.1. As an example, an n-type silicon in form of a thin disk (wafer) is shown. Silicon has a rectifying p^+n junction on the upper face and an nn^+ junction acting as a non-injecting ohmic contact on the lower face. The rectifying junction is reverse biased at a sufficiently high voltage to remove all free charges between the two surface electrodes (complete depletion of the silicon bulk). In the depleted bulk the space charge due to the ionized donors fixed in the silicon lattice is present. The electric field in the bulk is created by this space charge and the surface layer charges on the junctions.

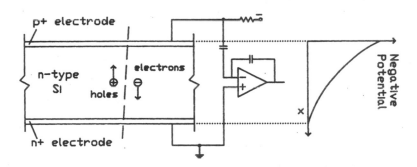

Figure 3.1: a) *Cross section of a classical semiconductor detector.* b) *Negative potential in the bulk of a classical detector.*

The presence of the field do not cause a strong current flow through the detector. Only a small reverse current of both junctions and a small current of carriers thermally generated in the bulk flows through the detector. The depletion and the electric field are essential for the functioning of the semiconductor detector. Electron-hole pairs generated by the radiation move apart and towards the two junctions in the electric field. The motion of the charges within the bulk of the detector induces signal current in the external circuit which includes a charge sensitive preamplifier. The total charge measured by the preamplifier is equal to the charge produced by the ionizing particles in the detector when all electrons and holes reach nn^+ and p^+n junctions respectively.

In the parallel plate geometry of Fig. 3.1 the field is normal to the plates. The negative potential or the potential energy of electrons in the bulk is also shown in Fig. 3.1. The motion of carriers in silicon can be easily visualized with the help of pictures of the negative potential. Electrons move down as small balls without inertia, while holes moves up as bubbles. The charges of both polarities spend minimal time within the detector bulk. The fast removal of all charges produced by ionization from the bulk of the detector is the main feature of classical semiconductor detectors.

Classical semiconductor diode detectors were the first semiconductor detectors developed in 1951 [1]. These detectors are widely use now as

1. X-ray detectors

2. Photodiodes

3. Microstrip Detectors

4. Pad Detectors

5. Some Pixel Detectors etc.

In spite of a very short carrier collection time in the classical semiconductor detectors the total read-out time is usually long. The same electrode geometry of classical semiconductor detectors which provides the fastest removal of charge from the bulk of the detector leads to a large anode capacitance of the detector. A large anode capacitance makes the noise of the preamplifier important. There is no signal amplification in semiconductor detectors and in order to obtain a sufficient signal to noise ratio, the bandwidth of the processing electronics must be limited. The rise time of the overall response is therefore much longer than the collection time of the carriers in the detector. The duration of the processing time rather than the carrier collection time limits the rate capability of the classical semiconductor detectors.

The position resolution in this kind of detectors is obtained simply by division of at least one junction into strips. Each strip is connected to its own preamplifier. Recently introduced double sided strip detectors have both junctions divided into parallel strips, strips on the upper junction running perpendicular to the strips on the lower junction. A very large number of preamplifiers is needed to read the position information from strip detectors.

4. - Semiconductor Memory Detectors

Charges produced by ionizing particle in semiconductor memory detectors are not removed immediately from the detector as was the case in classical detectors. In a memory detector at least one kind of charge carriers (usually electrons) is stored within the volume of the detector. The stored carries are transported in a controlled way in a direction parallel to the wafer surface to a read-out electrode. The position information is retained during the carrier transport. Moreover, the capacitance of the read-out electrode can be very small leading to a low noise contribution of the preamplifier.

The examples of semiconductor memory detectors are:

1. Surface and Buried Channel Charge Coupled Deviced (CCDs)

2. Fully Depleted CCDs

3. Semiconductor Drift Detectors

4. Small Capacitance X-ray Detectors and Photodiodes

5. Some Pixel Detectors, etc.

Two following subsections will describe principles of operation of fully depleted CCDs and semiconductor drift detectors.

4.1 - Fully Depleted Charge Coupled Devices

The electrode structure of any of semiconductor memory detectors is more complicated than the structure of classical detectors.

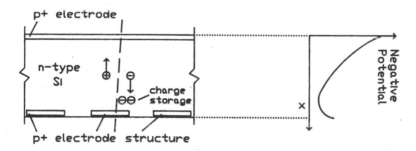

Figure 4.1: a) *Cross section of a Fully Depleted Charge Coupled Device illustrating general features of memory detectors.* b) *Negative potential in the bulk of a memory semiconductor detector.*

Fig. 4.1 shows a cross section of fully depleted Charge Coupled Device (CCD), a very good example of a memory semiconductor detector. There are rectifying junctions on both upper and lower faces of the wafer. The wafer is supposed again to be of n-type semiconductor and is completely depleted of mobile electrons. The space charge of the ionized donors in the silicon lattice together with the voltages applied to the upper and to the lower junctions produce an electric field. The negative potential of the electric field in a cross-section through the bulk of a fully depleted CCD is shown on the right hand side of Fig. 4.1.

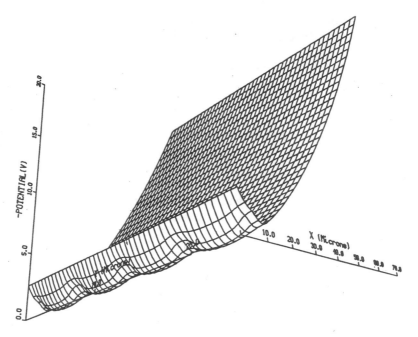

Figure 4.2: *Negative potential in a Fully Depleted CCD in two dimensions. Local minima confine signal electrons and define individual pixels.*

Let us assume that a minimum ionizing particle crossed the CCD and produced a column of electron-hole pairs. Holes move towards one of the rectifying junctions, but electrons move to the point of minimum negative potential where they are stored. The potential in a real fully depleted CCD is engineered in such a way that electrons are confined in this region and cannot diffuse in any direction. The picture of electron confinement in two dimension is shown in Fig. 4.2. A fully depleted CCD contains many (of the order of 10^6) such regions or elementary memory units (pixels) arranged into a two-dimensional matrix. Electrons produced by the mentioned minimum ionizing particle are stored in one or in a few pixels while the others are generally empty.

There is no external circuit and no preamplifier shown in Fig. 4.1. The signal produced by the motion of electrons falling in one pixel and holes moving to the surface junction is not read-out. There is an electrode structure, called a peristaltic pump, implemented on the lower face of a fully depleted CCD. When a sequence of appropriate voltages is applied to the shown electrode structure, the charges stored in the bulk are shifted from one pixel to the next one without any loss of electrons. Finally electrons are pumped into the last memory cell where their presence affect a read-out preamplifier. The pumping of electrons from the bulk to the reading electrode also cleans all pixels from any background electrons produced by thermal generation in the bulk or at junctions.

There is only one preamplifier for the whole CCD. The position of the passing particle is measured when the corresponding pixel is read. The capacitance of the ready-out cell of the CCD is very small so the preamplifier noise is kept at the minimum and CCD have a very good signal to noise ratio. These devices are ideally suited for the X-ray astronomy because they combine a good energy resolution of semiconductors implemented with a low noise read-out and the position resolution defined by the size of a pixel.

The CCDs are memory detectors which store the signal electrons for the longest amount of time. On the other extreme we have semiconductor drift detector where the storage time is the shortest.

4.2 - Silicon Drift Detectors

A perspective view of the drift detector [2] is shown in Fig. 4.3. The volume of the detecotr is fully depleted of mobile electrons. The field created by the remaining fixed charges confines electrons generated by a ionizing particle in a buried potential channel. An electrostatic field parallel to the surface is superimposed. This field transports electrons created by a particle passage along the buried channel toward a collecting electrode. The transit time of electrons inside the detector measures the distance of an incident particle from the anode.

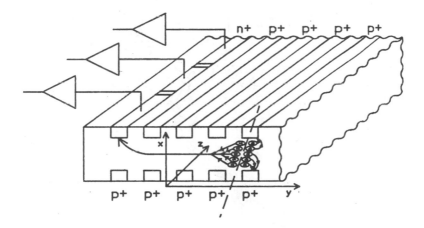

Figure 4.3: *Perspective view (not to scale) of a semiconductor drift detector. Electrons created by an ionizing particle are transported long distances parallel to the detector surface. The anode is divided into short segments to measure the second coordinate.*

The exact shape of the electric potential in drift detectors requires some explanation. The cross-section of the negative potential shown in Fig. 4.1 is the solution of Poisson's equation in one dimension

(4.1)
$$\frac{d^2\phi(x)}{dx^2} = \frac{\rho(x)}{\epsilon}$$

where $\phi(x)$ is the negative potential, x is coordinate perpendicular to the surface plates of the detector, ρ is space charge density and ϵ is the absolute dielectric constant of silicon. Potentials applied at the two faces determine the two integration constants. Negative potential showed in Fig. 3.1 and in Fig. 4.1 are parabolas corresponding to the solution of Eq. (4.1) in a simple case where $\rho(x) = \rho$ is independent of x.

While the electric field of a classical diode detector can be understood in one dimensional model, the electric field of drift (memory) detectors must be described at least in two dimensions as was already obvious from Fig. 4.2. A solution of Poisson's equation in

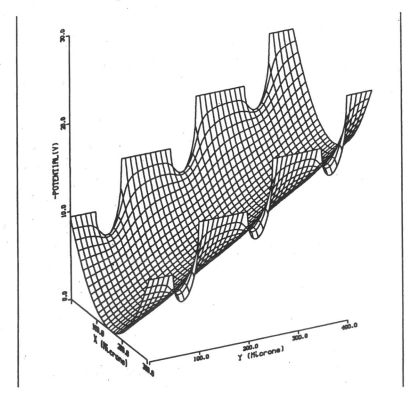

Figure 4.4: *Negative electric potential (potential energy of electrons) in a linear drift detector. Surfaces of detector are in planes* $x = 0$ *and* $x = 300\mu m$. *Reverse biased* p^+ *electrodes have the potential imposed by a voltage divider.*

two dimensions:

(4.2)
$$\frac{d^2\phi(x,y)}{dx^2} + \frac{d^2\phi(x,y)}{dy^2} = \frac{\rho(x,y)}{\epsilon}$$

is shown in Fig. 4.4. There the negative potential in the drift region of the detector is displayed. Poisson's equation in two dimensions Eq. (4.2) is satisfied by adding to the parabolic solution in x of Eq. (4.1) a linear term $U_y = E_d \times y$, where E_d is the drift field in the drift direction y. To realize the potential shown on Fig. 4.4 we have to impose the same linear y dependance of the potential on the surface of the detector.

Fig. 4.4, shows a relatively strong electric field at the surfaces of the detector. Any electron present on the surface would be injected into the valley. To prevent the electron injection, electrodes at both surfaces must be p^+ type. (p^+n rectifying junctions) This is a common requirement for all fully depleted memory detectors.

To prevent a large flow of current along the surfaces, the p^+ electrodes are segmented into strips. The inclined parabolic cylinder of potential energy for electrons (negative potential) contains electrons within the bulk and transports them to the anode.

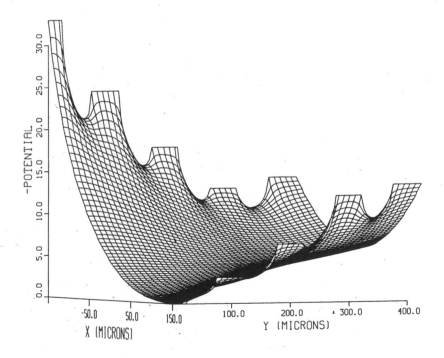

Figure 4.5: Negative electric potential in the anode region of a drift detector

The anode region of the detector is shown in Fig. 4.5. Potentials are applied at the surfaces in such a way to bend the bottom of the valley towards the surface collecting anode. The potential energy of the anode as shown is at the lowest potential energy in the active region of the drift detector. All electrons generated in the bulk arrive at the anode. Thus the detector is maintained in full depletion during the operation.

The electrodes, apart from imposing the desired surface potential on the drift detector, act as an electrostatic shield for drifting electrons. Electrons induce the signal charge only when they arrive to a close proximity of the anode. The drift time is the difference between the arrival time sensed by the anode and the time passage of the particle detected by other detectors of the experiment or provided by the accelerator. The information about the coordinate perpendicular to the drift direction is not lost during the transport in a linear drift detector. Fig. 4.3 shows that the anode divided into short segments provides also the position information in the direction perpendicular to the drift. The surface of individual anodes is small and independent of the size of the detector. Thus the anode capacitance is very small as is the case of all memory detectors.

Let us summarize differences and similarities between fully depleted CCDs and drift detectors. In semiconductor drift detectors the storage time is much shorter than in fully depleted CCDs. Drift detectors convert the position information into a drift time proportional to the distance between the crossing point of the particle and the anode location. In CCDs the time to position relationship is controlled by the clock frequency and by the chosen sequential order of reading pixels in the two-dimensional matrix.

5. - Performance of Silicon Drift Detectors

The amplifier noise is not the only limitation in position resolution of drift detectors [3]. The detectors leakage current and the statitical fluctuations of the pulse shape at the anode also limits the resolution. Electrons arriving at the anode spread because of diffusion and the electrostatic repulsion during the drift time. The sigma of diffusion for an individual electron is relatively large. ($150\mu m$ for a drift time of about $1\mu s$) The centroid of electron pulse has a sigma reduced by $\sqrt{20000}$. We can use the centroid of all electrons to measure the position of particles. The electric field in the drift detector is such that electrons created contemporary at a given y-coordinate, independently of the x-coordinate, arrive at the anode, on the average, at the same time. The achievable position resolution is a few μm for a drift distance of several mm [4].

5.1 - Position resolution of single anode drift detectors

Tests were done on a single anode linear drift detector. The length of the anode was $1cm$ and the maximum drift distance $4mm$. The thickness of the silicon wafer was about $300\mu m$ for all detectors reported here. The picture of the linear detector is in Fig. 2 of

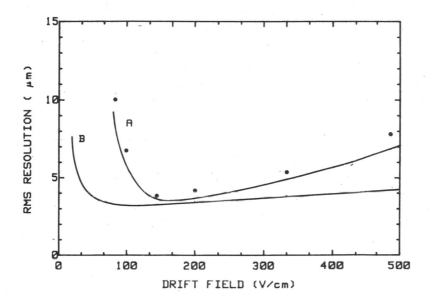

Figure 5.1: Position resolution of a single anode drift detector as a function of drift field. Drift distance is 3.85 mm. Measurement was done in a laboratory test apparatus using a light pulser.

Ref. 4. The position resolution as a function of drift field for a constant drift distance $d = 3.85mm$ is shown in Fig. 5.1. Experimental points are shown by black dots. Curve *A* shows the calculated resolution for the filter used in the measurement and optimized for a drift field $E_d \approx 150V/cm$. Curve *B* shows the best resolution obtainable if optimized filters were used for all values of the drift field E_d. The resolution shown in Fig. 5.1 was obtained under laboratory conditions with light pulses focused on the surface of the detector. The light intensity was adjusted to produce the same number of electrons as a minimum ionizing particle passing through the detector. The r.m.s. of the distribution of drift time for light pulses was measured for each value of E_d and multiplied by the drift velocity to obtain the position resolution.

5.2 - Position Resolution of a Multi-Anode Drift Detector

Results reported here were obtained with a drift detector where a $1cm$

anode was segmented into 41 anodes. Anode pitch was thus $250\mu m$ and the maximal drift distance 1.6cm. Each anode was read-out by its own preamplifier followed by a complete chain of

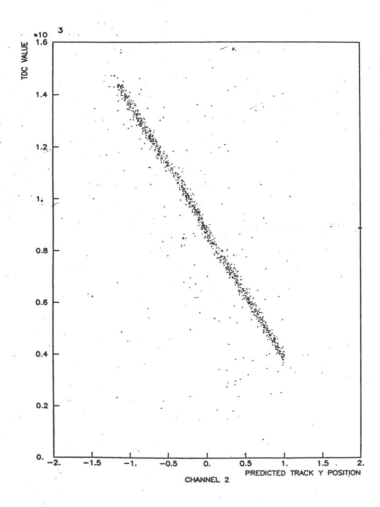

Figure 5.2: Position resolution along the drift coordinate of a multi-anode drift detector measured with a 100 $GeV/c\pi^-$ *beam.*

electronics. Drift time and total charge were measured. Drift time measures the y coordinate as in the case of a single anode detector, division of charge among individual anodes measures the coordinate (z) perpendicular to the drift direction. Multi-anode detector thus measures both coordinates of particles in an unambiguous way. A consequence of the larger number of read-out channels is an increase of the number of particles which can be measured within the detector for the same interaction event.

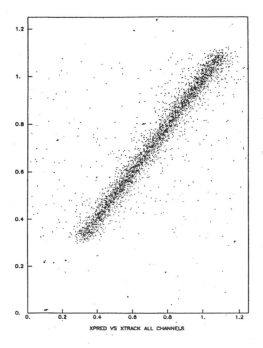

XPRED VS XTRACK ALL CHANNELS

Figure 5.3: Position resolution in the coordinate perpendicular to the drift direction of a multi-anode drift detector. The measurement is obtained from charge sharing among individual anodes.

The first test of multi-anode drift detector was performed at the CERN SPS with the help of equipment of the NA32 experiment. The position of $100\ GeV/c\pi^-$ was measured in the NA32 silicon microstrip detectors and the position of π^- was predicted in the plane of the drift detector. Fig. 5.2 and Fig. 5.3 show [5] the scatter plot of the position measured in the multi-anode detector versus position predicted from strip detectors in the drift coordinate and the coordinate perpendicular to the drift coordinate respectively.

The measured resolution is about 18 and $24\mu m$ in two respective coordinates. (After subtraction for the precision of the prediction from strip detectors). There is a visible kink on the scatter plot of Fig. 5.2. We believe it is due to the distorsion in the predicted position from strip detectors which were damaged by beam radiation. We were not aware of the damage of the strip detectors during the test.

Before discussing energy measurements in drift detectors we would like to point out that the achieved resolution of $18\mu m \times 24\mu m$ in the detector with the active area of $16mm \times 10mm$ corresponding to 4×10^5 resolution elements. This number of effective pixels was achieved with only 41 read-out channels.

5.3 - X-ray energy measurement

Results presented here were obtained from a small capacitance X-ray detector shown in Fig. 6 of Ref. 4. Reverse biased electrodes have the shape of concentric rings with a small point-like anode at the center. The drift field has a cylindrical symmetry and electrons drift radially toward the anode. The active volume of the detector has a diameter of $1cm$ and the anode capacitance is only $0.06pF$.

Results of room temperature and low temperature test were reported in Ref. 5. The room temperature test was done with the detector anode connected to a charge sensitive preamplifier realized in hybrid technology on a ceramic substrate from where the detector was mechanically supported. Special attention was paid to keep all stray input capacitances at a minimum. Total input capacitance was $5pF$ from which $3pF$ was the gate capacitance of the first transistor of the preamplifier (2N4416).

The equivalent noise charge of the cylindrical detector- preamplifier system at room temperature was 110 *electrons*, corresponding to a full width at half maximum of $930eV$ of energy. The shaping was pseudo-gaussian with a peaking time of $250ns$.

5.4 - Comparison of Classical and Memory Semiconductor Detectors

At first look one can have the following impression when trying to compare classical and memory semiconductor detector: Memory detectors can achieve a better performance due to their much lower anode capacitance, however, the classical detectors are faster. In reality, in almost all practical situations, measurement systems based on memory detectors are not only more precise but also faster than systems based on classical detectors. We will illustrate our point for both energy and position measurements with semiconductor detectors.

When the spectrum of energy of radiation is measured, the speed of the system is expressed by the rate of pulses the system can process without pileups. The pulse duration limits the processing rate. The duration of the pulse in classical detectors is not the carrier collection time, but the time required to decrease the amplifier noise down to an acceptable level. The processing time for systems based on classical semiconductor detectors is typically a few tens of μs with the corresponding maximal rate of a few tens of kHz.

For silicon drift detectors a longer drift (store) time does not limit the rate. Electrons drifting toward the anode have their charges screened by the electrode structure and do not interfere with the signal processing at the anode. The important time is the diffusion time, that is, the time it takes to collect full charge at the anode. It can be shown that the diffusion time t_{diff} is given by:

$$t_{diff} = 8 \times t_{drift} \left(\frac{U_{drift} \times q}{2 \times kT} \right)^{-1/2}$$

where t_{drift} is the drift time, U_{drift} is the absolute value of the drift voltage and kT/q is the thermal voltage equal to $0.025V$. For practical low capacitance X-ray drift detectors the diffusion time is about $100ns$. The shaping time is again about $100ns$ due to the extremely low capacitance of the detector anode. We see that the rate capability of low capacitance drift detectors is between a factor of 10 and 100 greater that the rate capability of classical detectors.

Comparing position sensing for classical and memory detectors gives similar results. We can compare the performance of a microstrip detector with that of a single anode linear drift detector. To obtain similar position resolutions a classical microstrip detector needs hundreds of read-out preamplifiers rather than a single preamplifier of a drift detector. Practical microstrip read-out system (microplexer or equivalent) has, after a stage of amplification, a parallel to serial conversion to keep the number of read-outs at a reasonable level. The read-out speed of a multiplexer is much slower than the drift time of electrons in the linear drift detector.

For very high rate environments (Superconducting Supercollider), the drift detector approach may lead to confusion since the maximum drift time in the detector is longer than the interaction rate. Signal electrons stored in the detector had not originated at the same time, and apart from the particle of interest we have several signals due to particles produced before or after the event of interest.

Doubling the number of detectors solves the problem. Placing two drift detectors parallel one to another in such a way that a particle has to cross both of them removes completely any possible confusion. It is sufficient to arrange drift fields in the two detectors in such a way that electrons drift in opposite directions in two detectors. If electrons were produced in both detectors at the same time the sum of the drift time in one detector plus the drift time in the second one is constant and equal the distance between two anodes over the drift velocity. If electrons were not produced at the time of interest the sum would differ exactly by the time difference of two interaction. Because the timing accuracy of the drift detector is better than $1ns$ it is possible to distinguish two interactions as close as a few ns apart and to use drift detectors in very high rate environments. (Superconducting Supercollider).

Multianode drift detectors and CCDs provide unambiguous two coordinates measurement with the position resolution in a few μm region. There are no other detectors to which we could compare them.

The negative side of memory detector is that their electrode structure is more complex than the structure of classical detectors. Their production is more complicated and the cost is higher than the cost of classical detectors.

6. - Future Developments

Semiconductor memory detectors are relatively new. There will be many new structures in the near future. The most important development for semiconductor memory detectors is the integration of the preamplifier directly on the silicon of the detector.

We have stressed very low values of anode capacitance for memory detectors (about $70fF$). This small value leads to a low noise performance which was observed and reported in the previous section. These tests were done, however, using electronics constructed from the best commercially available discrete transistors. The input capacitance of the smallest available FET 2N 4416 is about $4pF$. The total input capacitance achieved with a 2N4416 FET as the first transistor was $6pF$. Thus there is a factor of 100 mismatch between the low output capacitance detector and the input capacitance of commercially available FETs.

The realization of the first amplification stage directly on the wafer of the detector has two advantages

1.   the input transistor can be made small enough to match the small detector capacitance.

2.   stray capacitances due to the connection between the detector anode and the first transistor can be kept at a minimum.

The realization of matched preamplifiers with a minimum stray capacitance decreases the noise of the detector-preamplifier system, resulting in a substantial improvement of energy and position resolution of all silicon memory detectors. The first prototype of the integrated electronics is working [6] and in the very near future there will be memory detectors with better performance than reported here.

This paper focused for the most part on memory detectors in particle physics. However, we are also aware of the great interest of Luigi Radicati in astrophysics. Let us conclude by mentioning that the new fully depleted CCD will be used on the X- ray Multi-Mirror Satellite to bring down to Earth important data to astrophysics.

REFERENCES

[1]   K. McKay, Phys. Rev. **84**, 829 (1951).

[2]   E. Gatti - P. Rehak, Nucl. Instr. and Meth. **225** 608 (1984).

[3]   E. Gatti et al., Nucl. Instr. and Meth. **226**, 129 (1984).

[4]   P. Rehak et al., Nucl. Instr. and Meth. **A235**, 224 (1985).

[5]   P. Rehak et al., Nucl. Instr. and Meth. **A248**, 367 (1986).

[6]   V. Radeka et al. IEEE Trans. on Nuclear Science, **35**, 155 (1988).

Symmetries: An Experimentalist's
Point of View

GIORGIO SALVINI (*)

1. - The role of experiment and theory.

Physicists watch nature around us, and starting from phenomena and detectors they get rules and laws and symmetries.

I try to rivet with this note, also on the basis of my work as an experimental physicist, that our symmetries and conservation principles are extremely dependent on the capacity of our detectors, and that to make progress it will be necessary to improve in future their quality.

But let me start first with a bow to theory. Much of our recent story belongs to the theorists. They made a critical and subtle use of group theory, invariants, symmetries. The best of this effort has been the beautiful structure $SU(3) \times SU(2) \times U(1)$, which we refer to so casually as the "Standard Model" [1]. In particular the electroweak model $SU(2) \times U(1)$ incorporated Electrodynamics and Fermi theory of charged current weak interactions, successfully predicted the weak neutral current interactions and the existence and properties of the W and Z vector bosons. For all the Standard Model we can say that this mathematically renormalizable field theory can explain, or virtually agree with all known facts of elementary particle physics.

This capacity of control and prevision of the experimental results comes out, of course, after the necessary numerical input from the direct observation of nature arrived.

In fact the theory has succeeded in bridling and foreseeing the experimental facts with precise rules and symmetries (see for instance the existence of the W and the Z Boson and its radiative corrections) [2] [3], but the numerical values of the basic constants that we measure in our universe cannot be anticipated in any way by theory. Let me insist on this point.

Sometime it is said that the Standard Model is still incomplete, and cannot be the ultimate theory of nature, for it is too arbitrary, and it leaves too many questions opened. In particular theoreticians recall that there are at least eighteen

(*)   Dipartimento di Fisica, Università "La Sapienza", Roma

constants [4] of nature which are still unexplained (twentyone for others [5]). But this presentation is not yet the complete truth. The very fact is that we did not succeed (since the beginning of physics until today) in explaining or anticipating the numerical value of any fundamental constant.

Suppose that we prepare a complete detailed treatise of the Standard Model, our renormalizable quantum field theory $U(1) \times SU(2) \times SU(3)$. Suppose in this treatise are included all present developments, with radiative corrections, the Z° and W widths, with our hints on the number of families, the possible mass of the Higgs; the possible values of θ_W; the leptons and their decays, etc. This treatise is complete, and it quotes and indicates all the constants involved in the Standard Model and their relations.

But suppose that our treatise has one serious limitation: the electron mass, the muon mass, the quark masses, the coupling constants, the weak angle etc., are all indicated as usual by letters and symbols, $(m_e; m_\mu; \alpha_s; \theta_W; G_F;$ etc) but there is no number or experimental value attached to each of them. For instance it does not indicate any numerical ratio between the quantities $\frac{m_e}{m_\mu}, \alpha/\alpha_S, \frac{m_\tau}{m_H}$ etc.).

At this point, imagine that you go out of our Universe, in a place where intelligent people live, who never saw our Universe. You offer them our complete treatise of the Standard Model, in order to make them acquainted with our world.

They understand all of it, of course. Which picture or description of our Universe can they get from our Standard Model treatise? The correct reply is: Nothing. For instance, the image of the atoms is deeply different or inexistent if the mass of the electron and the proton were very near, and the strong and e.m. coupling were close in value. It would be a very different world, and we would probably look for different symmetries and conservations.

I think that from our S.M. treatise we cannot extract fundamental inequalities or limits: we cannot say if $2 < \frac{m_\mu}{m_e} < 2000$ or not, and if $.1 < sin^2\theta_W < .9$ or not. We could create a large variety of "worlds", equally coherent, and deeply different.

So, theory is blind. Theory in itself is only a topology or structure, and cannot anticipate the value of any fundamental constant. The day for instance we explain with certainty the why of the mass of the muon (or the observed ratio of the masses, m_e, mass of the electron, and m_μ, mass of the muon) or the value of the weak angle θ_W [6], this will be a great step forward respect to our present knowledge.

When trying to represent to me the relations between theory and experiment, I think of a tent that you can build and arise with tissue and pegs. The tissue (a very elastic one) with the proper knots and cuts is the theory, the topology. The pegs are our experimental constants.

Without the pegs, that is the numbers we have from experiment, the tent can have any form and size, and two tents with the same topology or structure (the Standard Model) could be largely different. So, in order to really inform our friends of the other Universe, our treatise on the Standard Model must be accompinied by a table of numerical values, and in case by a stick as a unit.

This numerical table today should give eighteen (or twenty one) numbers only. Yesterday, they would have been many more. We must not forget this progress, of course.

2. - How to observe symmetries. The detectors.

The mediators between nature and theory are our scientific eyes, the detectors, by which we measure. They are more than important to our knowledge; they are essential to our progress and understanding. Our capacity to discover symmetries and conservation laws depends on their time and space resolution, now and in future (§ 6).

When going through our recent history, we notice that in order to observe a symmetry we had to go out of it. I do not mean that one can never in principle establish a symmetry if he remains inside the field where it is valid: otherway it would be intriguing for me to discuss the case of CPT symmetry, within which we live. A symmetry appears when you have ways to observe the difference between the objects which constitute that symmetry [7] (for instance on octet or a decuplet in $SU(3)$). Otherwise it is an identity which cannot be disclosed, as shown in the clear analysis of L. Radicati [7]. Of course, the amount of the difference, the hyerarchy, is important: for instance, isospin concept (§ 3) would be much less significant if the e.m. forces and the nuclear forces were close in value.

It follows also that the qualities of detectors used in hunting for symmetries in elementary particles are of fundamental importance. Let's try to make this point clear in the next paragraph, starting from the possible Lagrangian of a physical system.

3. - Evidence of symmetries through the Lagrangian.

Let's take the simplest case of a Lagrangian describing a physical elementary system, and for which the variables of the system are the space coordinates and their time derivatives [8]:

$$L = L(q, q), \text{ with } q = dq/dt.$$

We can say that we have a symmetry when there is a transformation (in this case depending on a single parameter α)

$$Q = Q(q, \alpha), \quad \dot{Q} = \dot{Q}(q, \dot{q}, \alpha)$$

such that:

$$L(Q(q, a), \dot{Q}(q, \dot{q}, a)) = L(q, \dot{q}) \text{ for every } \alpha \quad (\forall \alpha)$$

It is on these lines that we can discover the main symmetries and conservation laws of classical physics. For instance we find the simplest of all: that the symmetry of translation (independence from space translation for the dynamical system we are considering) implies the conservation of momentum. In order to verify this invariant, or even think of it, we must have a stick and a square, to distinguish between two different frames of reference [7]. When we find that they are redundant, and we only need relative coordinates, then we deduce from this space symmetry the conservation of momentum.

This is possible, for nature was generous with us: we live at a temperature that allows solid states and rigid bodies, so that we can build a stik, and define lengths and angles. Without this length detector, our world would be empty or meaningless to our brains.

Let's consider a more significant case, which is at the start of nuclear and subnuclear physics: the discovery of the symmetry between protons and neutrons, which brought us to the invention of isotopic spin. As we shall see, we needed for this the use of electromagnetic detectors [8].

A system with hadronic particles, including protons and neutrons can be described by a proper Lagrangian function. This Lagrangian must contain the hadronic part, L_H, and the electromagnetic part L_E:

$$L = L_H + L_E$$

If the electromagnetic part is not included, then we cannot put in evidence the existence for instance of a proton neutron doublet, for these two particles appear as identical indistinguishable things.

Let's consider as an example the processes

(1) $$pn + pn \to ppn + n$$

(2) $$pn + pn \to pnn + p$$

where p is proton, n is neutron, pn is the nucleus of Deuterium, ppn is He^3, pnn is Tritium.

We assume that nuclear forces have charge symmetry: the interaction energy remains the same when in a nuclear system we substitute all protons by neutrons, and vice versa. When introducing the concept of isotopic spin T (1/2 to proton; $-1/2$ to neutron), this symmetry can be formulated by saying that the Lagrangian of the nuclear system remains invariant under the change of T_3:

$$L(T_3) = L(-T_3), \text{ or } L = L(|T_3|)$$

We can apply the operator that changes the sign of T_3 to the two members of (1): the first remains invariant, for it describes two particles with $T_3 = 0$. The second member of (1) goes to the second member of (2). We obtained reaction

(2). So with this operation, which is a symmetry of the system, we sent reaction (1) into reaction (2). The hadronic Lagrangian does not distinguish between the two final states of equations (1), (2). It follows that these reactions must be equal (apart the e.m. and weak differences) in all their physical characteristics.

As we already said in § 2, to discover the symmetry we went out of it, and observed with proper detectors the difference (between protons and neutrons). This is possible for we can use electromagnetic detectors which clearly separate the proton, a charged ionizing particle, from the neutron which has no charge.

Suppose we did not have electromagnetic detectors, magnetic field, drift chambers and all that. For instance we dispose only of nuclear detectors, like the nuclear emulsions of the old times, which could not see the track left by minimum ionizing particles: then to our "eyes"protons and neutrons of high energy could remain identical particles. Of course, thermodynamics and particle statistics could indicate the existence of two types of almost identical nucleons. But the isospin symmetry was introduced aniway by experiment: the nuclear difference between proton and neutron is small, so that we can distinguish p and n by electromagnetic probes or detectors, and at the same time we can approximate them as equal in their behaviour. This corresponds to establish symmetries (i.e. approximate identities), and it allows to anticipate the behaviour of many yet unmeasured processes, which are similar to those which we already discovered. So, symmetry in nature is an approximation, but it is a powerful guide toward discovery, it is esthetic and productive. Its existence is strictly bound to the actual values of our unexplained constants of nature.

We must add that the world appeared until now rather generous in helping our classifications and offering reasonable approximations. In fact we could profit of:

-    the large difference in scale among the forces of nature (e.m. to nuclear, for instance);

-    the existence of solid bodies at the temperature of our earth, which allow the construction of rigid frames of references;

-    the approximation of point-like bodies, for instance in astronomy, with an extended nothing between.

It is this structure which allowed us to build the basic detectors of elementary particle physics.

4. - The detectors, our eyes.

Our world is described by different forces, and we succeeded in qualifying them separately: symmetries will be discovered between particles of one realm of forces (f.i. strong forces), by using the clear difference between them in athor realm (f.i. e.m. forces). This difference is measured by the detectors, and their exactitude is important or decisive.

One could remark that our added classifications and quantum numbers constitute a long chain of unended complications in the theory of elementary particle physics. This is not necessarily true. For instance the "electromagnetic" part of our Lagrangian has brought us to identify different types of barions and pions, and then to the first synthesis of the vector meson model [9] [10], and from this to the $SU(3)$ symmetry. These classifications and symmetries allowed us to foresee many experimental results and gave us predictive power and synthesis. We could not have arrived to the helium content of the Universe and to the basic simplicity of the quarks without going through the agony of all e.m. differentiations between the particles.

It could be that there are specific differences (symmetries) among the elements described by our known fields of forces, which we are not able to observe yet. For instance we could have quantum numbers (and new symmetries) which differentiate the bosons and fermions according to their gravitational properties. But to do that we need gravitational detectors, which at present are not available. It could be that the attribution of proper new gravitational numbers to the existing particles bring us at the end to a larger sysnthesis.

5. - Some personal recollections.

Sometime experimental facts, that is signals from our detectors, came to us unexpected, to enlarge our view of the world, or to violate something which we belived. We shall call these serendipic results. (See the example of $J\Psi$ in this paragraph).

Sometime the information from the detectors came as a verification of expected symmetries, or clear consequence of well established results. We shall call this an experimental verification (confirmation) of a symmetry or conservation law. It is the case of the antiproton, or the discovery of the W, Z° bosons. (This paragraph).

I shall recall now some experimental results to which I took part. Some of them serendipic, other not:

   The $J\Psi$ discovery (I was not among the first discoverers, but I did follow its developments in Frascati, while working at the ADONE collider);

   The W and Z° discovery;

   The search for the top quark;

   The $B\overline{B}$ mixing;

   Hunting for Susy and new physics.

5.1 *On the $J\Psi$*.

This sharp intense resonance took us by surprise. I am among those who saw the first $J\Psi$'s in Europe, in Frascati, with ADONE [11].

ADONE had had the merit of observing multihadroninc production [12] a fact which gave almost direct evidence of quarks, real and confined. In the summer before $J\Psi$ came, we were looking for the ratio R [3], among hadronic and muon cross section:

$$R = \frac{\sigma(e^+ + e^- \rightarrow \text{ hadrons })}{\sigma(e^+ + e^- \rightarrow \mu^+ + \mu^-)},$$

in the region of energy $2 - 3$ GeV, and we were obtaining a reasonable value in agreement with $SU(3)$ and the three quarks hypothesis, while collecting with day and night work, a few events per day.

That night of November 1974 [11], the events came to us by hundreds. As we explained in our note [11], we found the resonance almost immediately, after having been informed by the discoverers, and we could contribute to the statistics. The only confort to us was that we did not loose the "glory" due to our heedlessness or negligence. ADONE was prepared for a maximum energy of three thousand MeV centre of mass, and to get that limit was already rather binding. No prophet in the world of physics could have come to tell us "pay attention, you are 100 MeV below the paradise: push a little more. The $J\Psi$ is at 3100 MeV". As I said, it was serendipic, in its numerical value at least.

The events came with the strength of those things who change your life. Our basic instruments were scintillation chambres, shining and loud with their sparks. It was like hail after a period of drought. Those days changed in part my attitude toward physics: new things may come unexpected, always, to change our scientific story. I have this feeling also today, in a period which suggests that the unexpected could still be tenths of years far from us.

The possible explanation of the $J\Psi$ came, beautiful and to stay: not the value of its mass, which is still unexplained today, but the structure and position of this resonance. Only for a moment it was thought that we were in presence of the longed for intermediate heavy boson. The lucidity of S.L. Glashow and A.de Rujula [13], on the basis of existing analysis, recognized from clear symptoms that it should have been the fourth quark. The story is well known [14] [3].

The discovery was possible due to the great development of the machines and detectors. The large detector of SPEAR [3] made possible observation of details which were absolutely necessary to theory and identification.

5.2 On the W, Z°.

In this case (I was among the authors since 1977) my recollection is different. These bosons came as a triumph of theory. Yes, we were curious at CERN in 1982-1983, after four years of intense operation, to see it or not, at the suspected place. Not existence of W, Z°, that would have been the great discovery!

It is beautiful, to go through the twenty years long experimental and theoretical chain which indicated the way to this success. It starts from those illuminated mathematicians and physicists who convinced us of the coherence of theory. First came the basic fundamental concept of gauge local invariance [15]. This has been the guiding powerful symmetry which produced the evidence and unicity of $U(1)$, and drove us to foresee a symmetrical unity in e.m. and weak forces. This brought to the discovery at CERN of W, Z° [16] in the expected mass region, of course after the weak angle had been measured, or at least guessed experimentally through neutrino interactions [17].

I remember walking up and down with Carlo Rubbia, in the hall close to the $UA1$ well, while taking data during the nights. It was a mixed feeling: we had confidence in our $UA1$ apparatus and in the calculations. Should W exist, it had to come out. On the other side, being (ambitious) experimentalists, the news "No $W^{\pm}$" would have been for us a tremendous serendipic success and we were ready to find in it a kind of happiness, while driving our car on the bones of some dead theory. But the W came, with all its characters in order; so in 1983 [16] we were (and still we are: nothing really new is happening) the notaries of the expected and legal facts of physics. The huge $UA1, UA2$ detectors were necessary to make out of question an identification which could have been otherway suspicious being so obvious.

Today, even before the imminent results from LEP [3] [18] (Large Electron Positron of CERN) and SLC (Stanford Linear Collider), the W and Z masses, the absence of flavor changing neutral currents, the charged and neutral current gauge couplings to fermions are well tested, and the CERN and Fermilab experiments [2] [19] place some clear constraints on the possible new physics into the $\leq 1\ TeV, GeV$ range.

5.3 The search for the top quark.

The top quark must exist [2] [3], given the clear simple assumption of the Standard Model. This reminds us of other theoretical bids which experimental physics respectfully obeyed: existence of antiproton and antineutron; the existence of W, Z we now recalled. But with a difference: the mass of the top is not predicted (see §1) and only the mass range $60 - 200\ GeV/c^2$ is strongly suggested.

This uncertainty on the mass of the top [20] is the best occasion to understand the importance and the serious limitations of our present detectors. The jets and leptons resulting from the decay of the top must be disentangled from a jungle of other possible jets and leptons coming from other quark and boson decays. The burden to our detectors was until now too heavy.

It took time to discover our experimental limitations, and in fact, in a moment of optimism there was among us at $UA1$ the feeling that the top had been observed with a mass around $40\ GeV/c^2$ [21]. Now the same group has made a more careful analysis, and gives a safe upper limit: $m_t \geq 50\ GeV$ with 95% c.l.

But the warning is clear: the quality of detectors is in future the serious problem of fundamental physics. And they must be precise and well thought. Think of those many times when one or more theories jumped out of the mathematical bed to explain and reasonably fit one experimental result which later was found to be wrong, due to the uncritical analysis of the data.

5.4 On the $B\overline{B}$ mixing.

$B\overline{B}$ came to us laboriously, and finally rather clear [22]. We knew that the mixing of B with $\overline{B}$ was a possibility, but the theory, especially being the mass of the top still unknown, could not foresee the percentages of our mixing (B_d and B_s mixing). We were, as well known [22] the first to find mixing. After us, e^+e^- colliders came with more complete results, and it was possible to separate mixing B_d from B_s [23].

We at $UA1$ had a well built central detector [24], imbedded in a .7 Tesla magnetic field. This expensive complicated detector was the main source of our information; it was the result of a great effort, something like 200 dedicated man-years. Around it we had an extended roof and floor and side walls of muon detectors. We measured pair production of muons having opposite or equal charge: you have mixing when $B_0(\overline{B}_0)$ of the $B_0\overline{B}_0$ pair changes to $\overline{B}_0(B_0)$. You notice it through a number of muon pairs with equal sign. Our measurements were rather elementary. The more precise ones came from the e^+e^- rings, with more specialized detectors [23].

I remember at the beginning a kind of cold interest on my side to this problem. It took some time for me (I should feel ashamed for this) to realize the potential interest of the problem, but when I fully realized it, I was quite impressed. The $B\overline{B}$ oscillations, the values of ϵ' and ϵ'/ϵ will become a severe test for the Standard Model [3] [25]: we must wait until the mass of the top, and the rate $b \rightarrow u$ are well established [3]. The interesting thing is to verify and measure CP conservation in $B\overline{B}$ mixing, and when we go to numbers, we discover that no accelerator can give us today as many $B\overline{B}$ pairs as it will be necessary [26] (many many millions).

A couple of comments are at this point in order with the aim of this paper. The first regards again the detectors, our eyes. At present only a small percentage of the $B\overline{B}$ decays ($r \sim 4\%$) [25] [26] can be completely reconstructed, with their charged and neutral components. To increase r we increase momentum resolution and wide angle range of the detectors. This is of overwhelming importance and it is not going to be easy. Part of the future of theoretical research shall depend on this.

The second point is that it will take many years to get m_t, m_H CP from $B\overline{B}$ [3], etc. Suppose, ten years or more. Which shall be our main interests, ten years from now? Will the physicists and the young generation have the temper for this journey? My reply is yes, notwithstanding I think that again something new could happen during the journey.

5.5 *On monojets, SUSY and the hopes for a new world.*

Susy was at the gates in the eighties, and after $W, Z°$ success we started to look for a new world in the 100 GeV area. Yes, I admit that we were psychologically too ready to it. It is in this period and atmosphere that we and $UA2$ started to observe some events in the $100 - 200$ GeV area that it was difficult to explain on the basis of known (Standard Model) structure [27]. The main evidences were jets and sometimes photons without charged particles on the opposite side. In this case we were not driven by a precise theoretical prevision or possibility; but an elegant symmetry was really anxious and ready to shelter our results: Supersymmetry [3] [28] (§6). The hope was really to find results which could shock, surprise the theoretical establishment. Of course, group of theorists were watching our results too: the hope to overthrow the existing system is since ever a strong inspiration to physicists. In § 6 we report the lower limits established by $UA1$ and other experiments to the Susy sparticles for their existence.

After four years, I must say that today little remains of our revolutionary hopes. Most of the monojets were explained by more refined $SU3$ theory and analysis of quantum chromodynamics, and some strange events with a single photon returned to the ranks (of the Standard Model).

The fact is that our detectors were not precise and critical enough. It is the same conclusion than at the end of § 5.3. This reminds me of the other attempts in the past to break the existing lines and symmetries. The old one [29] is the famous "break down" of electrodynamics. If you go through the literature, you find (wrong) experiments which were claiming the "breaking of e.d." [29]. We know today that no rough breaking exists, but rather the fusion into the electroweak theory.

So, today we admit that there is not yet any clear evidence of something new, Susy or not. The physicists strongly stick to their detectors: they learnt that to improve their calibration and resolving power is something important and fruitful, and carelessness does not pay. Still, the interest to Susy is a good example of our today's philosophy. The existence or not of this symmetry is a guiding line for our progress. To look for Susy through the experiments in the following years is a challenge to our capacity of induction and coherence. It is convenient to spend here a paragraph on the search for Susy particles, and its nature. This may be somewhat educational in drawing some (modest) conclusions on the relations among experimental physics and symmetries today.

6. - Looking for new symmetries. SUSY. The cost of the challenge.

We enter now a field whose bases are not in any experimental evidence or hint, but rather in a request of coherence in our theoretical considerations. Really we did not find yet any indication of resonances or state peeping out of the gates of the Standard Model. But powerful instruments are being prepared for the future Susy safari.

6.1 *What is it.*

I apologize for condensing in a few lines hundreds of articles and fine analysis. Supersymmetry [3] [30] refers to a symmetry between fermions and bosons in which each particle has a partner, alike in all quantum numbers, except that the spins differ by 1/2 unit. There appear to be powerful theoretical motivations for supersymmetry, mainly because it offers a way to understand the large mass ratios in physics, the so called hierachy problem. In any case, the idea has strongly captured the imagination of both theorists and experimentalists.

Table 1

Supersymmetric particles. Particle contenent of the Standard Model and their supersymmetric features

Particle	Spin	Sparticle	Spin
quark $_{L,R}$	1/2, 1/2	squark$_{L,R}$	0, 0
lepton$_{L,R}$	1/2, 1/2	slepton$_{L,R}$	0,0
Photon γ	1	photino $\tilde{\gamma}$	1/2
gluon	1	gluino $\tilde{g}$	1/2
$W^{\pm}$	1	Wino $\tilde{W}^{\pm}$	1/2
$Z^{\bullet}$	1	Zino $\tilde{Z}^{\bullet}$	1/2
Higgs H^{+}, H_2^0	0	Shiggs	1/2
H_1^0, H_2^-	0		1/2
chargino's			
neutralino's			

There are many versions of supersymmetry. The minimal particle content is shown in Table I. Susy particles are conventionally designated by a twiddle, for instance $\tilde{e}$ is a supersymmetric electron. The partners of leptons and quarks are called sleptons and squarks. Thus, $\tilde{e}$ is a selectron, $\tilde{\mu}$ is a smuon, and $\tilde{q}$ is a squark; gauge particles are designed by ending "ino" ("gauginos"). Thus $\tilde{\gamma}$ is a photino, $\tilde{W}$ is a Wino etc.

Theory does not offer guidance, until now, for the mass of the sparticles, but it tells us about topology and how to look for them. The crucial rules for (the simplest) Susy searches are:

1)   Supersymmetric theories are invariant under a new symmetry, R-parity. Particles have $R = 1$ and sparticles $R = -1$. R is a multiplicatively conserved quantum number. Consequently, sparticles are always produced in pairs [3] [30].

2)   Supersymmetric couplings are obtained from standard couplings by changing two of the particles at a vertex to their supersymmetric partners without changing the strenght of the couplings.

3)   There is a lightest Susy particle to which all the other decay rapidly. This particle is stable and interacts weakly so that it is not detected. The energy and momentum carried off from the lightest Susy particle is a crucial signature for their search.

Generally, the lightest particle is assumed to be the photino, $\tilde{\gamma}$. Thus $\tilde{e} \to e\tilde{\gamma}, \tilde{q} \to q\tilde{\gamma}$. Usually, one also assumes $M_{\tilde{\gamma}} = 0$. With these assumptions it is possible to calculate cross sections, and specify and foresee experimental signatures.

6.2 Recent researches for Susy sparticles, by UA1 and other experiments.

Search for sparticles was made with e^+e^- and $p\bar{p}$ colliders, until now unsuccessfully, as we already said. In the case of e^+e^- colliders, we recall looking for pair production of sleptons. Selectrons can be produced in pairs for instance through e^+e^- annihilation, like any pair of electrons or muons, or tao's.

The selectrons decay to $\tilde{e} \to e\tilde{\gamma}$, so that $e^+e^- \to \tilde{e}^+\tilde{e}^- \to e^+e^- \ \tilde{\gamma}\tilde{\gamma}$. The signature can be acoplanar pairs of electrons with large missing energy. Petra experiments succeeded in setting a lower limit for the selectron mass, $m_{\tilde{e}} > 23 \ GeV$ [31].

The best mass limit for any sparticle has been obtained from the reaction $e^+e^- \to \gamma\tilde{\gamma}\tilde{\gamma}$. The signature of this process is a single photon. The reaction $e^+e^- \to \gamma\tilde{\gamma}\tilde{\gamma}$ is sensitive to selectron masses much greater than the beam energy. The final result [32] obtained from a variety of experiments has produced a limit $m_{\tilde{e}} > 65 \ GeV$.

Research for sparticles with hadron colliders can be pushed to higher limits. But the real difficulty with the powerful hadron colliders is that the debris of the sparticle decays are imbedded in the tenths or hundreds of charged particles which are produced in each single proton-proton, or proton-antiproton interaction.

Searches for supersymmetric particles have been made by $UA1$ [33] and $UA2$ [34] at the CERN $p\bar{p}$ collider. Our analysis at $UA1$ has been extended to $715nb^{-1}, 546$ and 630 GeV $c.m.$ Precise cuts were imposed on jets energy and missing energy. The events were studied taking into account known sources of prompt neutrinos and jets, like $W \rightarrow e\nu, \mu\nu, t\nu, c\bar{s}$ plus jets; $Z^{0} \rightarrow \nu\bar{\nu}, c\bar{c}, b\bar{b}$, plus jets; and heavy quark production followed by semileptonic decay. The expectations from these sources were studied by Montecarlo analysis strictly applied to the $UA1$ apparatus [34], and no room was found for gluinos or squarks below some mass values.

In synthesis, the results are presented in Fig. 1. In this figure, the mass limits at 90% confidence level on squarks and gluinos are given. Assuming a massless photino, the $UA1$ results give

$$m_g > 53 \ GeV/c^2 \quad (90\% \ \text{c.l.})$$
$$m_q > 45 \ GeV/c^2, \quad \text{independently of the gluino mass.}$$

The analysis of $UA2$ [34], with a total luminosity of $910nb^{-1}$ at the same energies than $UA1$ has been extended to various exotic processes, involving excited electrons, supersymmetric particles and additional vector bosons. In regard to supersymmetric particles they have analyzed also the case of an unstable photino, which decays into $\tilde{\gamma}H$. In this hypothesis they can exclude squark masses between 9 and 46 GeV/c^2, and gluino masses between 15 and 50 GeV/c^2. They also exclude additional vector bosons W' and Z' with masses less than 200 and 180 GeV, respectively.

Similar researches than ours, and with similar experimental approach, are being developed at Fermi Lab with the Tevatron, which can arrive to a c.m. energy of 1800 GeV. Their limits are higher, and are given in the same figure 1.

6.3 Susy researches in future.

All the colliders in preparation and being planned have in their program searching for sparticles. Susy is only one hope, and of course each experimental program is larger, and openminded: to specialize on Susy only, would be an obvious mistake.

The method for the Susy safary is again as follows [35]. We antici-pate behaviour and topology of all known particles, once we know coupling and masses. We can also anticipate the behaviour of Higgs and quarks, if we make and hypothesis on their masses. This means that by Montecarlo (M.C.) distributions we can foresee in detail what should we observe (in a whole of

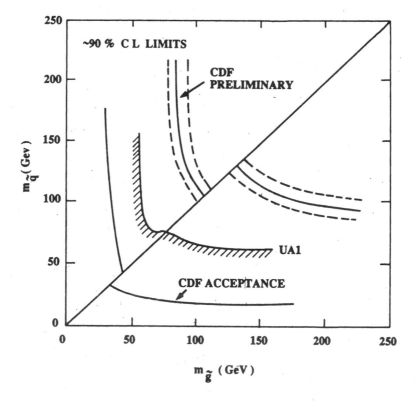

Fig. 1 - Limits on squark and gluino mass (90% C.L.) obtained by U A1 (see text). Also shown are preliminary results from CDF (Munich Proceedings, ref. (35)).

thousands of interactions) by assuming that all the particles foreseen by the Standard Model exist and only those.

The new events (sparticles, new families beyond the accepted three) should then remain by difference in our hands. There shall be (with proper normalization) a number of new events N given by·

$$N = Real\ distribution - MC\ prevision.$$

Looking into N, the experimentalist may hope to identify sparticles or other new bodies within the collected interactions. Let me say - being an experimentalist - that for us the hope is to have $N \neq 0$, irrespective of this being due to sparticles or other novelties; the real value of N being in its clear evidence and surprise.

To extract N from the future experiments will be a difficult and expensive investment, something like one thousand physicists for ten years, and ten thousand million dollars. I give now two simple examples of the dizzy heights we are going to climb.

6.4 *Two examples of sparticle production at super high energy and intensity.*

The first is a study of the expected cross sections for sparticles production in pp colliders [35]. In Fig. 2, we refer to processes

$$p + p \rightarrow \tilde{g}\tilde{g} + X; \tilde{q}\tilde{q} + X; \tilde{W}\tilde{W} + X$$

$$\tilde{e}\tilde{e} + X, \tilde{Z}\tilde{Z} + X; \tilde{\gamma}\tilde{\gamma} + X$$

The cross sections have been evaluated with the assumptions of § 6.1 and should be (if Susy exists) rather reliable.

The cross sections of Fig. 2 a) are given as a function of the sparticle mass. It is assumed for the pp collider a c.m. energy of 40 TeV ($40x10^3 GeV$). As we see, squarks and gluino's should be copiously produced in the LHC/SSC colliders. For example, for a total integrated luminosity of $10^{40} cm^{-2} = 10 fb^{-1}$, which correspond to about one year of running at $L = 10^{33} cm^{-2} s^{-1}$, we can expect $\sim 10^4$ gluino or squark pairs, each having a mass of 1.5 TeV/c^2 [36].

These numbers in themselves are rather encouraging, but still the devil is in the tail. As we see in Fig. 2a) again, the cross sections for the "ordinary" processes

$$pp \rightarrow t\bar{t} + x; W^{\pm} + X; Z^{\circ} + X$$

are $5 - 7$ orders of magnitude higher. The real problem will be to disantangle sparticle production from these events, which can occur with quark and gluon jets and leptons in a variety of forms. The way out will be precise measurements by excellent calorimetry (§ 7, detectors). This shall also help to localize neutrinos and other (f.i. photino's) non interacting particles.

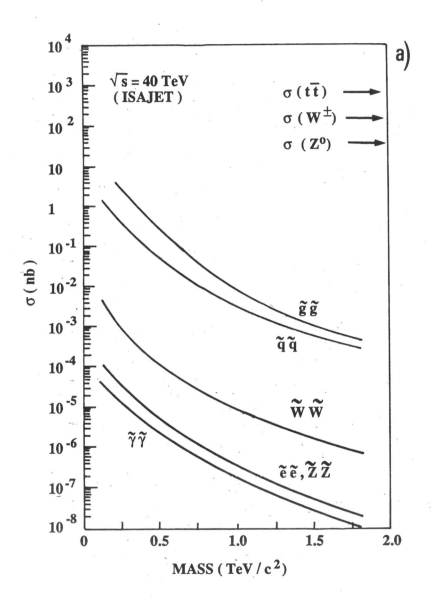

Fig. 2 a) Supersymmetric particle production cross sections, as a function of sparticle masses for pp collisions at $\sqrt{s} = 40 TeV$.

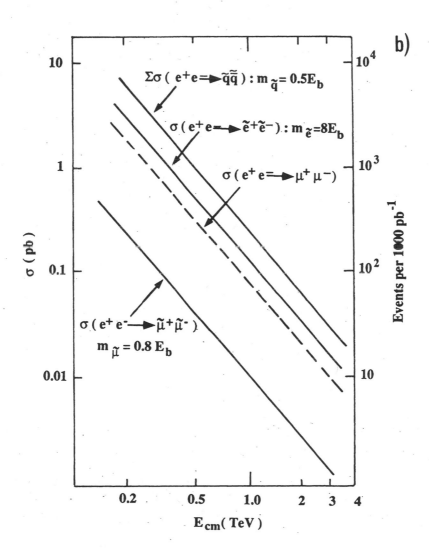

2 b) Supersymmetric sparticles cross sections as a function of center of mass energy in e^+e^- collisions (annihilation). Ref. (35).

The second example regards sparticles from giant linear accelerators e^+e^- colliders, and in particular from CLIC [37]. CLIC is thought to reach $1 - 2 \; TeV \, c.m.$

In Fig. 2b) we report the cross sections for processes

$$e^+e^- \rightarrow \bar{q}\bar{q}; \tilde{e}^+\tilde{e}^-; \tilde{\mu}^+\tilde{\mu}^-; \tilde{\mu}^+\tilde{\mu}^-.$$

as a function of $\sqrt{s}$, the e^+e^- beams centre of mass. For example, at $\sqrt{s} = 2TeV$ and again for a luminosity of $10fb^{-1}$, we can observe $30\tilde{\mu}^+\tilde{\mu}^-$ events, and 600 $\tilde{e}^+\tilde{e}^-$ pairs (with $m_{\tilde{e}} = .8TeV$). Those numbers can be compared with ~ 220 events per year expected from $e^+e^- \rightarrow \mu^+\mu^-$. Clearly, for e^+e^- colliders, high luminosity ($> 10^{33} cm^{-2}s^{-1}$) is required in order to maintain the physics reach for sparticles.

Let's remind that CLIC [3] [37] is going to be a difficult enterprise, and many technical problems have to be solved (two years, 10 years?) before starting its construction. Aniway it seems that disentangling particles from sparticles shall be somewhat more pellucid with e^+e^- colliders. Of course, appetite grows with what it feeds on. In a second (how far?) generation it will be a dainty bit to make measurements with beams having longitudinal polarization, for this would lead to very severe restrictions on the supersymmetry parameters [35]. But to perform such studies, a total of 300 fb^{-1} will be needed: something really too far in the future.

It is clear that we shall have to build new powerful detectors, and that without them our curiosity shall be unsatisfied. I insisted on the case of Susy safari for this case spans over all our natural philosophy. We need new experiments and observations: no physics without them. But we need the vision of theory, laws and symmetries: without it, our inspiration and motivation fail.

6.5 A lot of work ahead.

It comes out from our discussion that present apparatuses and detectors are not prepared yet to identify the components of an high energy interaction in such a way that a supersymmetric particle can be identified with certainty. Photons (γ rays), electrons, muons, even τ leptons can be identified with certainty; uncertainties remain with the other components, those particles which can be deduced only by missing energy and jets. Gluon jets and quark jets are expected to be different in structure, so that they can be distinguished statistically, but not one by one. The day we shall find new ways to anatomize jets of high energy, we'll have gone a big step forward.

Notice that supersymmetry is at the same time something which is worth to believe, and a beautiful false scope: all the experiments that we prepare are for something different, beyond the Standard Model. I cannot imagine any even absurd combination of quarks and leptons which could remain unobserved, being it a piece of Susy or not. Of course we should also be prepared to notice any possible deviation from present symmetries: including conservation of energy, momentum, CPT.

7. - Need for new detectors.

Let's consider briefly what we realistically need today in order to progress in our knowledge of symmetries and laws:

- High precision calorimetry [38]: this is due to the necessity of measuring the energy balance with high precision. We must go better than 1%; We saw (§ 5) that detailed energy balance is a precious, fundamental information: it is strictly bound to the precision of our calorimeters.

- High precision in jet identification and structure. This is a kind of integral property where charged particle detectors and calorimeters must strictly collaborate to measure jet total energy, particle numerosity and momentum distribution; it has not been established yet which confidence level we can achieve in distinguishing quark and gluon jets. Progress in this sector is certainly necessary.

- Quickness of the detectors in giving the quantitative signal and being ready again for a new one. This is particularly necessary for the apparatuses working with proton proton colliders. Fast detectors are the challenge for future SSC and LHD colliders [3] (§ 6). The problem is not as severe in case of e^+e^- colliders (CLIC).

- Polarization measurements. As we said, longitudinal polarization asymmetries would lead precise informations on the nature of the interactions (§ 6). This will require high intensity beams and fast detectors with precise geometries.

8. - Conclusions.

I tried to emphasize the decisive role of the quality of our detectors in the discovery, the shaping, and the selection of symmetries and conservation laws in physics (§ 1, 2). Without their data our theories are blind, we miss the landscape: we do not even succeed in imaging it (§ 2, 3).

This is getting even more clear in these years, let's say after 1984 when vector bosons were discovered (§ 4, 5). In fact a number of representations wider than the Standard Model is being carefully explored (§ 5, 6), from compositness of the fundamental fermions [39] to supersymmetries. But to get the reply and to go ahead, we need new colliders and new detectors (§ 7).

Our curiosity and waiting is rather dramatic: new very costly experiments are necessary, and the final green light for our activity will depend on tax payers, politicians and philosophers. They shall ask which is the real value of our findings. We shall give well thought and clear replies.

The present bases for our next progress ahead are sound: we know that the previsions of $SU(3) \times SU(2) \times U(1)$ gauge interactions are exact within the present experimental errors: whatever the next version may be, it is probable that the Standard Model will be the correct low energy limit.

But it is very improbable that we find nothing new when going from two to forty TeV in the centre of mass: two many symptoms indicate that we could be at the very start of a large representation of the world around us (§ 5, 6).

I insisted on Susy as a good example of the present relations between theory and experiment (§ 6): theory indicates rather precisely the structure of the new Superworld (the tent, § 1, 2). We need experiment to have the numerical values of the masses (the pegs). Without new data we are stuck. As we said, Susy is aniway a magnificent false scope. But who does not believe in Susy cannot be charged with misoneism.

As we saw, with theory and experiment we build a model and settle an order. This is good, and helps to guess what the world is. The world is not symmetric (it is not a pure crystal), it is not chaos. It lures and seduces us in many ways. It pushes us toward order and mathematics, but it constantly prooves to us the absolute necessity of experiment, and gives unexpected new facts.

The positive results we already reached tell us that the future of physics is today - more than ever - fascinating and unpredictable. We do not really know - we never knew - what we are going to discover and build in the coming years.

REFERENCES

[1]   S.M. BILENKY - J. HOSEK, Phys. Rep. **90** (1982) 73.
       Fifty Years of Weak Interaction Physics (Italian Physical Society, Bologna, April 25, 1984).

[2]   P. LANGACKER, Status of the Standard Electroweak Model, XXIV International Conference on High Energy Physics, Munich, Aug 4-10, 1988, p. 190, Springer Verlag Editors.

[3]   G. SALVINI - A. SILVERMAN, Physics Reports **171** (1988) 231-424.

[4]   G. ALTARELLI, in Fifty Years etc., quoted in Ref(1), p. 485.

[5]   See Ref(2). The minimal model, when including QCD and classical gravity has 21 free parameters.

[6]   Review of Particle Properties, Particle Data Group. Phys. Lett. B, **204** (1988) 1-486.

[7]   L. RADICATI, Remarks on the early developments of the notion of symmetry breaking, in: Symmetries in Physics (1600-1980), Seminari d'Historia de les Ciencies - Universitat Autonoma de Barcelona. Editors M. Doncel, A. Hermann, L. Michel, A. Pais, p. 197-206. R.M.F. HOUTAPPEL - H. VAN DAM - E.P. WIGNER, Rev. of Modern Physics, **37** (1965) 595. These two papers report the analysis of Pierre Curie in 1894

on the symmetry breaking as a necessary condition for the existence of phenomena.

[8] I am grateful to dr. Ugo Aglietti for frequent discussions on this point.

[9] J.J. SAKURAI, Vector Meson Dominance, Proceedings of 4th International Symposium on Electron and Photon Interactions, 1969, p. 91. Ed. D.W. Braben.

[10] N. KROLL - T.D. LEE - B. ZUMINO, Phys. Rev. **157** (1967) 1376.
J.P. PEREZ JORBA - F.M. RENARD, Phys. Rep. **31** (1977) 3.

[11] MK1 Collaboration, Phys. Rev. Lett. **33** (1974) 1406.
J.J. AUBERT et al., Phys. Rev. Lett. **33** (1974) 1404.
C. BACCI et al., Phys. Rev. Lett. **33** (1974) 1408.

[12] C. BERNARDINI in Proc. Int. Conf. on Electron and Photon Interactions at High Energies (Cornell Univ., Ithaca, N.Y. 1971) ed. N.B. Mistry, p. 37.

[13] A. DE RUJULA - S.L. GLASHOW, Phys. Rev. Lett. **34** (1975) 46.

[14] S.L. GLASHOW - J. ILIÓPOULOS - L. MAIANI, Phys. Rev. **D2** (1970) 1285.

[15] C.N. YANG - R.L. MILLS, Phys. Rev. **96** (1954) 191.

[16] UA1 Collaboration, Phys. Lett. B **122** (1983) 103; B **126** (1983) 398.
UA2 Collaboration, Phys. Lett. B **122** (1983) 476; B **129** (1983) 130.

[17] M. DIEMOZ - F. FERRONI - E. LONGO, Phys. Rep. 130 (1986) 294.

[18] Proc. LEP Summer Study Group, (Les Hauches and CERN 1978).

[19] CDF Collaboration, reported by S. Errede, Munich Conference as quoted in (2), p. 689.

[20] K. EGGERT, Proc. of *Les Rencontres de Physique de la Vallée d'Aoste*, p. 583, edited by M. Greco, 1988.

[21] UA1 Collaboration, Phys. Lett. B **147** (1984) 493.

[22] UA1 Collaboration, Phys. Rev. Lett. B **160** (1985) 188.

[23] Argus Collaboration, H. Albrecht et al., Phys. Lett. B **192** (1987) 245.

[24] M. CALVETTI et al., Nucl. Instr. Methods **176** (1980) 255.

[25] NA31 Collaboration reported by L. Gatignon, Munich Conf. as quoted in Ref(2), p. 528.

[26] Workshop on Heavy-Quark Factory, 14-18 Dec 1987, SIF, Editrice Compositori, Bologna, p. 3-1021.

[27] UA1 Collaboration, Phys. Lett. B **139** (1984) 115.
UA2 Collaboration, Phys. Lett. B **139** (1984) 105.

[28] J. WESS - B. ZUMINO, Nucl. Phys. B **70** (1974) 39.

[29] See for instance the Proceedings quoted in Ref(12) and references therein.
C. BERNARDINI, High Energy experiments in QED. Presented at the Theoretical Physical Institute, Unoversity of Colorado, Summer 1968, p. 465-491.

[30] H.E. HABER - G.L. HANE, Phys. Rev. 117 (1985) 76.

[31] CELLO Collaboration, Phys. Lett. B **176** (1986) 247.

[32] M. DAVIER in Proc. XXIII Int. Conf. on High Energy Physics, (Berkeley, 1986, Ed. S.C. Loken), p. 25.

[33] UA1 Collaboration, Phys. Lett. B **198** (1988) 261.

[34] UA2 Collaboration, Phys. Lett. B **195** (1987) 613.

[35] F. PAUSS in Munich Proc. as quoted in Ref(2), p. 1275.

[36] E. EICHTEN - I. HINCHLIFFE - K. LANE - C. QUIGG, Rev. Mod. Phys. **56** (1984) 579-707.

[37] U. AMALDI, Nucl. Instr. Meth. A243 (1986) 312.

U. AMALDI in Proc. Workshop on Physics of Future Accelerators (La Thuile, 1987) p. 323 and references therein.

[38] M. ALBROW et al., (Part of $UA1$ Collaboration), submitted to Nucl. Instr. Meth., 1988.

[39] A clear synthesis of the (too many) roads in front of us can be found for instance in the Plenary Section Reports in the already quoted Proceedings of the Munich XXIV International Conference on High Energy Physics, p. 190-386.

Luminous and Dark Matter in the Universe

The most secure observational evidence on dark halos of galaxies and in general on dark matter in the universe comes from the study of the kinematics of systems in dynamical equilibrium. The following is a short account of the main observational results obtained for the various systems.

The existence of large discrepancies between the dynamically computed mass and the luminous mass of *spiral galaxies* seems firmly established (cf. review by van Albada and Sancisi 1986). The total masses and mass distributions of these systems are derived from optical and HI rotation curves, which are now available for a large number of objects (Rubin et al. 1985, Bosma 1981, Sancisi and van Albada 1987a). The most reliable evidence to date, however, comes from the study of a small number of systems which have exponential luminosity profiles and HI layers extending far beyond the optical images - out to $10-12$ disk scale lengths. Their velocity fields are regular and do not show signs of large deviations from axial symmetry or flows in the radial direction: in other words, there are no indications that the assumption of gas moving in circular orbits is not valid. The estimated total masses, i.e. the mass inside the last measured point on the rotation curve, lead to values of M/L_B in the range of 10 to 30. The corresponding values of the cosmological density parameter Ω, 0.01 to 0.02, are, however, negligible with respect to the closure density of the universe.

The rotation curves for three of these galaxies are shown in Fig. 1. The curves extend to $10-12$ disk scale lengths and are flat out to the last measured point. The uncertainties in the circular velocities are approximately 2 to 5 km/s. Circular velocities have been calculated from the luminosity profile (see Fig. 2 top) by assuming a constant M/L ratio. If the value of M/L is chosen such as to maximize the contribution by the luminous disk, one obtains an estimate of the

(*)   Kapteyn Astronomical Institute, Groningen University, Postbus 800, 9700 AV Groningen

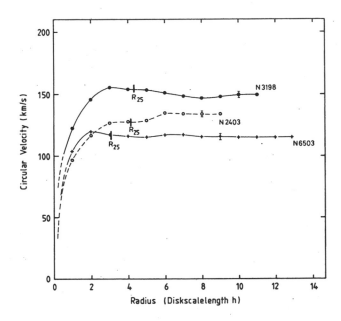

Fig. 1 - HI rotation curves for three spiral galaxies with extended, symmetrical HI disks and exponential luminosity profiles (from Begeman 1987; see also Sancisi and van Albada 1987b).

minimum amount of dark matter required to explain the observed rotation curve (see Fig. 2 bottom). Clearly there is a large discrepancy between the observed and predicted curves, starting around 2 − 3 scale lengths and increasing in the outer parts. The local M/L ratio increases from about 1 − 4 in the inner parts to several thousands in the outer parts (assuming that the dark matter lies in a disk; see Albada and Sancisi 1986). The amount of dark matter inside the last measured point is about 4 times as large as the mass of the disk (assumed to have constant M/L). This should be regarded as the minimum amount of dark matter required as, in principle, the disk contribution could be much less. It is likely, however, that the estimated maximum value of the M/L ratio for the disk is close to the true value (cf. Kent 1986, and van Albada and Sancisi 1986).

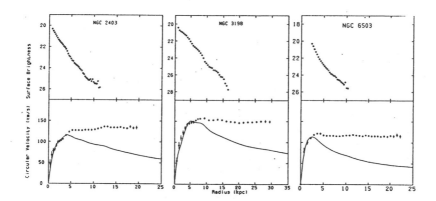

Fig. 2 - Light profiles and rotation curves for the three galaxies of Fig. 1. Upper panels: observed luminosity profiles (Wevers 1984). Lower panels: observed HI rotation curves (dots with error bars) and model curves (solid line) representing the circular velocity of stars and gas (Begeman 1987). The large discrepancy between observed and predicted (constant M/L) velocities is the "signature" of the dark halo.

The conclusion then is that dark matter is concentrated in the outer parts of spiral galaxies; in the inner parts, although not entirely ruled out, its presence is not required by the observations. The dark material amounts to at least 4 times the luminous material inside 2 Holmberg radii (about 10 disc scale lengths). The local *M/L* values are of order 5000 or larger.

Several question remain unanswered at present:

1.   What is the total mass of a galaxy? Where is the boundary of the dark halo? No convincing case has been found as yet of declining rotation curves.

2.   What is the explanation of the "conspiracy" of disk and halo to produce a flat rotation curve through the combination of a declining rotation curve of the disk and a rising one of the halo?

3.   What is the distribution of dark matter: in a spherical halo, or a disk which becomes darker (and perhaps thicker) in the outer parts? Although the halo hypothesis is generally favoured there is no compelling observational evidence as yet.

4.   Do the amount and distribution of dark matter depend on luminosity, morphological type, or environment?

5.   What is the nature of the dark matter associated with galaxies? Is it all baryonic? What are its constituents?

The case for dark halos associated with *dwarf galaxies* (spheroidal and irregular) is still in doubt (cf. Kormendy 1987, Sancisi and van Albada 1987b).

The presence of massive dark halos around *elliptical galaxies* has been inferred from X-ray data showing extended, hot gaseous coronae (Forman et al. 1985). Values of M/L up to 100 are derived (see also Fabian et. al. 1986). The best studied objects are M 87 (Fabricant and Gorenstein 1983) and NGC 4472 (Forman et al. 1985) for which the density and the temperature profiles are both available, although the latter are quite uncertain. The M/L values are about 200 at $r = 10re$ for M 87, and 20 at $r = 3re$ for NGC 4472. A recent discussion of the large uncertainties in all these estimates is given by Trinchieri, Fabbiano and Canizares (1986), who find that the mass within the observed region could be as low as that indicated by the optical data. This is true also for NGC 4472. The evidence of a dark halo around M 87 seems quite compelling, but it is not clear whether the halo belongs to M 87 itself or to the Virgo cluster of galaxies. The X-ray results, although certainly consistent with the presence of extended and massive halos around elliptical galaxies, do not seem, therefore, to offer incontrovertible evidence of their existence.

Since the total mass and extent of dark halos around individual galaxies are not known, it is important to investigate multiple systems such as binaries, groups and clusters.

The studies of *binary galaxies* and *groups* present serious difficulties (cf. Sancisi and van Albada 1987b and references therein). They do indicate, however, that dark halos are needed to explain the observations, with typical values of $M(\text{dark})/M(\text{lum}) \approx 3$ and suggest that they do not extend much beyond 5 optical radii (Van Albada 1988).

The presence of dark matter in *clusters* of galaxies, already advocated fifty years ago by Zwicky (1933), is strongly indicated by the virial analysis of radial velocities of the cluster members on the assumption that the clusters are bound and in equilibrium. The mass-to-blue luminosity ratios found for the Coma (Kent and Gunn 1982), Perseus (Kent and Sargent 1983) and Virgo (Huchra 1985) clusters are about 300 to 500 indicating values of Ω between 0.2 and 0.3, ten times larger than for spirals. In these analyses the distribution of dark matter is assumed to follow that of the light. If the dark material is more concentrated than the galaxies to the cluster centre the mass estimate becomes a factor three smaller; if, on the other hand, it is more extended to the outer parts the mass required is about a factor three larger (The and White 1986). Even in the former case, a large amount of dark matter is still needed.

The known baryonic matter consists of stars ($M/L \sim 10$) and hot intergalactic gas. The total mass of the latter as estimated from the observations of diffuse X-ray emission in clusters is rather uncertain, but it seems to be at least as large as the stellar component (Henriksen and Mushotzky 1985), and may be represented, therefore, with M/L values ranging from 10 to 30. If we compare the M/L values of clusters with those of stars and gas together we find a factor 10 in the cluster M/L unaccounted for. Some of this dark matter is located in the halos of individual galaxies. This is, however, only a small fraction. It is not clear where the remaining part is, whether it is all associated with galaxies and located in much more massive and extended halos than observed so far or whether it is spread out in intracluster space. The former possibility seems unlikely considering the small separation between galaxies in the dense cluster regions. It is interesting to note that the dark matter in spirals, although closely coupled to the light, does not follow its distribution, as it is more extended. One may wonder whether such a segregation might also occur on the scale of clusters and whether dark matter might be generally more extended than luminous matter and perhaps not be traced at all by the light.

The nature and composition of dark matter are still a complete mystery. The amounts strictly required from observations would not be inconsistent with the possibility of it being exclusively baryonic material. Any hypothesis on the nature and origin of dark matter, at least for the dark halos around spiral galaxies, should account for the large local M/L values mentioned above and for its close "coupling" with the luminous material (i.e. the disk-halo conspiracy).

Alternative explanations, less conventional than the presence of large quantities of dark matter, have been proposed for the observed mass discrepancies. Finzi (1963), and more recently Milgrom (1983), Bekenstein (1987 and references therein) and Sanders (1984, 1986), have made radical suggestions involving modifications of Newtonian dynamics or gravity and have compared their predictions with the observations. So far these comparisons do not seem to have led to inconsistencies which would rule out such alternative.

Acknowledgement to Luigi Radicati

Several times during the past ten years I have been a guest at the Scuola Normale, invited by Luigi Radicati. My memories of these visits are very warm. I have greatly enjoyed the generous hospitality at the Scuola and the scientific collaboration with friends and collegues. But above all, the friendship and kindness of Luigi Radicati and his interest and sympathy for my research have been very important to me. It is indeed a privilege for me to present this short contribution in his honour.

REFERENCES

[1] T.S. VAN ALBADA 1988, in IAU Symposium 130, Large Scale Structures, ed. J. AUDOUZE - M.-C. PELLETAN - A. SZALAY (Dordrecht: Kluwer Academic Publishers), p. 401.

[2] T.S. VAN ALBADA - R. SANCISI 1986, in Material Content of the Universe, Phil. Trans. R. Soc. Lond., A. 320, 447.

[3] K. BEGEMAN 1987, Ph. D. THESIS, University of Groningen.

[4] J.D. BEKENSTEIN 1987, in Proceedings of the Second Canadian Conference on General Relativity and Relativistic Astrophysics, Ed. Dyer. C., World Scientific, Singapore.

[5] A. BOSMA 1981, Astron, J., 86, 1825.

[6] A.C. FABIAN - P.A. THOMAS - S.M. FALL - R.E. WHITE III 1986, Mon. Not. Roy. astr. Soc., 221, 1045.

[7] D. FABRICANT - P. GORENSTEIN 1983, Astrophys, J. 267, 535.

[8] A. FINZI 1963, Mon. Not. Roy astr. Soc., 127, 21.

[9] W. FORMAN - C. JONES - W. TUCKER 1985, Astrophys J., 293, 102.

[10] M.J. HENRIKSEN - R.F. MUSHOTZKY 1985, Astropys. J., 292, 441.

[11] J.P. HUCHRA 1985, in The Virgo Cluster of Galaxies, ESO Workshop, ed. O.-G. RICHTER - B. BINGGELI, p. 181.

[12] S.M. KENT 1986, Astron. J. 91, 1301.

[13] S.M. KENT - J.E. GUNN 1982, Astron. J., 87, 945.

[14] S.M. KENT - W.L.W. SARGENT 1983, Astron. J., 88, 697.

[15] J. KORMENDY 1987, in IAU Symposium 117, *Dark Matter in the Universe*, ed. J. KORMENDY - G.R. KNAPP (Dordrecht: Reidel), p. 139.

[16] M. MILGROM 1983, Astrophys. J., 270, 365.

[17] V.C. RUBIN - D. BURSTEIN - W.K. FORD Jr. - N. THONNARD, 1985, Astrophys. J. 289, 81.

[18] R. SANCISI - T.S. VAN ALBADA 1987a, in IAU Symposium 117, *Dark Matter in the Universe*, ed. J. KORMENDY - G.R. KNAPP (Dordrecht: Reidel), p. 67.

[19] R. SANCISI - T.S. VAN ALBADA 1987b, in IAU Symposium 124, Observational Cosmology, ed. G. BURBIDGE - A. HEWITT (Dordrecht: Reidel), p. 699.

[20] R.H. SANDERS 1984, Astron. Astrophys., 136. L21.

[21] R.H. SANDERS 1986, Mon. Not. Roy. astr. Soc., 223, 539.

[22] L.S. THE, S.D.M. WHITE 1986, Astron. J., 92, 1248.

[23 G. TRINCHIERI - G. FABBIANO - C.R. CANIZARES 1986, Astrophys. J., 310, 637.

[24] B.M.H.R. WEVERS 1984, Ph. D. Thesis, University of Groningen.

[25] F. ZWICKY 1933, Helv. Phys. Acta., 6, 110.

Euler's *"Anleitung zur Naturlehre"* [1]

D. SPEISER (*)

For Gianna and Luigi A. Radicati

It is a great honour and even more so a special pleasure to be invited to contribute to a volume dedicated to Luigi Radicati. Our friendship (or should I say my discipleship?) dates from fall 1960 when we shared an office at the Institute for Advanced Study. There we discussed many problems connected with symmetries, internal and other ones, and we even struggled bravely for several days and evenings with $SU(3)$-Clebsch-Gordan coefficients.

Since 1962, I had the privilege to spend many weeks in Pisa, first at Piazza Torricelli, then at the even more illustrious Piazza dei Cavalieri. At both places, I spent only delightful hours and in addition, I spent many more at Via del Risorgimento and at Via del Capannone!

Euler attempted to formulate in one, as we would say, unified system, the whole science of inorganic nature as it was known in his time. In his great *"Anleitung zur Naturlehre"*, or: "Guide to the Science of Nature", he indicated how he thought that such as ambitious goal could be realized. Unfortunately, the book was never published during his lifetime, but only in 1862, and by then parts of two sections had been lost. In the Op. Omnia the *Anleitung* is contained in Vol. III 1 edited by F. Rudio et al..

For an understanding of the ideas that motivated and guided Euler's theories of light, electricity, magnetism and gravitation the *Anleitung* is indispensable, but so far it has been examined only very little [2]. For this reason we try to give a few indications here concerning the notion of field as it had evolved from Euler's hydrodynamical investigations. This idea was to guide him in his views on the subject mentioned above. As we shall see, some of Euler's ideas again have a certain topicality today. There can be no question of giving a detailed account of the whole work here, let alone a detailed analysis, since we are concerned here only with some of its 21 sections, but a short view over the

(*)   Institut de Physique Théorique, Université Catholique de Louvain, B-1348 Louvain-la-Neuve.

whole work is necessary for clarifying the significance of what Euler is driving at.

The work in its present form fills 163 pages of vol. III 1 of the Op. Omnia. Since the end of Chapter 5 plus the beginning of Chapter 6 may have filled another 7 - 9, (there are 9 missing Sections), we now find us a work of about 170 pages. But there are reasons to think that the work as it stands is not complete and at the end of the paper we shall ask the question: why did Euler not finish it?

The book is divided in 21 Chapters, that may be grouped as follows.

Chapters 1 - 6: Introduction and discussion of the four fundamental properties of all bodies (chap. 1):

> extension (chap. 2)
>
> mobility (chap. 3)
>
> persistence ("*Standhaftigkeit*") (chap. 4)
>
> impenetrability, (chap. 5)

Of this last Chapter the end is missing and so are the title and beginning of Chapter 6. Rudio's guess was that this section may have been called "*Von den Kräften im Allgemeinen*". But the title may have been also "*Von den Veränderungen der Körper und ihren Ursachen*". At any rate, it looks like the introduction to the next part.

Chapters 7 - 11: General introduction to the principles of mechanics; the introduction includes a thorough discussion of what call today the *principle of relativity* [3].

Chapters 12 - 14: Euler states his main hypothesis concerning the structure of matter: there are only two kinds of matter "*die grobe*" (gross, coarse, heavy) and "*die subtile*" (subtle, fine). Of those two elementary matters only all materials that we observe are built up!

Chapters 15 - 18: contain a discussion of the different kinds of bodies found in nature: fluid, solid, elastic.

Chapter 19: is a speculative digression on gravity.

Chapters 20 - 21: are a thorough exposition of the principles of hydrostatics and of hydrodynamics; this theory is at the basis of his speculation.

Symmetries, the subject of this volume, play, we may add, a double role in the *Anleitung*. In sections 7 - 11, as mentioned above, Euler discusses explicitly what we call today the principle of relativity. Thereby he settled the dispute between Newton and Leibniz concerning the role played by space and time in nature. I wrote about this question elswhere and I shall not return to it [4]. But, throughout the *Anleitung*, symmetries play a hidden but none the less decisive role. Euler had since 1750 [5] slowly begun to grasp the most important consequence of the principle of relativity and (in modern terms) the action of a group on space and time, namely the fact that the laws of nature are relations between tensors. Euler was the first to see this, and for this reason after 1750 he used the "manifestly covariant form" in most of his books and papers in mechanics i.e. used Cartesian coordinates which is the closest substitute to the use of vectors and tensors. The first fruit of this new understanding was the equations of hydrodynamics based on his own discovery of the internal pressure, (a scalar density) which one to formulate the effect through contact of one part of a fluid on anothet without the explicit a priori knowledge of the nature of the forces between these parts.

Without it the view of nature proposed in the *Anleitung* would be meaningless; this is really at the bottom of his idea.

This short survey shows that the *Anleitung* is not a *Cosmology* [6] in the sense of Leibniz and Wolff and of some of their followers, that is, a deduction of the fundamental laws of Physics from metaphysical principles, and we shall presently what distinguishes it also from Descartes speculations to whom in some respect is ressembles. While Sects. 6,7 and 11 contain a clear exposition of the basic notions following a metaphysical tendency that once prevailed in the scientific world, and a thorough discussion of it, the *Anleitung* itself is rather an easily understandable presentation and discussion of the fundamental laws of Mechanics, especially hydrodynamics for the professional and nonprofessional reader alike, followed by a grandiose speculation about the structure and composition of all matter. But this speculation is a strictly scientific one. Euler's hope is to show that from these dynamical laws and from his simple hypothesis concerning matter, all other of physics can be derived. More about this speculation will be said later.

The introduction in the first sections, of the four properties ("*Eigenschaften*") common to all bodies: extension, mobility, persistence and impenetrability, may strike the reader who glances through the book for the first time as a purely philosophically minded and even as a slightly scholastic procedure. But if we read these sections carefully and if we observe what Euler is driving at in the later ones, we see that he presents and discusses simply the various elements that enter into the Euler equations of hydrodynamics

(1)
$$\rho \frac{d\vec{u}}{dt} + \vec{\nabla}p = \vec{f}$$

$$\partial_t \, \rho + \vec{\nabla}(\rho\vec{u}) = 0$$

Obviously:

extension ("*Ausdehnung*") corresponds to space: $\vec{x}, d\vec{x}$

mobility ("*Beweglichkeit*") to time and velocity:

$$t, dt, \vec{u}, d\vec{u}, \frac{d\vec{u}}{dt}$$

persistence ("*Standhaftigkeit*") [or inertia ("*Trägheit*")] corresponds to mass m and mass density ρ

impenetrability ("*Undurchdringlichkeit*") to Euler's own creation: the internal pressure p.

Indeed out of these elements the left hand side of the Euler equation [1] is constructed. So we may see in the importance attributed to these "properties" a mere concession to the style of his time. What about the right hand side, that is the outer forces (or force-densities to be precise)? In most of the usual applications of these equations $\vec{f}$ represents the gravitational field or e.g. the constraint due to a tube, etc... This, as we shall see presently, is where Euler's speculation enters.

For understanding Euler's idea the reader must recall two limitations of classical physics.

First: classical mechanics speaks only of the propagation of bodies through space and time. What these bodies are, what their structure is, classical physics cannot tell without making some further adhoc assumptions: today we know that Quantum Mechanics is here of essence.

Second: classical mechanics, hydrodynamics for instance, assumes the existence of forces, e.g. with Newton it introduces into his computations gravitational forces, without, however, having an answer as to the mechanism that transmits them. Euler tries to fill both gaps.

We beginn with the second point. According to Euler the force densities $\vec{f}$ in eq.(1) result, in the last analysis always from the *impenetrability* of some matter, i.e. from, the pressure that this matter exerts on some other bodies. I.o.w. according to Euler's Naturlehre, what we observe in nature, is *a system of dynamically coupled* (as we would say today) *fluids*. The coupling is the result of the impenetrability of each matter and all observed phenomena can be resolved into such interactions.

The big question is now: how many different materials are there in nature? Infinitely many as it would seem to the ordinary observer or only a few basic oncs, out of which all other ones are built up? To being with, surely some fluids or at least one must be responsible for optical, electric, magnetic and gravitational phenomena. This fluid is the ether: we shall come back to it and to the question whether there is one ether only or various ones.

But what about the ordinary materials: e.g. the metals, the gases, the insulators to say nothing of organic matter? Is there really an endless variety of them or can we hope to bring some mathematical order into what we find through the known laws of mechanics? This is the question that Euler tries to answer with his speculation. But before presenting it to the reader we must recall, that chemistry, as we know it today hardly existed at that time; almost no law of chemistry was known and but a few elements had been identified. Thus there was not much that could guide or stimulate Euler, at least nothing comparable in importance and general applicability with his own equations for fluids.

From Euler's paper E 109 [7] and especially also from the *Lettres* on Electricity and on Magnetism [8] we know that Euler conceived all ordinary matter as permeated by "pores" and "canals" through which the ether flows. Obviously size and form of these canals and pores are characteristic for any material and may vary considerably from one to the other. Euler conjectures now that it is really only by the configuration of these canals and pores, their size and form etc. that the various materials as we know them, differ from each other, but that the **matter** itself, the **stuff** of which they are built, is in all of them the same!

This one fundamental matter fills only a very small part of a body, it consists of tiny but very heavy specks and droplet whose density is always the same. Between these heavy droplets run the canals through which the ether flows. How and according to what laws this structure that caracterizes a material is formed, he cannot say, of course. However, this basic heavy matter, like the fine one, the ether, obeys the equations of hydrodynamics (1).

Lest we should give the reader who does not knows the *Anleitung* the impression that Euler's ideas are pure verbiage, we recall here that his verbal explanations are underpinned throughout with a thorough presentation of the fundamentals of mechanics, especially of hydrodynamics. It is indeed his long and enduring occupation with this continuum theory that guided his speculation. These underpinnings show also that his ether was not just one new matter more, like those which so many of his predecessors had proposed or even the same: it was a matter *that obeyed established laws of continuum mechanics.* Thus its properties were not ad hoc assumptions but could be deduced from these laws. Knowledge and use of these laws separates Euler's speculation from older ones it may recall.

Thus his speculation, at least, is not lacking consistency and coherence. We may sum up these ideas as follows.

There are only two kinds of matter in the universe: the gross ("*grobe Materie*") and the fine ("*feine Materie*"). The structure of all bodies and their qualities are determined by them. The heavy matter is the same in all bodies; it is porous, made up of *drops*. All drops have the same density and if the *observed* weights of bodies are different, this is due to the varying number of drops per unit volume as well as the different number and size of drops of heavy fluid contained in these bodies. The constitution of bodies is fixed by their stereometric configuration, and the distribution of the drops in the body is determined also by the fine matter, the ether. That is: the same ether which we met in the theory of heat, in optics, in the theory of electricity and of magnetism, is partially responsible for the structure and for some of the macroscopic properties of all bodies [9].

Ether, according to Euler, is responsible for gravitation too. Thus only the gross matter i.e. the heavy drops, contribute to the body's weight. He leaves undecided (Sect. 96) whether there is only one ether or several. Empirically we cannot say, but he states that it would be against the principles of sound research to suppose the existence of more than one, unless such an assumption is made unavoidable by experience (Sect. 96, p. 103: *"Ob es von dieser subtilen Materie weiter verschiedene Arten gibt, von welchen eine dichter sei als die andere, müssen wir hier an seinen Ort gestellt sein lassen und wollen zum wenigsten alle diese Arten, wenn je mehrere vorhanden wären, unter dem allgemeinen Namen der subtilen Materie begreifen. Den so lange die Erklärung der Begebenheiten der Natur nicht mehrere solche Arten erheischet, so würde es verwegen sein und gegen die Regeln einer gesunden Naturlehre laufen, wenn wir bloss aus unserer Einbildung die Anzahl der subtilen Materien vermehren wollten"*.

Before discussing the most interesting feature of the *Anleitung* we try to evaluate what Euler achieved by asking: where did he go in the right direction, and where did he go wrong? What, with hindsight, must be judged as hopeful, even fruitful, and finally, which of his ideas were eventually followed up by successors?

That all ordinary materials are made up of one and the same matter and that these materials differ only by their stereometric structure must have seemed at the time a fantastic speculation. Even the 92 elements in Mayer's and Mendeleev's table, the crowing achievement of a long line of reseach, more than 100 years later are a far cry from one single heavy matter! Yet Euler's heavy droplets bear a curious resemblance to today's nuclear matter; the tiny nuclei too are, after all, responsible for over ninety-nine percent of the weight of all bodies. And even though it would be silly to claim a tangible connection between Euler's drops and the nuclei this resemblance shows, at least, that Euler's vision was not absurd.

But then: what determines the specific structure of ordinary bodies and what is the interplay between matter and the ether? Euler does not say, and of course he cannot. Manifestly there is a huge gap here. The whole world of the atom, of chemistry and of crystallography and what determines it, namely the electron, are, of course, entirely missing. Modern chemistry, as we said, was hardly born; though crystals may have occupied Euler, they could not lead him very far. The electron, to say nothing of quantum mechanics, lay in a far distance beyond any dream. No wonder then that nothing of all this can be found here.

But it is a very different story with the ether, *"the subtle air of heaven"*. Here, as Faraday's testimony [10] shows, Euler was definitely on the right track and his ideas remained not without influence. We have to ask how Euler had come to such a profound as well as powerful insight. Indeed optics, electricity and magnetism are today united by the same electromagnetic field, as we call it, of which the ether was a forerunner. Perhaps Euler went wrong when he hoped to make this ether responsible also for gravitation. After the innumerable attempts by Weyl, Einstein and many others which all failed too, we doubt now that there is a direct way to bring gravity and electromagnetism together. Maybe this goal can be reached if the theory also includes the *heavy matter*, as is the case for instance with the *supersymmetric* field theories that comprise also spinor fields.

But what does all this amount to? These single ideas, interesting and fruitful as they were, are in fact not the most interesting aspects of the *Anleitung*. What strikes us most today is the totality of his ambition and the way he tries to realize it. Euler attemps here neither more nor less than to conceive a coherent system that is able to describe the whole of inorganic nature starting from a few principles. Newton in the famous last queries 28-31 of the *Opticks*, had sketched a few ideas in this direction, yet they remained queries [11] Newton was lacking the new powerful tool that is at basis of fieldtheories: partial differential equations. Here in the *Anleitung* the ambition is greater, and the realisation goes further. It shows how more powerful mathematics and physics had become when they had learned to master the mechanics of the continua. Especially this new branch of analysis had permitted d'Alembert and Euler [12] to create the field concept.

Thus Euler was the first to conceive and to imagine a *unified field theory*, an idea that stands today in the center of the physicist' aim. Indeed according to Euler we can reduce all phenomena to the interplay of four scalar and of two vector fields: the densities and pressure of the heavy matter and of the ether and of their two respective velocity fields, i.e. their streamlines, and to the partial differential equations that connect and govern them. As an attempt, Euler's speculation, in spite of the enormous additional amount of accumulated empirical knowledge that we possess today as well as our much more penetrating insight into the structure of matter is in principle not all too different from present day speculation that aims at unified field theories: a well established differential equation, basic in one domain of mechanics, is declared the foundamental principle for all other observed processess too, and this principle is combined with a radical hypothesis that must complete it. Here, as we said, the basic principle, the field equations are the Euler equations of hydrodynamics.

This is another one of the many instances that allow and even compel us to say that the physics of the 20th century, where mathematics again plays a dominant role, is much closer to the physics of Newton, of the Bernoullis, of Euler and of Lagrange than the physics of the 19th century by and large was, and therefore we are often much more apt to appreciate the former period than the scientists of the 19th century were. For today we have learned again that such speculations are not idle but can be the motor to progress.

We conclude this report on the *Anleitung* by pursuing the subsequent work of Euler a bit further. When after 1759 Euler again took up his work on acoustics [13] he discovered the linear differential equation, today named after d'Alembert [14]:

$$(2) \qquad \mathbf{u} \equiv (\frac{1}{c^2}\partial_t^2 - \Delta)\; u = 0$$

by deriving it from the equations (1) of hydrodynamics. Euler had first dispaired of solving the wave equation in three dimensions since in two it had been shown to be intractable. But then, as he says in the following paper E 307, p. 485: "C'est précisement ici, comme je l'ai remarqué depuis, que les difficultés ne sont pas invicibles, et c'est là qu'a lieu un cas semblable à ceux que la Comte Riccati a proposé autrefois, où une certaine équation devient intégrable, pendant qu'en général elle ne l'est pas [15].

For Euler, the theory of sound had always been the model for the theory of light [16]. Now finally after 1759, he had in his hands all the tools for explaining and formulating almost all optical phenomena, especially interference and diffraction, but with one significant exception to which we shall return. But then after that date other subjects attracted him more strongly, and in his last years he did return to only twice the questions discussed in this volume. We may regret this but we cannot quarrel with Euler.

However that may be, shortly after this discovery Euler explained his ideas concerning the propagation of light in a letter sent to Turino. (Indeed: what better place could he have chosen?) To M. de la Grange he wrote:

"*Monsieur,*

Je suis bien flatté de l'approbation dont votre illustre Academie, et vous en particulier, aves bien voulu honorer mon essai sur les ébranlemens dans un milieu elastique; l'honneur de ces profondes recherches est uniquement dû à votre sagacité, et je n'y ai rien fait que profiter des lumières que votre excellent Memoire [17] m'a fournies. Vous y aves ouvert une carrière tout a fait nouvelle, où tous les géomètres qui viendront après nous trouveront toujours abondamment de quoi occuper leur addresse, et, à mesure qu'ils y reussissent, l'Analyse en acquerra des accroissements très considerables. Or la matière même est sans doute la plus importante dans la Physique: non seulement tous les phenomènes de la propagation du son en dépendent, mais je suis assuré que la propagation de la lumière suit les mêmes loix. On n'a qu'à substituer l'ether au lieu de l'air, et les ébranlemens qui y sont répandus nous donneront la propagation de la lumière. Maintenant, il serait à souhaiter qu'on pût déterminer les alterations que les ébranlemens excités dans un milieu souffrent lorsqu'ils passent dans un autre milieu dont la densité et [l'] élasticité sont différentes. Je ne sais pas si l'on peut esperer la solution de ce probleme, mais je suis très convaincu qu'on y decouvriroit infalliblement non seulement les veritables loix de la refraction, mais aussi la plus complette explication de la reflexion dont la refraction est toujours accompagnée: on verroit qu'il est impossible que les rayons passent d'un milieu dans un autre, sans qu'une partie rebrousse chemin. Peut-être que cette consideration pourroit faciliter le developpement de l'analyse et fournir au moins quelques solutions particulieres;

mais on rencontrera ici une nouvelle difficulté. Comme il faut estimer tant la densité que l'elasticité des autres milieurx transparens comme par exemple le verre, la densité étant si grande par rapport à celle de l'ether sans qu'on puisse supposer plus grande son elasticité [, ainsi] que la vitesse des rayons dans le verre, deviendront extrêmement petite; cependant, je crois que la refraction même prouve suffisamment qua la vitesse des rayons dans le verre à celle dans l'ether doit être dans la raison de 2 à 3 [18]".

But this is not all: there is a second "*Lettre de M. Euler de la Grange* [19]" destinated this time for publication in which Euler says "*Depuis ma derniere lettre j'ai réussi à ramener au calcul la propagation du Son, en supposant à l'air toutes les trois dimensions, et quoique je ne doute pas que Vous n'y soyés parvenu plus heureusement, je en crois pouvoir mieux témoigner mon attachement envers Votre Illustre Société qu'en lui présentant mes Recherches sur ce même sujet*".

But then this second letter is *not* called: "*De la propagation du son*" as the first mémoire had been but: "*Recherches sur la propagation des ebranlemens dans un milieu elastique*". In view of what Euler had written to M. de la Grange earlier and what we find stated all over Euler's work, it is clear that "*milieu élastique*" refers here not only to air but also to *the ether, in other words to light*. The derivation of the wave equation is now shorter and again we find the solutions that describe spherical waves [20].

So we may justly consider this letter to Torino printed in the Mélanges de philosophie et de mathématique de la société royale de Turin [21] *as the first paper on mathematical wave optics* [22]!

What we said just now allows us to put the finger more precisely on one limitation of Euler's physics. Equation (2) is a *scalar* equation: thus while interference and diffraction phenomena can easily be derived from it, one optical phenomenon cannot, and therefore resists Euler's attempts: polarization. Indeed Euler never speaks of polarization, although it is possible that he knew this phenomenon. In any case there is no room for polarization in a theory based on a *fluid* i.e. scalar ether: what was needed here is an *elastic ether*. From the point of view of field theory: a scalar field like pressure cannot produce transverse effects but such effects (shearing) are obviously needed for making a material cohesive!

The modern theory of elasticity as Truesdell has shown in his history of elasticity, Vol. II, 11 of the Opera, goes back to Jacob Bernoulli. Daniel Bernoulli [23] created the linearized theory and discussed oscillations and his brother Johann II Bernoulli had discussed later transverse vibrations [24]. Euler had taken up the subject independently of Daniel Bernoulli, exchanging his ideas with him, and his numerous papers must be considered as one of the greatest contributions to the subject in the 18th century. Yet they were almost all limited to one-dimensional problems: only once did Euler overcome this restriction successfully when he found the equation of the oscillating membrane i.e. the drum. To formulate a true three-dimensional theory of elasticity was impossible at the time due, as Truesdell showed, to the lack of sufficiently powerful mathematical, especially differential geometrical, tools. This gap in mechanics was closed only in the twenties of the 19th century by Cauchy through his creation of the stress tensor δ_{ik}. Thereby Cauchy laid the general foundations of the mechanics of continua. Euler's internal pressure p is a special case of Cauchy's stress tensor valid for a non viscous fluid

$$(3) \qquad \sigma_{ik} = p \; \delta_{ik}$$

Through the creation of this concept the ground was opened for further progress, and the fluid ether was now superseded by its successor the elastic ether. From Cauchy to Kelvin many worked on these theories, yet polarization again proved to A be major obstacle, even the stumbling block; while the fluid ether could not accomodate it, the elastic ether proved to be too generous: the longitudinal polarization, predicted by every elastic theory, failed to show up. Eventually Maxwell's theory gave the correct solution. But even Maxwell conceived his theory of the elastic ether [25] and this view was certainly also held by H.A. Lorentz. This may seem surprising since in the vacuum, at least, Maxwell's theory is a linear one, while an ether theory necessarily must contain non linear terms. At any rate the modern field concept emerged only after the investigations of Poincaré and Einstein [26].

Furthermore this new electromagnetic field, although a classical concept, is already very close to the photon, which is a true quantum notion, far outside anything classical mechanics could think of.

Is the *Anleitung*, leaving aside the missing chapters five and six, a complete work and if so, why was it not published, or if not, what stopped Euler from pursuing the subject?

Trivial reasons, like: Euler could not find a publisher, or: he was interrupted by more urgent business, e.g. an order by Frederick II, who in those years was involved in the Seven Years War, cannot be excluded. Yet there may have been a deeper reason. Euler must have become conscious somehow, that without a theory of elastic materials he could not achieve his goal. With a scalar pressure only, it is not possible to generate shear forces, etc., and those are badly needed for a solid world!

True, most of Euler's work on two and three dimensional elasticity [27] especially E 410, 1771 and E 481, 1774, were written long after 1760. But his one successful work E 302 (on the drum) dates from 1759 too and the subject certainly must have occupied him earlier. It may well have been the lack of a 3-dimensional theory of elasticity which stopped Euler from pursuing his speculation.

Did the *Anleitung* find an echo?

I have mentioned already (Cf. note 9) that Faraday had read and quoted repeatedly Euler's Lettres à une Princesse d'Allemagne. Once he says speaking about magnetism *"There are at present two, or rather three general hypotheses of the physical nature of magnetic action. First, that of aethers, carrying with it the idea of fluxes or currents, and this of Euler has set forth in a simple manner to the unmathematical philosopher in his Lettre; - in that hypothesis the magnetic fluid or aether is supposed to move in streams through magnets, and also the space and substances around them"*... *"My physicohypothetical notion does not go so far in assumption as the second and third of these ideas, for it does not profess to say how the magnetic force is originated or sustained in a magnet; it falls in rather the first view [i.e. Euler's], yet does not assume so much."* But Faraday could not know the *Anleitung*.

There is a puzzling sentence in the posthumous work of B. Riemann Concerning his own philosophical investigations, a note in his papers says: *"Meine Hauptarbeit betrifft eine neue Auffassung der bekannten Naturgesetze - Ausdruck derselben mittelst anderer Grundbegriffe - wodurch die Benutzung der experimentellen Data über die Wechselwirkung zwischen Wärme, Licht, Magnetismus und Electricität zur Erforschung ihres Zusammenhangs möglich wurde. Ich wurde dazu hauptsächlich durch das Studium der Werke Newton's, Euler's und anderseits Herbart's geführt [28]."* This note is not dated but it is in a close relation to what he writes to his brother in a letter written December 28th 1853 [29].

What did Riemann allude to in Euler's work? One thinks first of the Lettres a une Princesse d'Allemagne. The second and especially the third of these essays on natural philosophy stand in a close relation to Euler's papers E 305, 306, 307 quoted above. In the third one, Riemann discusses Euler's scalar wave equation and applies it to motions that cause optical and gravitational phenomena [30].

These notes are then probably written before Euler's *Anleitung* eventually appeared in 1862. The Riemann Archives in Göttingen seem not to show traces of this book [31].

But given Riemann's interest in Euler's work and in his ideas it is not unlikely that he may have read this book. He may even have done so in Pisa where he went twice in 1863 and in 1864, since, as is well known, he had many friends among the Pisan mathematicians and physicists [32]. His biographer R. Dedekind calls these stays *"einen wahren Lichtpunkt in seinem Leben"*.

No one who knows Pisa, Pisan scientists and especially the Directors and above all their wives will doubt this statement!

Mrs P. Radelet helped me with the manuscript and the references. My wife Ruth herself a sometimes guest in the Timpano, and my friend David Finkelstein provided good and much needed linguistic advice. To all of them go my best thanks.

REFERENCES

[1] Op. Omnia III 1, p. 16 - 178. This paper is a slightly extended version of a section of the Introduction for the forthcoming vol. of Leonhardi Euleri Opera Omnia III 10.

[2] One exception is H. Weyl who expressed his high regard for the Anleitung in his book "Philosophy of Mathematics and Natural Science"; rev. and augm. edition based on a translation by O. Helmer, Princeton NJ, Princeton Univ. Press 1949.

[3] Cf. D. Speiser, *The principle of relativity in Euler's work*, First International Meeting on the History of Scientific Ideas: *Symmetries in Physics 1600 - 1980*, Barcelone 1983, Gràficas Ferba 1987.

[4] Cf. note 2.

[5] Cf. E 177, Decouverte d'un nouveau principle de mécanique, 1752 and note 2.

[6] Cf. the book by M. Schramm, *Natur ohne Sinn? Das Ende des teleologischen Weltbildes.*, Reihe Herkunft und Zukunft, Styria, Graz Wiem Köln 1985.

[7] "*Dissertatio de magnete*", Prix de l'Académie de Paris pour 1746 (1748).

[8] Leonhardi Euleri Opera Omnia, ser. III, vol. 11 et 12.

[9] Whittaker, Ed. "*A History of the Theories of Aether and Electricity*", London 1951.

[10] Faraday read Euler's "Lettres à une Princesse d'Allemagne" and refers to them repeatedly in his Diary. For a full list of these quotations cf. D. Speiser, *L'oeuvre de L. Euler en Optique physique*, Actes du Colloque "Roemer et le vitesse de la lumière", Vrin, Paris 1978.

[11] Cf. *Sir Isaac Newton Opticks* Dover Publications, New York 1952, pp. 362 - 406.

[12] Cf. C.A. Truesdell's introduction to Vol. II, 12 especially p. LXXXI.

[13] See E 306, Vol. III 1, p. 478, 480.

[14] D'Alembert had discovered in 1746, the 1-dimensional d'Alembert equation valid for the vibrating string.

[15] Also at that period everybody and especially people from Basel took great profit from Italy (So many thingh are possible in Italy but not so, alas, elsewhere!)

[16] E 88, Vol. III 5, p. 1.

[17] Lagrange, "Recherches sur la nature et la propagation du son", Misc. Taurin. I., 1759, p. 1 - 112 Ouvres 1, 39 - 148. Concerning Euler's and Lagrange's work, cf. C.A. Truesdell's introduction to Vol. II 13 of L. Euleri Op. Omnia pp. XXXV-IL.

[18] Lettre de L. Euler à Lagrange, Berlin le 24 juin 1760, Euler, Briefwechsel Bd. V. p. 441.

[19] L. Euleri Opera Omnia II 10 pp. 255 - 263. The letter is also reproduced in the oeuvres de M. de la Grange t. 14, pp. 178 - 188.

[20] Cf. also M. Fierz: Zur Geschichte der Optik, in Vierteljahrsschrift der Naturforschenden Gesellschaft in Zuerich.

[21] Vol. 2, 1760/61 (1762), p. 1 - 10.

[22] Huygens', *Traité de la lumière*, a book that cannot be praised enough, deals with geometric not with wave - (= mechanical) optics.

[23] Die Werke von Daniel Bernoulli, Bd. 6 to appear.

[24] Johann II Bernoulli, *Recherches Physiques et Géométriques sur la question: comment se fait la propagation de la lumière*, Prix de l'Academie de Paris pour 1736.

[25] J.C. Maxwell, *A treatise on electricity and magnetism*, Third edition, Oxford 1904.

[26] A. Pais, *"Subtle is the Lord, The Science and the life of Albert Einstein"*, Oxford, New York 1982.

[27] See C.A. Truesdell's introduction to II 11 p. 320 ff. See also Truesdell's article, *Whence the law of moment of momentum*, in Mélanges A. Koyré; l'aventure de la science; Paris Vrin 1964.

[28] B. Riemann's gesammelte mathematische Werke und wissenschaftlicher Nachlass, H. Weber, Dover New York 1953, p. 507.

[29] Loc. cit., p. 547.

[30] Loc. Cit., p. 536 - 538 sects. b and c.

[31] I am indebted to Dr. U. Neuenschwander for this information.

[32] Exactly as it is pleasing to think of Euler's connection to Turin it is pleasing to think of a possible one through Riemann to Pisa.

Symmetry and Symmetry-breaking in Astrophysics

M. STIAVELLI (*)

Abstract

We briefly review several aspects of the applicability of symmetry arguments to astrophysical problems by discussing the well known classical example of the rotating figures of equilibrium and their relations with other more complex and physically interesting problems.

1. - Introduction

In this short review we plan to briefly discuss three different ways in which symmetry and symmetry-related arguments may appear in astrophysical problems. The first and, probably, the most fruitful way in which symmetry arguments can be applied, is in cases where the equations describing a given phenomenon are left unchanged by some symmetry group G. The solutions to these equations will have in general lower symmetry, represented by G_1, a subgroup of G. Such a phenomenon is often referred to as *spontaneous symmetry breaking*. In general it is easy to understand which symmetry breakings are *kinematically*, i.e. mathematically, allowed by using group theoretical arguments. In cases where there are no (or only a few) internal degrees of freedom in addition to those specified by the symmetry group, the symmetry breakings kinematically allowed can be simply related to those dynamically, i.e. physically, allowed. Unfortunately, in realistic systems of astrophysical interest there are many degrees of freedom, which cannot be easily dealt with in terms of group theoretical arguments. Hence, it is usually very hard to understand which breakings can actually occur and these sorts of considerations are of smaller applicability. In simple systems this can, however, be done and an explicit example will be discussed in the following Section. Note that when the equations of the motion admit a continuous symmetry group G, this can often be related to the existence of a conserved quantity, e.g., by application of Noether's theorem

(*)   Dept. of Physics and Astronomy, Rutgers University - Piscataway, NJ 08855.

when the equations can be derived from a Lagrangian function.

Another kind of symmetry which has some relevance in astrophysics is that relating the form of the equations themselves to those describing some entirely different physical system. Here the interest is of a more mathematical nature, but such a relation can give some insight into problem at hand. Finally, in some cases, one can find a discrete symmetry between different solutions of the same set of equations. In this case too, the interest is of a more mathematical nature, but, sometimes, the physical implications may be interesting.

Since a rigorous and exhaustive treatment of the subject would be far beyond scope of this short article, we will instead illustrate the three above mentioned aspects with a classical example: that of the rotating figures of equilibrium. Whenever possible, we will show how the physical features of the described phenomena are of more general astrophysical application.

2. - Classical figures of equilibrium

The study of the equilibrium of a rotating, self-gravitating fluid mass is a very old subject and it has been reviewed by several authors. We refer to Chandrasekhar (1969) for a comprehensive review of all the classical results. Here we will consider the problem from a different point of view (Stiavelli 1983), which is a generalization of a method originally introduced by Dyson (1968) to study interstellar gas clouds in cases where self-gravity could be neglected.

The idea is to describe the motion of the system by using linear operators and by deriving an equation for them. This is possible since the classical Dedekind problem (Chandrasekhar 1969) deals with linear fluid motion, hence the relations between the eulerian coordinates, $\mathbf{X}(t)$, and the lagrangian coordinates, ξ, of a fluid element are:

$$(1) \qquad\qquad \mathbf{X}(t) = L(t)\xi,$$

where $L(t)$ is a linear, time-dependent, operator. We consider the case where the figure of equilibrium is an ellipsoid, thus the axes of the ellipsoid allow us to identify a reference frame. If $\mathbf{x}(t)$ is an orthonormal reference frame comoving with the figure, i.e. the fluid surface, we have the following relation:

$$(2) \qquad\qquad \mathbf{x}(t) = T(t)\mathbf{X},$$

between the inertial and the comoving frame. T is a $SO(3)$ matrix describing the motion of the figure of equilibrium.

Now let ϵ_i be the semi-axis squared of the ellipsoidal figure of equilibrium, along the axis x_i. We define a matrix $A(t)$ as:

$$(3) \qquad A(t) = \begin{pmatrix} \sqrt{\epsilon_1} & 0 & 0 \\ 0 & \sqrt{\epsilon_2} & 0 \\ 0 & 0 & \sqrt{\epsilon_3} \end{pmatrix}.$$

A is a function of time, since in principle the ellipsoid can have non constant shape. We also define $A_0^{-1} \equiv A^{-1}(t = 0)$. Finally, define $F(t)$ by:

$$(4) \qquad L(t) = F(t) A_0^{-1}.$$

We want to find an expression for F in terms of quantities with clear physical meaning. We can do so by making use of the definition of T and by recalling that $\mathbf{x}(0) = \xi$ are the lagrangian coordinates of a fluid element.

$$(5) \qquad A^{-1}\mathbf{x} = A^{-1}T\mathbf{X} = A^{-1}TFA_0^{-1}\mathbf{x}(0) \equiv SA_0^{-1}\mathbf{x}(0).$$

The matrix S characterizes internal motions. If the shape and size of the ellipsoidal figure of equilibrium are constant, i.e. $\dot{A} = 0$, so that $\dot{\epsilon}_i = 0$, and $\epsilon_1\epsilon_2\epsilon_3 = 1$, we find that $S \in SO(3)$, since on the system surface $A^{-1}(= A_0^{-1}\mathbf{x})$ and $A_0^{-1}\mathbf{x}(0)$ can be seen as two generic vectors of length 1 related to one another by the application of S following Eq. (5).

The system motions are described by the linear operator:

$$(6) \qquad F = T^t A S,$$

where T and $S \in SO(3)$ and A is diagonal and independent of time. The Euler equation for the system is:

$$(7) \qquad \rho\left[\frac{\partial \mathbf{v}}{\partial t} + (\mathbf{v} \cdot \nabla)\mathbf{v}\right] = \rho\nabla\Phi - \nabla P,$$

where $\mathbf{v}$ are the fluid velocities, Φ is the gravitational potential, P the pressure. By making use of the classical results for the potential inside a constant density ellipsoid, after a straightforward algebraic manipulation, Eq. (7) can be cast in the following matrix form:

$$(8) \qquad \ddot{F} = T^t BAS,$$

where B is a diagonal matrix dependent on the central density, the central pressure and the values of ϵ_i. Note that, since T and S are specified by three parameters and A by two, F contains 8 free parameters in total. In the Appendix we show how a few solutions of interest for Eq. (8) can be derived. Here we note that Eq. (8) is formally equivalent to the equation of motion of a point mass constrained to move on a 8 dimensional manifold. Hence, we have explicitly found a one to one correspondence, i.e. a symmetry in our broad sense, in the equations describing a mechanical and a very simple fluid problem. Pegoraro (1972) has been able to show this equivalence in an alternative way by writing down a Lagrangian function for the system from which Eq. (8) could, in principle, be derived.

The correspondence between classical mechanics and fluid dynamics is however more general. For example, there is a relation between the motion of a solid body and the hydrodynamics of a perfect fluid, this is illustrated by, e.g., Arnold (1978).

3. - Dedekind's theorem

The operator F describing the system motion has been expressed in term of two $SO(3)$ matrices. It is easy to see that the quantities:

$$(9) \qquad\qquad K = F^t \dot{F} - \dot{F}^t F,$$

and

$$(10) \qquad\qquad J = \dot{F} F^t - F \dot{F}^t,$$

are conserved. It is also possible to introduce the two invariants $K^2 - J^2$ and KJ of the $SO(3) \times SO(3)$ symmetry. The physical meaning of the two conserved quantities can be easily identified. J is the total angular momentum, while K is related to the vorticity tensor ς by the relation $K = -F^t \varsigma F$. The solutions of Eq. (8) are thus characterized by given values of the angular momentum and the vorticity. Since the angular momentum is a measure of the amount of rotation, represented by the matrix T, and the vorticity is related to the internal motions as given by S, one might wonder what happens if we exchange the role of T and S. It is readily seen that this transformation exchanges F with F^t and K with J. F^t remains a solution of Eq. (8) since:

$$\ddot{F}^t = S^t ABT = S^t BAT,$$

as both A and B are diagonal. This result represents a generalization of Dedekind's theorem which refers to the case where one of the two matrices T and S is the identity. Two solutions which can be obtained from one another by exchanging the roles of T and S, are called reciprocal. The reciprocal of the triaxial Jacobi ellipsoids (with $S = 1$) are the non-rotating Dedekind ellipsoids (with $T = 1$). Reciprocity between solutions is another kind of (discrete) symmetry and it is, in this example, caused by the $SO(3) \times SO(3)$ symmetry which allows us to exchange the roles of the groups.

4. - Spontaneous symmetry breaking

It is well known that when the amount of rotation energy as measured with the Ostriker and Peebles (1973) parameter, $t \equiv T/|W|$, of a Maclaurin spheroid exceeds the value $t_c \simeq 0.137$, the system becomes secularly unstable and bifurcates to the Jacobi triaxial sequence. The symmetry group of a Maclaurin spheroid is $D_{\infty h}$ and that of a Jacobi ellipsoid D_{2h}. This spontaneous (minimal) symmetry breaking, which corresponds to a second order phase transition and is associated with an increase in the entropy (Bertin and Radicati, 1976), is not the only one known to appear in the framework of the classical figures of equilibrium. Jacobi ellipsoids are known to bifurcate to the pear-shaped Poincare's figures (Chandrasekhar 1969) and many more bifurcations are known to be kinematically possible (Constantinescu, Michel, and Radicati 1979). Note that a spontaneous breaking of the equatorial symmetry in the classical figures of equilibrium would be kinematically allowed but it is nevertheless forbidden by dynamical reasons. It is, in fact, possible to prove that symmetry with respect to the equatorial plane is always enforced.

The breaking from $D_{\infty h}$ to D_{2h} is not restricted to classical figures of equilibrium with constant density. The breaking of axial symmetry is also present in polytropic fluid configurations and their stellar dynamical analogues with index $n \leq 0.808$ (see, e.g., Vandervoort 1980), and in more realistic stellar dynamical models, suitable to describe elliptical galaxies (see, Ostriker and Peebles 1973). Stellar dynamical systems are in fact unstable, in a way comparable to the Maclaurin-Jacobi sequence, when they rotate so rapidly that the parameter t exceeds the value t_c.

In addition, a symmetry-breaking instability, leading to bar-formation, is present when the pressure anisotropy is large enough (Polyachenkho and Shukhman 1981; Bertin and Stiavelli 1989). The relevant parameter is now $2T_r/T_t$, where T_r and T_t are, respectively, the radial and tangential kinetic energies, the parameter being different from unity since stellar dynamical systems need not to have isotropic velocity dispersion. The instability is observed when $2T_r/T_t \geq 2$. During dissipationless formation of elliptical galaxies (van Albada 1982; Aguilar and Merritt 1989) it is easy to violate this constraint, so that elliptical galaxies are good candidates for being physical systems where the breaking of axial symmetry is not only possible, but also rather common. The symmetry breaking instability would explain the observed shapes of many elliptical galaxies. Unfortunately, given the complexity of the systems, the forementioned results are essentially based on numerical experiments and often lack a proper *analytical* understanding, even though work on this subject is in progress (Bertin and Pegoraro 1989). In Fig. 1 we show the initial and final (after 25 half-mass crossing times) isodensity contours, projected onto two orthogonal planes, of a model unstable with respect to bar formation due to its sizeable radial anisotropy (model $\Psi = 0.86$ of the f_∞ sequence, Bertin and Stiavelli 1989).

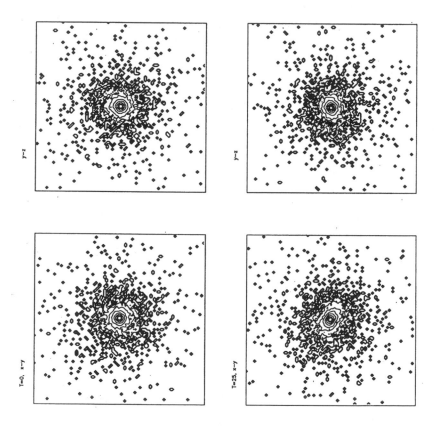

Fig. 1: projected density contours for the model $\Psi = 0.86$ of the f_∞ sequence
(Bertin and Stiavelli 1989). The top plots refer to the initial equilibrium system,
the bottom plot to the final system (after 25 half-masss crossing times). The
plots on the left hand side are projections on the $x - y$ plane, those on the
right hand side projections on the $y - z$ plane. The ratio of the box side length
to the half-mass radius is 70. Contours are logarithmic and, at large radii,
are dominated by the noise due to the limited number of particles used in the
simulation (2×10^4).

Finally we note an analogy between the two processes of symmetry breaking by rotation and by pressure anisotropy. The last has been interpreted as being caused by too small a velocity dispersion along the tangential direction, so that the system is tangentially Jeans unstable. It is thus convenient to write both criteria in terms of the minimum amount of tangential velocity dispersion needed for stability or, alternatively, in terms of the minimum amount of tangential, random kinetic energy U_t. The virial theorem ensures that $U_t + U_r + T + W/2 = 0$, i.e. $u_t + u_r + t = 1/2$, having defined $u_t = U_t/|W|$ and $u_r = U_r/|W|$. It turns out that in terms of u_t both the criteria can be conveniently expressed as follows:

$$(11) \qquad\qquad u_t \geq 0.24.$$

It is remarkable that the two criteria assume the same form and it is worthwhile to study the stability of systems that are both rotating and radially anisotropic. Preliminary results (Stiavelli 1989, in preparation) indicate that their behaviour is not simply described by Eq. (11). One should, however, keep in mind that the two global criteria, even when taken separately, are only approximate and do present some scatter in the marginal stability value, which seems to depend on the details of the model considered.

5. - Conclusions

Symmetries and symmetry breaking arguments have been shown to play an important role in astrophysics, although the intrinsic complexity of astrophysical systems makes these arguments alone often insufficient to draw firm conclusions on the physical processes actually taking place. When supported by *external* physical arguments, group theoretical considerations can, however, be of great help.

Acknowledgements

The author wish to thank all those who taught him physics at the Scuola Normale Superiore and, especially, Prof. Luigi Radicati who has also introduced him to the study of the subject here presented and to astrophysics in general.

REFERENCES

[1]   L.A. AGUILAR - D. MERRIT, 1989, Rutgers University astrophysics preprint 103.

[2]   V.I. ARNOLD, 1981, *Mathematical Methods of Classical Mechanics*, New York: Springer-Verlag.

[3]   G. BERTIN - F. PEGORARO, 1989, in *Plasma Astrophysics*, ed. T.D. Guyenne, ESA Sp-285, p.329.

[4] G. BERTIN - L.A. RADICATI, 1976, Astrophys. J. **206**, 815.

[5] G. BERTIN - M. STIAVELLI, 1989, Astrophys. J. **338**, 723.

[6] S. CHANDRASEKHAR, 1969, *Ellipsoidal Figures of Equilibrium*, Dover.

[7] D.H. CONSTANTINESCU - L. MICHEL - L.A. RADICATI, 1979, J. Physique **40**, 147.

[8] F.J. DYSON, 1968, J. Math. Mech. **18**, 91.

[9] J.P.O. OSTRIKER - P.J.E. PEEBLES, 1973, Astrophys. J. **186**, 467.

[10] F. PEGORARO, 1972, Rend. Cl. Scienze Mat. Fis. Nat. Accad. Naz. Lincei, serie VIII, **LII**, 73.

[11] V.L. POLYACHENKO - I.G. SHUKHMAN, I.G., 1981, Soviet Astron. **25**, 533.

[12] M. STIAVELLI, 1983, *Tesi di laurea*, University of Pisa.

[13] T.S. VAN ALBADA, 1982, Mon. Not. Roy. astron. Soc., **201**, 939.

[14] P.O. VANDERVOORT, 1980, Astrophys. J. **240**, 478.

Appendix

When a solution of Eq. (7) is sought, one finds that the pressure must be quadratic in the coordinates. By imposing this condition Eq. (7) reduces to Eq. (8), where the matrix B is:

$$
(A1) \qquad B = \begin{pmatrix} B_1 & 0 & 0 \\ 0 & B_2 & 0 \\ 0 & 0 & B_3 \end{pmatrix},
$$

with the diagonal elements B_k given by:

$$
(A2) \qquad B_k = 2 \left(\frac{p_c}{\epsilon_k} - \pi G \rho_0 \alpha_k \right).
$$

Here p_c is the pressure at the center, ρ_0 is the (constant) density, and α_k is given by:

$$
(A3) \qquad \alpha_k = \int_0^\infty \frac{du}{\Delta(\epsilon_k + u)},
$$

with

$$
(A4) \qquad \Delta^2 = (\epsilon_1 + u)(\epsilon_2 + u)(\epsilon_3 + u).
$$

Maclaurin spheroids and Jacobi ellipsoids are solutions characterized by rotation without internal streaming. They are obtained when $S = 1$, so that $F = T^t$ and $\breve{F} = T^t B$. By taking T to be a rotation with angular velocity Ω with respect to x_3 one has:

(A5)
$$T = \begin{pmatrix} \cos\Omega t & \sin\Omega t & 0 \\ -\sin\Omega t & \cos\Omega t & 0 \\ 0 & 0 & 1 \end{pmatrix}.$$

The equations of the motion reduce to:

(A6)
$$p_c = \pi G\rho_0\alpha_3\epsilon_3,$$

and

(A7)
$$\Omega^2 = 2(\pi G\rho_0\alpha_1 - p_c/\epsilon_1) = 2(\pi G\rho_0\alpha_2 - p_c/\epsilon_2).$$

Maclaurin spheroids are the axisymmetrix solutions of Eqs. (A6) and (A7), characterized by $\epsilon_1 = \epsilon_2$ and $\alpha_1 = \alpha_2$. Jacobi ellipsoids are triaxial, i.e. $\epsilon_1 \neq \epsilon_2$.

If we now consider internal motions aligned with the rotation, i.e. S to be of the form:

(A8)
$$S = \begin{pmatrix} \cos\varsigma t & \sin\varsigma t & 0 \\ -\sin\varsigma t & \cos\varsigma t & 0 \\ 0 & 0 & 1 \end{pmatrix},$$

we find the following equations of the motion:

(A9)
$$-(\Omega^2 + \varsigma^2)\sqrt{\epsilon_1} - B_1\sqrt{\epsilon_2} + 2\Omega\varsigma\sqrt{\epsilon_2} = 0,$$

(A10)
$$-(\Omega^2 + \varsigma^2)\sqrt{\epsilon_2} - B_2\sqrt{\epsilon_2} + 2\Omega\varsigma\sqrt{\epsilon_1} = 0,$$

(A11)
$$B_3\sqrt{\epsilon_3} = 0,$$

whose solutions are Riemann's type S ellipsoids.

If T and S are taken to be of the most general rotations with constant angular velocity, one can find the Riemann's ellipsoids of types I, II, and III. All these classical solutions are described in detail by Chandrasekhar (1969), although he derives the relevant equations in a less transparent way. It turns out that the *rotation* axes of T and S have to lie on a principal plane of the ellipsoid. This result, known as Riemann's theorem, is a direct consequence of the conservation of the two invariants K and J (Pegoraro 1972). A more general result can be found in the cases where the angular velocities are not taken to be constant.

Extension of the Bloch-Floquet Theorem to Crystalline Point Groups

G. STRINATI ([1,2]) - U. FANO ([2])

Abstract

The Bloch-Floquet theorem, which underlies band structure calculations for perfect crystals, is generalized to point group symmetries which are relevant to molecules and to crystals in the presence of isolated impurities.

Calculations of single-particle (or quasiparticle) energy levels in perfect crystals are considerably simplified by exploiting the translational symmetry of their Hamiltonian. This symmetry enables one to solve the Schrödinger equation only within a unit cell of the crystal, the rest being taken care of by suitable boundary conditions at the cell boundary. Mathematically, these conditions are spelled out by classifying the energy eigenfunctions $\Psi(r)$ through a wave vector $\mathbf{k}$ restricted to the first Brillouin zone. Translational symmetry then requires that

$$(1) \qquad \Psi_{\mathbf{k}}(\mathbf{r} + \mathbf{R}_n) = \exp(i\mathbf{k} \cdot \mathbf{R}_n)\Psi_{\mathbf{k}}(\mathbf{r})$$

for all crystal translations $\mathbf{R}_n$. A convenient way to enforce the condition (1) expresses the wave function $\Psi_{\mathbf{k}}(\mathbf{r})$ in the so-called Bloch form [1]

$$(2) \qquad \Psi_{\mathbf{k}}(\mathbf{r}) = \exp(i\mathbf{k} \cdot \mathbf{r})u_{\mathbf{k}}(\mathbf{r}),$$

where $u_{\mathbf{k}}(\mathbf{r} + \mathbf{R}_n) = u_{\mathbf{k}}(\mathbf{r})$ is an *invariant* of the translational group while the plane wave $\exp(i\mathbf{k} \cdot \mathbf{r})$ alone bears the symmetry requirement (1).

Specifically, $\exp(i\mathbf{k} \cdot \mathbf{r})$ is *the simplest function* that transforms according to the irreducible representation of the group of crystalline translations specified by the wave vector $\mathbf{k}$. The simple form (2) of the Bloch-Floquet theorem rests

([1]) Scuola Normale Superiore, 56100 Pisa, Italy
([2]) Department of Physics and J. Franck Institute, University of Chicago, Chicago, IL 60637, U.S.A.

on the abelian property of the translational group, whose elements commute
with one another. The computational simplification lies thus in reducing the
Schrödinger equation for $\Psi_{\mathbf{k}}(\mathbf{r})$ to an equation for its invariant factor $u_{\mathbf{k}}(\mathbf{r})$
only.

This result extends easily to certain potentials relevant to molecular
problems. Thus a potential with a ternary axis (C_3 symmetry), $V(r, \theta, \varphi + 2\pi/3) = V(r, \theta, \varphi)$, yields eigenfunctions of the form

(3) $$\exp(im\varphi) I_m(r, \theta, \varphi) \quad (m = 0, \pm 1)$$

in which the factor $I_m(r, \theta, \varphi + 2\pi/3) = I_m(r, \theta, \varphi)$ has the same periodicity as
V, i.e., is an invariant of the abelian goup C_3.

One may wonder whether the Bloch-Floquet theorem is peculiar to periodic
potentials and abelian groups or may instead extend to non-abelian groups such
as the groups of symmetry elements of molecules and the crystalline point
groups. (The latter constitute the only symmetry that remains in the presence
of an isolated impurity in an otherwise perfect crystal).

Consider then a symmetry-adapted wave function $\Psi_{\Gamma i}(\mathbf{r})$ which transforms
under point group operations according to the i-th row of the irreducible
representation Γ of the group. The factorization on the right-hand side of (2)
suggests its representation:

(4) $$\Psi_{\Gamma i}(\mathbf{r}) = \sum_{\lambda=1}^{\dim\Gamma} \xi_{\Gamma i}^{\lambda}(\hat{r}) I_{\lambda}^{\Gamma}(\mathbf{r}).$$

Here, the functions $I(\mathbf{r})$ are *invariants* of the point group while the $\xi_{\Gamma i}(\hat{r})$
are meant to be *the simplest functions* that tranform according to the i-th row
of the irreducible representation Γ. Notice that the non-abelian nature of the
point group manifests itself in (4) through the occurrence of multi-dimensional
irreducible representations with $\dim\Gamma > 1$. *More than one basic function* $\xi_{\Gamma i}^{\lambda}(\hat{r})$
$(\lambda = 1, \ldots, \dim\Gamma)$ *is accordingly required to bear the symmetry character of
the wave function* $\Psi_{\Gamma i}(\mathbf{r})$.

Given the form (4), specification of the symmetry-adapted function $\Psi_{\Gamma i}(\mathbf{r})$
reduces to the knowledge of its invariants $I_{\lambda}(\mathbf{r})$, which should be more readily
constructed than the full function $\Psi_{\Gamma i}(\mathbf{r})$ itself. Thereby one would calculate
numerically only wave function factors having the *full symmetry* of the potential,
much as one calculates only radial wave functions in potentials with central
symmetry. More precisely, the Schrödinger equation for $\Psi_{\Gamma i}(\mathbf{r})$ would reduce
to a coupled system for the invariants $I_{\lambda}(\mathbf{r})$ only, as the equation for the $\Psi(\mathbf{r})$
of a *non central* potential $V(\mathbf{r})$ reduces to a coupled system for radial $f_{\ell}(\mathbf{r})$
through expansion of Ψ and V in spherical harmonics.

The conjecture (4) occured to us earlier while bridging from the atomic-
like description of electronic properties about a localized crystal impurity
onto the band theory applicable far from the impurity [2-5]. Specifically, we
had classified homogeneous polynomials of the form $\sum_{p,q=0}^{\ell} a_{p,q} x^p y^q z^{\ell-p-q}$ as

symmetry-adapted harmonics of crystalline point groups. The conjecture of an extension of the Bloch-Floquet theorem for point groups in the form (4) was explicitly verified in that context. We appreciate the opportunity of the volume in honor of Luigi Radicati to bring now our findings to his attention. Our extension of the Bloch-Floquet theorem to point groups appears here as a search for symmetries within symmetries.

For purposes of illustration, we outline our main concepts with reference to an NH_3-type potential, whose symmetry operations include both a ternary axis and symmetry planes, thus forming a C_{3v} group:

$$(5) \qquad V(r,\theta,\varphi) = V(r,\theta,\varphi + 2\pi/3) = V(r,\theta,-\varphi).$$

Additional examples for three major point groups are listed in the Tables at the end of the paper.

Eigenfunctions of the potential (5) can be expressed in terms of functions $I(r,\theta,\varphi)$ having the *full symmetry* of the potential as follows:
(i) Eigenfunctions invariant under the subgroup C_3 (i.e., with $m = 0, 3, \cdots$ in (3)) occur again here but may be either even or odd under the reflections as indicated by

$$(6a) \qquad I_0(r,\theta,\varphi), \quad \sin 3\varphi I_3(r,\theta,\varphi).$$

(ii) The wave functions with $m = \pm 1$ in (3), which are degenerate under C_3, combine now into a pair of functions which are even or odd under reflection through $\varphi = 0$ and belong to a two-dimensional (i.e., an E) representation of C_{3v}. A pair of E wave functions may be indicated by

$$(6b) \qquad \{\cos\varphi I_1(r,\theta,\varphi), \sin\varphi I_1(r,\theta,\varphi)\}.$$

Additional pairs of eigenfunctions of the Schrödinger equation beloging to the E representation are, however, cast as more general superpositions of two terms according to (4), one with the form (6b) and another with the form

$$(6c) \qquad \{\cos 2\varphi I_2(r,\theta,\varphi), -\sin 2\varphi I_2(r,\theta,\varphi)\}.$$

The functions (6c) result upon considering the pair $\exp(\pm 2i\varphi)$ which is equivalent under C_3 to the pair $\exp(\pm i\varphi)$ of (3). The Schrödinger equation for states of the E representation can then be shown to reduce to a *system* of equations for I_1 and I_2, coupled through kinetic rather than potential energy terms.

To this end, we recall that the factors $\exp(i\mathbf{k}\cdot\mathbf{r})$ of Bloch functions form a complete orthogonal base set in the following sense. The $\sum_n \exp[i(\mathbf{k}-\mathbf{k})\cdot(\mathbf{r}+\mathbf{R}_n)]$, summed over all points $\mathbf{r}+\mathbf{R}_n$ obtained from $\mathbf{r}$ by lattice translations, vanishes for any pair of different vectors $(\mathbf{k},\mathbf{k}')$ of the first Brillouin zone; the set $\{\exp(i\mathbf{k}\cdot\mathbf{R}_n)\}$ is complete in that the orthogonality property holds for no additional functions. The analog of the set of lattice points $\{\mathbf{R}_n\}$ for the group C_{3v} of an NH_3-like potential consists of "stars" of six position vectors $\{\mathbf{r}_n\} = \{(r, \theta, \varphi_n)\}$ for which φ_n takes the six values

(7) $$\{\pm\varphi + 2m\pi/3 \ (m = 0, \pm 1)\}.$$

[The arbitrary values of (r, θ, φ) correspond to those of $\mathbf{r}$ in a crystal cell.] One verifies readily that the six functions of φ in (6), namely

(8) $$\{\xi_p(\varphi)\} = \{1, \ \sin 3\varphi, \ \cos \varphi, \ \sin \varphi, \ \cos 2\varphi, \ -\sin 2\varphi\}$$

are linearly independent, though not orthogonal over the six values of φ_n in (7). This set of six functions is complete in that no additional function could be linearly independent of them over the φ_n. Note that stars $\{\mathbf{r}_n\}$ with $\varphi = 2m\pi/3$ are "special" inasmuch as the set in (8) reduces for them to only three distinct nonzero elements.

Consider now a wave function $\Psi(\mathbf{r})$ that depends on φ through a superposition of the six functions in (8). The Schrödinger equation $(H - E)\Psi(\mathbf{r}) = 0$ with the symmetric potential (5) will reduce to a system in the totally symmetric $I(\mathbf{r})$ by projection onto a set of functions $\{\xi_p^+(\hat{r})\}$ orthogonal to the set $\{\xi_p(\hat{r})\}$:

(9) $$\sum_n \xi_p^+(\hat{r}_n)[H(\mathbf{r}_n) - E]\Psi(\mathbf{r}_n) = 0 \quad (p = 1, \cdots, 6).$$

Owing to this orthogonality, the equations with $p = 1, 2$ will separate out, each of them governing a single function I_0 or I_3, respectively. The pair with $p = 3, 5$ and the pair with $p = 4, 6$ will instead remain coupled through the kinetic terms of H.

Projection onto functions $\xi_p^+(\hat{r})$, as in (9), also serves the more limited purpose of factoring any symmetry-adapted spherical harmonic $X_{\Gamma i}^{(\ell)}(\hat{r})$ into a standard $\xi_{\Gamma i}(\hat{r})$ and a residue $\overline{X}_\Gamma^{(\ell)}(\hat{r})$ invariant under all group operations. Both this factorization and the applications of (9) rest on the fact that functions belonging to different representations are orthogonal over the group operations, meaning that

(10a) $$\sum_n X_{\Gamma i}^{(\ell)}(\hat{r}_n)^* \ X_{\Gamma' i'}^{(\ell')}(\hat{r}_n) = 0,$$

(10b) $$\sum_n \Psi_{\Gamma i}(\hat{r}_n)^* \ \Psi_{\Gamma' i'}(\hat{r}_n) = 0,$$

whenever $\Gamma \neq \Gamma'$ and $i \neq i'$. Such orthogonality relations underlie the applications of (9), which we shall describe initially for the simpler case of one-dimensional representations.

In the one-dimensional case one may set simply

(11a)
$$\xi_\Gamma^+(\hat{r}) = \xi_\Gamma^*(\hat{r})/\Xi_\Gamma(\hat{r}),$$

where the normalization coefficient

(11b)
$$\Xi_\Gamma(\hat{r}) = \sum_n |\xi_\Gamma(\hat{r}_n)|^2$$

is a group invariant. Factorization of the wave function $\Psi_\Gamma(\mathbf{r})$ of (9) is now achieved by projecting it onto (11a) to yield a group invariant $\overline{\Psi}_\Gamma(\mathbf{r})$

(12a)
$$\overline{\Psi}_\Gamma(\mathbf{r}) = \sum_n \xi_\Gamma^*(\hat{r}_n)\Psi_\Gamma(\mathbf{r}_n)/\Xi_\Gamma(\hat{r}),$$

whence follows

(12b)
$$\Psi_\Gamma(\mathbf{r}) = \xi_\Gamma(\hat{r})\overline{\Psi}_\Gamma(\mathbf{r}).$$

Entering (11a) and (12b) in (9) yields now its more explicit form

(13)
$$\sum_n \xi_\Gamma^*(\hat{r}_n)[H(\mathbf{r}_n) - E]\xi_\Gamma(\hat{r}_n)\overline{\Psi}_\Gamma(\mathbf{r}_n)/\Xi_\Gamma(\hat{r}).$$

Invariance of the factor $\overline{\Psi}_\Gamma(\mathbf{r}_n)$ means that it is independent of the index n of its argument, whereby $\mathbf{r}_n \to \mathbf{r}$. The same holds for the potential $V(\mathbf{r}_n)$ and for the radial derivative term of $H(\mathbf{r}_n)$. The Hamiltonian itself is thus indeed a group invariant,

(14)
$$H(\mathbf{r}) = -\frac{\hbar^2}{2mr^2}\left[\frac{\partial}{\partial r}\left(r^2\frac{\partial}{\partial r}\right) - \mathbf{l}^2\right] + V(\mathbf{r}),$$

l being the angular momentum operator. For these terms and for the energy term E, the $\sum_n$ in (13) bears then only on the ξ and cancels the Ξ^{-1} factor. The only nontrivial quantity to be evaluated is the orbital momentum term $\sum_n \xi_\Gamma^*(\hat{r}_n)\mathbf{l}^2\xi_\Gamma(\hat{r}_n)$, in which $\mathbf{l}^2$ operates both on the following ξ and on the subsequent factor $\overline{\Psi}$. Because l is a linear differential operator, this term yields three contributions

(15)
$$\sum_n \xi_\Gamma^*(\hat{r}_n)\{[\mathbf{l}^2\xi_\Gamma(\hat{r}_n)] + 2[\mathbf{l}\xi_\Gamma(\hat{r}_n)]\cdot\mathbf{l} + \xi_\Gamma(\hat{r}_n)\mathbf{l}^2\},$$

where the square brackets limit the range of application of each operator. The last term of (15) yields trivially $\Xi_\Gamma(\hat{r})\ell^2$. For the first term, recall that $\xi_\Gamma(\hat{r})$ is chosen to coincide with the symmetry-adapted spherical harmonic $X_\Gamma^{(\ell_\Gamma)}(\hat{r})$ of lowest degree ℓ_Γ for the given irreducible representations Γ. Hence this term yields $\Xi_\Gamma(\hat{r})\ell_\Gamma(\ell_\Gamma + 1)$. In the second term, the factor $\rho_\Gamma(\hat{r}) = \sum_n \xi_\Gamma^*(\hat{r}_n)\mathbf{l}\xi_\Gamma(\hat{r}_n)$

transform as the vector $\mathbf{l}$ under general rotations and is invariant under the group operations; accordingly, the whole term will be indicated by $\rho_\Gamma(\hat{r}) \cdot \mathbf{l}$. Entering in (15) this expression and those of the first and last terms, and then (14) and (15) itself into (13) yields the desired equation for the group invariant wave function $\overline{\Psi}_\Gamma(\mathbf{r})$.

The application of (9) to multi-dimensional representation extends (11a) by constructing sets $\{\xi_p^+(\hat{r})\}$ orthogonal to the sets $\{\xi_p(\hat{r})\}$ over the group operations. We illustrate this rather elementary construction for the example of the pair of two-dimensional irreducible representation of the group C_{3v}. From (6b) and (6c) we identify the sets $\{\xi_p(\hat{r})\} = \{\cos\,\varphi,\ \sin\,\varphi\}$ and $\{\xi_p(\hat{r})\} = \{\cos\,2\varphi, -\,\sin\,2\varphi\}$, respectively. Sets orthogonal to them are readily obtained by multiplication with the element $\sin 3\varphi$ of (8) and adjustment of a sign, yielding the pair

(16a) $$\{\xi_p^+(\hat{r})\} = \{\sin\,3\varphi\,\sin\,\varphi, -\,\sin\,3\varphi\,\cos\,\varphi\},$$

(16b) $$\{\xi_p^+(\hat{r})\} = \{\sin\,3\varphi\,\sin\,2\varphi,\ \sin\,3\varphi\,\cos\,2\varphi\},$$

to within invariant normalization factors analogous to (11b). It is then possible to cast the Schrödinger equation into a coupled system in which the potential terms remain separate while the kinetic terms couple the invariant parts $I_\lambda^\Gamma(\mathbf{r})$ of (4). In this way, one simplifies the treatment of the potential energy which is generally more complicated and depends on the system under consideration. On the other hand, the kinetic energy terms which couple the equations are generally simpler and of standard form for all systems of the same symmetry.

The identification of the set $\{\xi_p(\hat{r})\}$ for non-abelian groups can be standardized in notable cases, namely, when the point symmetry group can be generated starting from one of its abelian invariant subgroups by sequential accretion of the operations of one or more cyclic groups. This accretion procedure, called a "semi-direct product", has been used by Altmann to identify the symmetry-adapted spherical harmonics [8]. For example, the C_{3v} group considered above results from the semi-direct product of the abelian C_3 and of the C_s group consisting of the identity and of the reflection σ_y, the remaining reflections resulting as products of σ_y by rotations of C_3. Similarly, the tetrahedral group T results from the semi-direct product of its D_2 subgroup (the group of an octahedron with unequally stretched axes) and of the C_3 operations about an axis in the (1,1,1) direction with respect to the D_2 axes.

Altmann's method relies on the following fact. Any invariant of the subgroup N splits, under the action of a cyclic group C, into a sum of terms that belong to different one-dimensional representations of the semi-direct group $N \wedge C$. In the example where $N \equiv C_3$ and $C \equiv C_s = \{E, \sigma_y\}$, invariants of C_3 are superpositions of spherical harmonics $Y_{\ell,3m}(\hat{r})$. The operation σ_y changes $\hat{r} \equiv (\theta, \varphi)$ into $(\theta, -\varphi)$, and hence each $Y_{\ell,3m}(\hat{r})$ into $Y_{\ell,3m}(\hat{r})^*$. Accordingly, $\mathrm{Re}[Y_{\ell,3m}(\hat{r})]$ and $\mathrm{Im}[Y_{\ell,3m}(\hat{r})]$ are even and odd under σ_y, respectively, while remaining invariant under C_3. They thus belong to the two one-dimensional irreducible representations of C_{3v}, i.e., A_1 and A_2, respectively. We select then

$\xi_{A_1}(\hat{r})$ and $\xi_{A_2}(\hat{r})$ as proportional to the nonzero functions $\mathrm{Re}[Y_{\ell,3m}(\hat{r})]$ and $\mathrm{Im}[Y_{\ell,3m}(\hat{r})]$ of lowest degree, that is

(17) $$\xi_{A_1}(\hat{r}) = 1, \quad \xi_{A_2}(\hat{r}) = \sin^3\theta \, \sin 3\varphi.$$

[The invariant factor $\sin^3\theta$ of ξ_{A_2} had been previously omitted in (6a) for simplicity but it is part of a spherical harmonic.]

Noninvariant functions belonging to irreducible representations of C_3 can be represented by superposition of terms

(18a) $$\xi_{1E}(\hat{r}) = \sin\theta e^{i\varphi} Y_{\ell,3m}(\hat{r})$$

(18b) $$\xi_{2E}(\hat{r}) = \sin\theta e^{-i\varphi} Y_{\ell,3m}(\hat{r}).$$

The operation σ_y changes φ into $-\varphi$ and thus interchanges these ^1E and ^2E representations of C_3, combining them into the E representation of C_{3v}. Standard practice identifies the two rows ($i = 1, 2$) of this representation as even and odd under σ_y. Hence the real and imaginary parts of the functions (18) belong to these two rows. Since these functions consist of two factors, their real parts can be formed with either the real parts of both factors or with their corresponding imaginary parts; imaginary parts can be constructed with the analogous alternatives. We obtain the base functions $\xi_{Ei}^{(\lambda)}(\hat{r})$ and $\xi_{Ei}^{+(\lambda)}(\hat{r})$ with $\lambda = 1$ by this procedure, utilizing in (18a) the lowest values of ℓ and m that provide nonzero results, namely $\ell = m = 0$ and $\ell = 3m = 3$:

(19a) $\{\xi_{E1}^{(1)}(\hat{r}), \ \xi_{E2}^{(1)}(\hat{r})\} = \{\sin\theta \, \cos\varphi, \ \sin\theta \, \sin\varphi\}$

(19b) $\{\xi_{E1}^{+(1)}(\hat{r}), \ \xi_{E2}^{+(1)}(\hat{r})\} = \{-\sin^4\theta \, \sin 3\varphi \, \sin\varphi, \ \sin^4\theta \, \sin 3\varphi \, \cos\varphi\}.$

The sets $\xi_{Ei}^{(\lambda)}(\hat{r})$ and $\xi_{Ei}^{+(\lambda)}(\hat{r})$ with $\lambda = 2$ are similarly obtained from the above procedure by taking $\ell = 3m = 3$ and $\ell = 3m = 6$ in (18b), respectively.

The procedure followed above for C_{3v} can be repeated step by step for any group generated as a semi-direct group $N \wedge C_s$. The procedure becomes trivial when the additional symmetry element is the inversion C_i at the center. In this case N combines with C_i into the direct product $N \times C_i$. The set $\{\xi_p(\hat{r})\}$ of N is expected to double its size in this case by multiplying each ξ_p by the lowest-degree odd representation of C_i which is invariant under N.

We note finally that additional applications of the Bloch-Floquet theorem to point groups in the form (4) are provided by the groups C_{4v}, T_d, and O_h which are relevant to the two-dimensional square lattice, the diamond and zinc blende structures, and the cubic lattices, in the order. The harmonics $\xi_{\Gamma i}^{(\lambda)}(\hat{r})$ for these groups are listed in Tables 1-3. The validity of the expansion (4) can be explicitly verified by inspection of the symmetry-adapted spherical harmonics of any ℓ, as listed in Ref. (6).

REFERENCES

[1]   See, for intance, F. BASSANI - G. PASTORI PARRAVICINI, *Electronic States and Optical Transitions in Solids* (Pergamon, Oxford, 1975).

[2]   U. FANO, Phys. Rev. Lett. **31**, 234 (1973).

[3]   G. STRINATI - U. FANO, J. Math. Phys. **17**, 434 (1976).

[4]   G. STRINATI, Phys. Rev. B**18**, 4096 (1978); *ibid.*, 4104 (1978).

[5]   G. STRINATI, J. Math. Phys. **20**, 188 (1979).

[6]   Symmetry-adapted spherical harmonics for the 32 crystal point groups are conveniently listed in the book by C.J. Bradley and A.P. Cracknell, *The Mathematical Theory of Symmetry in Solids* (Clarendon, Oxford, 1972).

[7]   Note that the invariant (12a) remains nonsingular in the "special" directions (where $\hat{r}$ belongs to a symmetry axis or a reflection plane so that its star $\{\hat{r}_n\}$ consists of fewer elements), because both the numerator and the denominator of the ratio on its right-hand side vanish quadratically. Recall, in fact, that any function that transforms according to an irreducible representation and is *not* invariant under a particular group operation (rotation about an axis $\hat{u}$ or reflection on a plane p) vanishes when $\hat{r}$ coincides with $\hat{u}$ or lies on p, respectively (e.g., any spherical harmonics $Y_{\ell m}(\hat{r})$ with $m \neq 0$ vanishes for $\hat{r} \equiv \hat{z}$). Therefore, $\xi_\Gamma(\hat{r})$ vanishes on various symmetry elements and its norm $\Xi_\Gamma(\hat{r})$ vanishes quadratically there.

[8]   S.L. ALTMANN, Rev. Mod. Phys. **35**, 641 (1963).

TABLE 1

A_1	1
A_2	$\sin 4\varphi$
B_1	$\cos 2\varphi$
B_2	$\sin 2\varphi$
$E \ (\lambda = 1)$	$\{\cos \varphi, \ \sin \varphi\}$
$E \ (\lambda = 2)$	$\{\cos 3\varphi, \ -\sin 3\varphi\}$

Table 1 - Basic harmonics $\xi_{\Gamma i}^{(\lambda)}(\hat{r})$ for the group C_{4v}. Their normalization and dependence on the angle θ are omitted.

TABLE 2

A_1	1
A_2	$(x^2 - y^2)(z^4 - x^2 z^2 - y^2 z^2 + x^2 y^2)$
$E \ (\lambda = 1)$	$\{2z^2 - x^2 - y^2, x^2 - y^2\}$
$E \ (\lambda = 2)$	$\{2z^4 - x^4 - y^4 - 6(z^2 x^2 + z^2 y^2 - 2x^2 y^2),$
	$x^4 - y^4 + 6z^2(y^2 - x^2)\}$
$F_1 \ (\lambda = 1)$	$z(x^2 - y^2)$
$F_1 \ (\lambda = 2)$	$xy(x^2 - y^2)$
$F_1 \ (\lambda = 3)$	$z(2z^2 - x^2 - y^2)(x^2 - y^2)$
$F_2 \ (\lambda = 1)$	z
$F_2 \ (\lambda = 2)$	xy
$F_2 \ (\lambda = 3)$	$z(2z^2 - 3x^2 - 3y^2)$

Table 2 - Basic harmonics $\xi_{\Gamma i}^{(\lambda)}(\hat{r})$ for the group T_d, listed in the form of (unnormalized) polynomials of x, y, and z. For the three-dimensional representations F_1 and F_2 only the third row ($i = 3$) is given, since the other rows with $i = 1, 2$ are obtained by cyclic permutations of x, y, and z.

TABLE 3

A_{1g}	1
A_{1u}	$xyz[x^4(y^2 - z^2) + y^4(z^2 - x^2) + z^4(x^2 - y^2)]$
A_{2g}	$x^4(y^2 - z^2) + y^4(z^2 - x^2) + z^4(x^2 - y^2)$
A_{2u}	xyz
E_g $(\lambda = 1)$	$\{2z^2 - x^2 - y^2, x^2 - y^2\}$
E_g $(\lambda = 2)$	$\{2z^4 - x^4 - y^4 - 6(z^2x^2 + z^2y^2 - 2x^2y^2),$ $x^4 - y^4 + 6z^2(y^2 - x^2)\}$
E_u $(\lambda = 1)$	$\{xyz(x^2 - y^2), -xyz(2z^2 - x^2 - y^2)\}$
E_u $(\lambda = 2)$	$\{xyz[13(x^4 - y^4) - 10(x^2 - y^2)(x^2 + y^2 + z^2)],$ $-xyz[13(2z^4 - x^4 - y^4) - 10(2z^2 - x^2 - y^2)]\}$
F_{1g} $(\lambda = 1)$	$xy(x^2 - y^2)$
F_{1g} $(\lambda = 2)$	$xy(x^2 - y^2)z^2$
F_{1g} $(\lambda = 3)$	$x^3y^3(x^2 - y^2)$
F_{1u} $(\lambda = 1)$	z
F_{1u} $(\lambda = 2)$	z^3
F_{1u} $(\lambda = 3)$	zx^2y^2
F_{2g} $(\lambda = 1)$	xy
F_{2g} $(\lambda = 2)$	xyz^2
F_{2g} $(\lambda = 3)$	x^3y^3
F_{2u} $(\lambda = 1)$	$z(x^2 - y^2)$
F_{2u} $(\lambda = 2)$	$z^3(x^2 - y^2)$
F_{2u} $(\lambda = 3)$	$zx^2y^2(x^2 - y^2)$

Table 3 - Same as Table 2 for the group O_h.

Non-perturbative Approach to the Infrared Problem and Confinement. Breaking of the Poincaré Group

F. STROCCHI (*)

Foreward

I wish to dedicate this work to Prof. Luigi A. Radicati who taught me the relevance of spontaneously broken symmetries in elementary particle physics and in theoretical physics in general.

1. - Infrared problem

Gauge theories in general and quantum electrodynamics (QED) in particular exhibit infrared singularities whose control goes under the name of infrared problem. Actually, the removal of the infrared cutoff (e.g. a fictitious photon mass) in the Green functions is not a serious difficulty and the real problem occurs when such removal is done in the scattering amplitudes. The pragmatic prescription of effectively replacing the infrared cutoff by the energy resolution ΔE of the experimental apparatus (by summing over all processes in which soft photons with energy $E < \Delta E$ are present in the initial and final states) yields finite transition probabilities for finite ΔE, but it leave unsolved the general problem of whether there exists an underlying theory independent of ΔE, rather than many theories labelled by ΔE. Such question is not a merely academic one, since it is at the roots of problems like leading order summations, exponentiation, effective expansion parameter $\alpha ln(\Delta E/m)$, construction of charged states, Bloch-Nordsieck ansatz etc. All these problems become crucial in quantum cromodynamics (QCD), where α is not small [1]. All this indicates that a non perturbative approach to the infrared problem also for QED is a necessary step.

Much of the non-perturbative wisdom on the infrared problem comes from non-relativistic soluble models like the Bloch-Nordsieck (BN) model and the Blanchard-Pauli-Fierz (BPF) model, the features of which have been

(*)   International School for Advanced Studies, Trieste, Italy.

extrapolated to QED [2] (and sometimes even to QCD). A rigorous control of the infrared problem in non-trivial models has been pioneered by J. Fröhlich [3], who provided a general framework and mathematical structures for the non-perturbative approach to the infrared problem. Fröhlich's ideas combined with the deep results by Buchholz on the existence of the asymptotic limit in the case of massless particles (a remarkable extension of the Haag-Ruelle scattering theory) [4] were then successfully applied to QED and led to a non-perturbative treatment of the infrared problem in QED [5], [6]. As a result one obtained: i) a characterization of the charged states in terms of asymptotic electromagnetic (e.m.) radiation; ii) a solution of the infraparticle problem, namely the identification of the charged particle mass which makes possible an approach to scattering theory similar to the Haag-Ruelle (or LSZ) theory; iii) the proof that non-Fock coherent charged states lead to the breaking of the Lorentz group.

The exploitation of locality and of local gauge invariance at the level of field algebra (by using a local and covariant gauge) then allowed [7], [8] an explicit construction of (physical) charged states and the proof of the Bloch-Nordsieck ansatz, namely the standard BN mean values of the asymptotic e.m. field on charged particle states. Such sharp information in turn allowed [9] to explore situations in which the infrared singularities are stronger than in the standard QED case and in some sense mimic the QCD case, as the case of massless charged particles in QED in four dimensions, (QED_4), and of charged particles in QED_3. In both cases one can prove that if the asymptotic photon field exists, then (not only the Lorentz group but also) the space-time translations are broken in charged sectors, i.e. *massless* charged particles are confined in QED_4 and charged particles are confined in QED_3, whenever asymptotic photons exist.

2. - Charged scattering states. Gauss' law. Infraparticles

The non-perturbative construction of scattering amplitudes in QED requires the solution of the following problems

a)   the asymptotic limit for the photon field, $F_{\mu\nu}^{as}$, $as = $ in/out

b)   the characterization of charged states as representations of $F_{\mu\nu}^{as}$

c)   the asymptotic limit of the charged fields.

The solution of a) has been given by Buchholz [4], [10], by exploiting Huyghen's principle and locality in a general framework. The solution of b) requires a careful analysis of the physical characterization of a charged state. The main point is that the identification of a charged particle state involves not only the measurement of particle observables like momentum, spin etc. but also (and crucially) of the asymptotic e.m. radiation associated to it. More precisely a charged state Ψ is defined by the expectation values of the asymptotic e.m. field $< \Psi, F^{as}_{\mu\nu}\Psi >$. In the massive case, by suitable experimental detectors one can obtain states consisting of a "charged" particle and no "massive" photon, whereas in the massless case such elimination is not possible since an infinite number of soft photons always escape the finite energy resolution of the counters. In the pragmatic treatment of the infrared problem in QED, this difficulty is somewhat overcome by adopting a non sharp description of a charged state, leaving the associated e.m. radiation of energy less than ΔE unspecified. In view of the above considerations it is more satisfactory [5], [6] to characterize a charged state as a representation of the algebra A^{as} of the asymptotic e.m. field $F^{as}_{\mu\nu}$, i.e. a state described in terms of charged particle observables *and* of the properties of the associated asymptotic e.m. radiation, as for example the energy spectrum, but not (in general) the photon number. The general properties of charged (scattering) states (like the energy spectrum, the breaking of symmetries etc.) are then analyzable in terms of properties of representations of A^{as} (charged representations).

The *detectability of a charged state* requires that its energy and momentum be finite, i.e. that space-time translations are not broken in the corresponding representation of A^{as}. The above considerations lead to the following general structure as a basis of the discussion of the infrared problem [5], [6].

I. The asymptotic limits $F^{as}_{\mu\nu}$, $as = $ in/out, of the electromagnetic field exist as free fields and they satisfy the free fields canonical commutation relations.

II. The space-time translations

$$F^{as}_{\mu\nu}(x) \rightarrow F^{as}_{\mu\nu}(x+a)$$

are described by unitary operators in each (charged) sector describing detectable charged states.

To avoid technical domain problems, in the following we will regards A^{as} as the Weyl algebra generated by $F^{as}_{\mu\nu}$ (a property guaranteed by general conditions [3], [5]). Representations of A^{as} satisfying I and II will be briefly called (charged) *scattering representations*.

The first problem one faces in the analysis of the scattering representations is the identification of the *mass of a charged particle*. Experience with soluble models (in particular the Schroer model and the BN model with Dirac equation [1]) as well as QED results obtained through a leading order summation of the perturbative series, show that in the charged one particle sector the energy-momentum spectrum does not have a sharp hyperboloid, i.e. one particle states are not eigenstates of P^2 with definite mass (*infraparticle*). Typically, the propagator behaves like $(p^2 - m^2)^{-\beta}$ rather than having a pole. The values of β depends on the gauge, but the phenomenon is a real physical one, since in the Coulomb gauge, where no unphysical degree of freedom occurs to affect the mass-shell singularity of the propagator, $\beta \neq 1$. Recently, Buchholz has proven quite generally [11] that the infraparticle phenomenon is a consequence of the state being labelled by a superselected charge obeying a local Gauss' law. This shows that one of the crucial assumptions of the Haag-Ruelle (or LSZ) theory, namely the existence of one particle states with definite mass, fails for charged particles. It is worthwhile to stress that such infraparticle phenomenon crucially affects the S-matrix at the perturbative level (at least if no and hoc or clever summation is done) and its solution requires a genuine non-perturbative analysis [5].

The general framework mentioned above provides a solution of this difficulty [3], [5].

THEOREM 1 [5]. *It π is a representation of $\dot{A}^{as}$, satisfying I and II, then the energy and momentum can be separated into an asymptotic photon part and a charged particle part*

$$(2.1) \qquad\qquad H = H^{ph} + H^c, \qquad P = P^{ph} + P^c$$

where (H^{ph}, P^{ph}) are self-adjoint operators affiliated to $\pi (A^{as})''$, (i.e. they are functions of $F^{as}_{\mu\nu}$, as = in/out), spectrum $(H^{ph}, P^{ph}) = V_+$, $H^{ph} \leq H$, and (H^c, P^c) commute with A^{as} and spectrum $(H^c, P^c) \subseteq V_+$, (V_+ = the forward cone).

Thus, roughly speaking, the separation of the asymptotic e.m. radiation from the charges, which cannot be done at the level of states, can be done at the level of energy and momentum and it is therefore natural [5], [6], to identify a concept of charged particle mass in terms of the spectrum P^c_μ. The idea is that the charged particle mass is related to the asyptotic dynamics of the charge ad it is this dynamics which is expected to display a sharp hyperboloid $(P^c)^2 = m^2$. Mass singularities corresponding to charged particles should then appear only after the energy of the asymptotic e.m. radiation has been subtracted out. This provides a clear cut definition of charged particle mass and it makes possible an approach to scattering theory à la Haag-Ruelle (this programme has been worked out in a non-relativistic model by J. Fröhlich [3]).

3. - Proof of Bloch-Nordsieck ansatz. Breaking of the Lorentz group and its physical implications

A further structural question about the charged states is whether they give rise to coherent state representations of the asymptotic e.m. field. The indications from soluble models, like the BN and BPF models, do not provide a clear cut extrapolation to the full QED case (not to speak of the non-abelian case). In particular, the need of considering an uncountable set of inequivalent scattering representations incoherently labelled by the particle momentum (which is therefore superselected) and a non-separable Hilbert space, as advocated by several authors [12], [13], on the basis of the above models, does not seem compelling. Such features, when extrapolated to QED would have so strong consequences at the level of general principles (above all the momentum superselection, but also the non-existence of the asymptotic limit of the Heisenberg (picture) fields in the Hilbert space in which the theory is defined for finite times [14]) that a careful analysis is needed.

On the basis of semiclassical considerations and/or by invoking a classical correspondence "principle" [3], [5] one can argue that charged states are generalized coherent states and one can prove that if such coherent state representations are not equivalent to the Fock representation then the Lorentz group is broken [5], [6].

THEOREM 2 [6]. *In the framework discussed above, let $H^{(1)}$ denote the one charged particle space corresponding to $(p^c)^2 = m^2$*

$$(3.1) \qquad H^{(1)} = \int_{p^2 = m^2} d\mu(p) \; H_p^{(1)}$$

with $H_p^{(1)}$ a separable space, $\forall p$, (countable number of quantum numbers besides momentum), then, if $H^{(1)}$ contains a non-Fock coherent representation in the above decomposition, the Lorentz group is broken in $H^{(1)}$.
(For a detailed proof see also Ref. [1]).

The strong physical implications of such result and the apparent conflict between this result and the Lorentz covariance of the Wightman functions in the local Gupta-Bleuler formulation (at each perturbative order) requires a careful analysis of the hypothesis of the non-Fock coherent property of the charged states. Actually, at this stage, one cannot exclude the point of view according to which it is this assumption (i.e. the Bloch-Nordsieck ansatz) which has to be questioned rather than Lorentz covariance, in the line with the conventional wisdom of local (covariant) renormalizable formulations of QED.

Our next point is therefore a non-perturbative analysis of local and covariant formulations of QED, based on the following general structures [7], [8]:

1. A local field algebra F generated by the Wightman fields $A_\mu, \overline{\psi}, \psi$ and a cyclic vacuum Ψ_0

2. A gauge group G acting on F (leaving the observable algebra A invariant), with the property that the gauge transformations of the first kind (i.e. the so-called rigid ones) are generated by the electric current j_μ (the source of A_μ)

$$(3.2) \qquad \Box\, A_\mu = j_\mu, \qquad \partial^\mu j_\mu = 0$$

and that the gauge transformations of second kind (the so-called local ones) are generated by

(3.3) $$Q^\Lambda \equiv \int d^3x \Lambda(x, x_0) \overset{\leftrightarrow}{\partial_0} \partial^\sigma A_\sigma(x, x_0)$$

with the gauge function $\Lambda(x, x_0) \in S(R^3)$ obeying $\square \Lambda = 0$.

3.   The asymptotic limit A_μ^{as} exists as free field satisfying the free field commutation relations, and covariant under the Poincarè transformations (on a dense domain stable under the time translations and the Hamiltonian).

By exploiting the above local structure (including the restrictions coming from local gauge transformations) one may get rather strong information on the infrared properties of the theory [7,8].

THEOREM 3 [7]. *Let Ψ_p denote a local charged (improper) state of (charge) momentum p, (obtained by decomposing a local charged state Ψ over the spectrum of P^c):*

(3.4) $$\Psi = \int d^3p\, \Psi_p, \quad p^2 = m^2,$$

then

(3.5) $$< \Psi_p, A_\mu^{as}(k)\Psi_{p'} > = -(2\pi)^{-3/2}\delta^3(p - p')\delta(k^2) \in (k)q$$
$$\left(\frac{p_\mu}{p \cdot k}\, \rho(k) + gauge\ terms \right),$$

*with $\rho(k) \to 1$, as $k \to 0$ (**Bloch-Nordsieck covariant factors**), and q denotes the charge of Ψ_p.*

THEOREM 4 [7]. *Given a local charged (improper) state Ψ_p one can construct a physical charged state $\hat{\Psi}_p = U\Psi_p$, satisfying the Gauss' law constraint, where U is a unitary operator and one has that $< \hat{\Psi}_p, F_{\mu\nu}^{as}(k)\, \hat{\Psi}_{p'} >$ gives a **non-covariant Bloch-Nordsieck factor***

(3.6) $$i(2\pi)^{-3/2}\delta(p - p')\delta(k^2) \in (k)q\ [(k_\mu p_\nu - k_\nu p_\mu)/k \cdot p$$
$$-(k_\mu v_\nu - k_\nu p_\mu)/k \cdot v]\ \rho(k),$$

where v is a fixed time-like vector. (The standard BN form is obtained by taking $v = (p_0, 0)$).

The above result can be regarded as a *proof of the BN ansatz, namely the non-Fock coherent character of the charged states.* It also explicitly displays how the Lorentz covariance is broken in going from local states to physical states. The infrared dressing transformation U, needed for such mapping in order to satisfy the Gauss' law constraint, maps the local state representation, in which the Lorentz symmetry is unbroken into one in which it is broken. As we will see below such mechanism is relevant also in the breaking of the space-time translations as a mechanism of confinement.

Before closing this section it is worthwhile to comment on the physical implications of *the Lorentz breaking*, which *is not a gauge artifact*, as sometimes incorrectly stated in the subsequent literature.

First of all it implies that observable matrix elements, like the expectation value of $F_{\mu\nu}^{as}$ on a charged state, do not transform covariantly under the Lorentz group. For example, the e.m. radiation associated to an electron vanishes if the electron is at rest $\mathbf{p} = 0$ and obviously it also vanishes in the reference frame in which the electron has momentum $\mathbf{p} \neq 0$. On the other hand, if the electron is accelerated from $\mathbf{p} = 0$ to $\mathbf{p} \neq 0$ the associated e.m. radiation is given by

$$F_{oi}(k) \sim q \, \frac{k_0 p_i}{p \cdot k} \left(\delta_{ij} - \frac{k_i k_j}{|k|^2} \right) \, \delta(k^2) \in (k) \neq 0.$$

Thus, the two electron states both of momentum $\mathbf{p}$ (one obtained by accelerating the electron and the other by changing reference frame) yield different matrix elements of $F_{\mu\nu}^{as}$, i.e. Lorentz covariance is lost.

It should be remarked that no breaking occurs in the zero charge sector, (a distinguishing feature with respect to standard spontaneous breaking of symmetries), as it can also be seen by considering the asymptotic e.m. radiation associated to a pair of oppositely charged particles $(+q, -q)$, with momenta (p_1, p_2)

$$F_{\mu\nu}^{as} \sim q \left[k_\mu \left(\frac{p_1}{p_1 \cdot k} - \frac{p_2}{p_2 \cdot k} \right)_\nu - (\mu \leftrightarrow \nu) \right] \delta(k^2) \in (k).$$

Thus, a covariant description of physical processes could be maintained at the price of considering together with a charged particle $(+q)$, the oppositely charged one $(-q)$ from which it has been separated in the experimental preparation process. The definition of charged one particle state, labelled only by the one particle observables and by its associated e.m. radiation, without keeping trace of the e.m. radiation of the opposite charged particle "removed behind the moon" leads to the breaking of the Lorentz group.

[†] Clearly, if the Lorentz group were unbroken

$$\Psi_{p=(p_0,\mathbf{p})} = \Psi_{\Lambda(p_0=m, \, \mathbf{p}=0)} = U(\Lambda) \, \Psi_{m, \, \mathbf{p}=0}$$

(Λ being a Lorentz transformation such that $p = \Lambda(p_0 = m, \, \mathbf{p} = 0)$) and

$$<\Psi_p, F_{\mu\nu}^{as}(k)\Psi_p> = <\Psi_{m,\mathbf{p}=0}, \, U(\Lambda)^{-1} \, F_{\mu\nu}^{as}(k) U(\Lambda) \Psi_{m,\mathbf{p}=0}> =$$
$$= (\Lambda^{-1})_\mu^\rho \, (\Lambda^{-1})_\nu^\sigma \, <\Psi_{m,\mathbf{p}=0}, F_{\rho\sigma}^{as} \, (\Lambda^{-1}k)\Psi_{m,\mathbf{p}=0}> = 0$$

Such *Lorentz breaking can* in principle *be detected experimentally*, e.g. by measuring the e.m. radiation at large distances associated to an electron of momentum p coming out of an "accelerator" like a cathode tube. Changing the momentum $p \rightarrow p'$ of a particle can be done in two in principle very different ways: i) by some local operation i.e. by the action of a local observable on the state; ii) by a Lorentz boost, which is, in principle, a global non-local "operation". If the particle state is a local state then the two ways are equivalent, since the global character of the Lorentz boost is not seen by such state. In this case, the result of a Lorentz boost can be equivalently obtained by acting on $|p>$ with a local observable $A \in \mathcal{A} : |p'> = A|p>$, i.e. it leaves stable the "sector" $\mathcal{H}^{(1)} = \{\overline{A\Psi_p}\}$. This is no longer true if the particle is charged: whereas the state $|p'> = A|p>$, i.e. the state obtained in the way i), belongs to $\mathcal{H}^{(1)}$, the Lorentz boosted one does not, i.e. the *Lorentz boost leads to an inequivalent representation*. Thus, one gets two electron states both with momentum p' (and with the same one particle observable like spin etc.) yielding inequivalent representations of the algebra of observables, characterized by different Gauss fluxes at ∞.

From this point of view, the breaking of Lorentz boosts in QED in *not different* from all other symmetry-breaking phenomena. The main point is that charged sectors consist of non-local states labelled by "observables" at infinity (like the Gauss flux at ∞ per unit solid angle) which are not scalar under Lorentz boosts [10].

This implies that the breaking cannot be seen in the vacuum sector (as it happens in all other symmetry breakings) and it may explain why it is not easy to detect it. One must in fact measure quantities which distinguish different Gauss fluxes at infinity, like e.g. inequivalent coherent factors.

Due to the smallness of the fine structure constant, such coherent effects are extremely small at the microscopic scale and sizable effects related to the non-Fock coherent character of the charged states can be expected only when macroscopic currents (10^{20} electron/sec.) are present (for a more detailed discussion see Ref. [1]). To further explain how subtle is the breaking of the Lorentz boosts in the charged sectors, one should also add that its occurrence does not spoil the relativistic relation between energy, momentum and mass for a charged particle, as proved in Ref. [14].

Finally, it is worthwhile to remark that the simplest characterization of the Lorentz breaking (and also its more direct proof) is in terms of inequivalent coherent representations of the asymptotic e.m. algebra (see Refs. [5-7] and [1]); however its validity has a much more general basis ([10], [1]). In particular, its proof does not require special assumptions on the observable algebra (like that of being generated by $F_{\mu\nu}$). The main point is that the Lorentz boosts intertwine between inequivalent representations of the asymptotic algebra (as a consequence of the non-Fock coherent property of the charged states) as well as between inequivalent representations of the local observable algebra (as a mere consequence of its being local with respect to $F_{\mu\nu}$); now, both on experimental as well as on theoretical grounds one cannot accept an uncountable number of superselection rules labelled by the Lorentz boost parameters, (this would lead in particular to a non-separable Hilbert space), and this leads *inevitably* to the Lorentz breaking in the charged sectors. The Lorentz breaking is also necessary to avoid the superselection of the charge momentum p^{ch} (see Ref. [1]).

4. - Confinement as a breaking of space time translations. Confinement of massless charged particles in QED$_4$.

As discussed in Sect. 2 a charged state is physically realizable provided its energy is finite (see condition 2 in the definition of scattering representation). The breaking of space time translations in a charged sector means that the energy is infinite and this can be regarded as a precise formulation of confinement. Since no such breaking is expected in the zero charge sector, the phenomena must be related to the impossibility of separating two oppositely charged particles, i.e. to the large distance (or infrared) behaviour of the correlation functions and as we will see this possibility occurs also in the abelian case, when the infrared singularities are worse than in the standard QED case. This is the case of quantum electrodynamics of massless charged particles, which exhibits dangerous collinear infrared singularities similar to those occurring in quantum cromodynamics. The confinement mechanism which will be discussed below may therefore shed light on the (more difficult) non-abelian case.

It appears that the only massless particles occurring in nature have zero electric charge and it is reasonable to ask whether this fact is not an accident and it can be understood in terms of general arguments. It is known that massless charged particles would give rise to worse infrared singularities [15] and a discussion of this problem with the aid of the renormalization group indicates difficulties of the theory. The renormalization group is used to get a leading term summation of the perturbative expansion in ordinary QED$_4$, in the region $k^2 \gg m^2, k^2 > \lambda^2 \geq m^2$, where λ^2 is the renormalization point and m is the electron mass. One then takes the limit $m \to 0$ and obtains the following expression [16] for the transverse photon propagator $d(k^2)/k^2$

$$(4.1) \qquad d(k^2) = \left(1 - \frac{\alpha(\lambda^2)}{3\pi} \; ln \; \frac{k^2}{\lambda^2}\right)^{-1}$$

and for the renormalized charge (defined through the vertex function)

$$(4.2) \qquad \alpha(k^2) = \alpha(\lambda^2) \left(1 - \frac{\alpha(\lambda^2)}{3\pi} \; ln \; \frac{k^2}{\lambda^2}\right)^{-1}$$

$(k^2 \ll k_L^2 = \lambda^2 \; exp \; [3\pi/\alpha(\lambda^2)])$.
In this standard analysis however

i)   the interchange of the limit $m \to 0$ and the leading order summation is questionable.

ii)   The problems related to the infraparticle structure of the charged states
        as well as the nedd of an infrared dressing transformation for going from
        local to physical states [7] have not been taken into account.

iii)   The standard relation between renormalized charge and photon propagator,
        relying on the cancellation of radiative corrections to the proper vertex
        and electron propagator, is not under control for $m = 0$.

iv)   The derived form of the transverse photon propagator (4.1) violation
        positivity of the two-point function of $F_{\mu\nu}$ since it implies that the
        corresponding Wightman function has a singularity for low k^2, which
        is not a measure and therefore it is not compatible with positivity [9].
        This pathology shows that eq. (4.1) is not reliable and it cannot be used
        to discuss the renormalized electric charge. Since such difficulty appears
        to have gone unnoticed in the literature, we reproduce the details of the
        argument in the Appendix.

Furthermore, one should remark that previous statements on the
renormalized charge in massless QED$_4$ are crucially affected by the questionable
ways in which the limit $m \to 0$ is taken. For example, if one keeps $\alpha(m^2)$ fixed
one gets

$$\alpha(k^2) = \alpha(m^2) \left(1 - \frac{\alpha(m^2)}{3\pi} \ln \frac{k^2}{m^2}\right)^{-1},$$

$$k_L^2 = m^2 \exp\left[3\pi/\alpha(m^2)\right] \to 0 \text{ as } m \to 0$$

and the range of validity of eq. (4.2) becomes zero. With this choice, λ^2 has
been eliminated by putting it equal to m^2, the only surviving scale parameter,
and therefore the limit $k^2 \to \infty$ is equivalent to $m^2 \to 0$. The behaviour of the
charge in the limit $m^2 \to 0$ is then given by the high k^2 behaviour, in particular
by whether the theory is asymptotically free or not.

Alternatively, in order to have a non-trivial range of validity of eq. (4.2),
(see [17], [18]), one makes the choice of keeping $\alpha(\lambda^2)$ fixed and one then
extrapolates eq. (4.2) to $k^2 \geq m^2$ so that

$$\alpha(m^2) = \alpha(\lambda^2) \left(1 - \frac{\alpha(\lambda^2)}{3\pi} \ln \frac{k^2}{\lambda^2}\right)^{-1} \sim (\ln m^2)^{-1} \to 0$$

when $m \to 0$. With this choice, the only charge renormalization compatible with
$\alpha(k^2) \neq 0$ is

$$\lim_{k \to 0} \alpha(k^2) = 0$$

and this makes the global gauge group trivial. As remarked above, this approach
also meets non trivial problems of positivity.

To get non-perturbative information on the properties of massless charged
states it is convenient to use the general framework discussed in Sect. II. The
crucial issue is whether

i)   massless charged states yield Bloch-Nordsieck type mean values of the asymptotic e.m. field $F_{\mu\nu}^{as}$ (here the wisdom from soluble modles and/or from classical considerations cannot be used!)

ii) The space time translations are unbroken in the corresponding sectors, i.e. massless charged states are physically detectable.

One can answer these questions by using a local and covariant formulation as in Sect. III, based on

1)   a local gauge group (see properties 1), 2) in Sect. III)

2)   the Gauss law constraint to select the physical states

3)   the existence of A_μ^{as}, as = in/out, as free massless fields obeying canonical commutation relations and transforming covariantly under the Poincaré group.

We will also assume that

4)   the space time translations are described by unitary operators $U(a)$, whose P_μ satisfy the relativistic spectrum condition $(P^2 \geq 0,\ P_0 \geq 0)$ and investigate the compatibility of this assumption in the sectors defined by massless charged states.

It is not difficult to see [9] that the arguments and results of Ref. [7] can be extended to the case in which the mass of the charged particle, i.e. the eigenvalue of the charge momentum squared $(P^c)^2$, is zero:

PROPOSITION 5. Let $\hat{\Psi}_p$ be a physical charged *massless* state then the infrared behaviour of the matrix elements $< \hat{\Psi}_p, F_{\mu\nu}^{as}(k)\hat{\Psi}_{p'} >$ is that of **Bloch-Nordsieck type mean values** (see eq. (3.6)).

In the following analysis we will only need the above information on the infrared behaviour of $< \hat{\Psi}_p, F_{\mu\nu}^{as}(k)\hat{\Psi}_{p'} >$, no commitment being made about $\hat{\Psi}_p$ being a Bloch-Nordsieck coherent state. Under such stronger assumption the analysis would simplify a lot (along the lines of Ref. [19])[†], but this would make the conclusion questionable, since the infrared behaviour of massless charged states is expected to be worse than in the massive QED case and the wisdom from soluble models, classical considerations, perturbative expansion etc. is not available and in any case not convincing in this case.

Within the above framework we have:

THEOREM 6 [9]. *If charged massless particle states with finite mean energy exist, then the Hamiltonian cannot be bounded from below.*

Hence, if the energy is (re) normalized in such a way that the vacuum is the lowest energy state, charged massless particle states get infinite energy, i.e.

[†]   When Ref. [9] was in print, we were informed by J. Fröhlich that he was aware of the difficulties related to massless charged states being standard Bloch-Nordsieck states.

they are confined.

PROOF. Let us assume that the Hamiltonian H is bounded below; then by the proof of Theor. 3.2 in Ref. [5], for any infrared cutoff ρ, one has the decomposition

$$H = H_\rho^{ph} + H_\rho^c$$

with

$$H_\rho^{ph} = \frac{1}{2} \int_{|k| \geq \rho} d^3k : \vec{E}(k)^2 + \vec{B}(k)^2 :$$

(E, B denote the asymptotic electric and magnetic fields and the double dots denote Wick ordering) and

$$[H_\rho^{ph}, H_\rho^c] = 0$$

Thus, since the spectrum of H_ρ^{ph} is bounded below by zero, Inf $\sigma(H_\rho^{ph}) = 0$, one has

$$\text{Inf } \sigma(H_\rho^c) = \text{Inf } \sigma(H) \equiv C$$

i.e. H_ρ^c is bounded below independently of ρ.

Now, if Ψ is a (physical) charged state with finite mean energy $< H >_\Psi$ one has

(4.3) $< H_\rho^c >_\Psi = < H >_\Psi - < H_\rho^{ph} >_\Psi$

and

$$< H_\rho^{ph} >_\Psi = < \int_{|k| \geq \rho} d^3k \, [\vec{E}(k)^+ \, \vec{E}(k)^- + \vec{B}(k)^+ \, \vec{B}(k)^-] >_\Psi$$

(4.4) $$\geq \int_{|k| \geq \rho} d^3k \, [| < \vec{E}(k)^+ >_\Psi |^2 + | < \vec{B}(k)^+ >_\Psi |^2]$$

$$< E_i(k)^+ > = \int_{k_0 \geq 0} dk_0 \, < F_{oi}(k_0, k)$$

The infrared singularity of the Bloch-Nordsieck mean values (eq. (3.6)) imply in turn that the integral on the right hand side of eq. (4.4) is divergent like [9]

$$\lim_{\varepsilon \to 0} \int_0^{\pi - \varepsilon} (1 + \cos \theta)^{-1} \, d \cos \theta \sim - \lim_{\varepsilon \to 0} \log \varepsilon$$

Hence, by eq. (4.3), H_ρ^c cannot be bounded below independently of ρ and therefore the original assumption that H is bounded below cannot hold.

The above result can be summarized by the statement that *the existence of (massless) asymptotic photons and of charged massless particles is incompatible with the relativistic spectral condition.*

This does not exclude, in principle, a "phase" in which the photons are massive or do not even exist as asymptotic states and "charged" particles are massless. However, if $F_{\mu\nu}$ is observable and the electric charge is unbroken, then the two point function of $F_{\mu\nu}$ has a $\delta(k^2)$ singularity [20] and by Buchholz results [4] [10] asymptotic massless photons exist. Thus, *in our world* one cannot have massless charged particles.

5. - Confinement of charged particles in QED$_3$

The argument of the previous section can be applied to the case of charged particles in QED$_3$ [9]. Again the analysis relies on the exploitation of the local structure associated to a local formulation: properties 1)-3). Actually in this case the Bloch-Nordsieck type mean values of $F_{\mu\nu}^{as}(k)$ can be proved under more general conditions [9]; it should be stressed on the other side that the existence of the massless asymptotic fields A_μ^{as}, i.e. of massless asymptotic photons, is not a harmless assumption in $2+1$ dimensions and there may well be "phases" of the theory in which it does not hold. One can prove, similarly to the previous Section, that the existence of massless asymptotic photons, and of charged states is incompatible with the space time translations being unbroken:

THEOREM 7 [9]. *Under the above assumptions the space time translations cannot be implementable with a Hamiltonian which is bounded from below in the charged sectors.*

Acknowledgements

I wish to thank J. Fröhlich for all I have learned from him on the infrared problem and for the pleasure and fortune of collaborating with him. I am also deeply indebted to G. Morchio whose contributions, collaboration and help have been crucial for me. The subject of this note was presented as an invited talk at the annual meeting of the Società Italiana di Fisica and I wish to thank the organizing committee.

APPENDIX
Non-positivity of the two point function of $F_{\mu\nu}$

We want to show that the transverse photon propagator (4.1) leads to a two point (Wightman) function of $F_{\mu\nu}$ with an infrared singularity incompatible with positivity [9]. We will split the detailed argument into various steps.

LEMMA 1. As a consequence of posititivity the two point Scwinger function $S(k)$ is analytic for Re $k_0 \neq 0$ and it vanishes for $|\text{Re } k_0| \to \infty$.

PROOF. From the definition of the Schwinger function in terms of the Wightman function

$$S(x_0, x) = W(ix_0, x)$$

we have

$$S(k) = S_+(k) + S_-(k)$$

$$S_\pm(k) = \text{Fourier transform of } W(\pm ix_0, x)\theta(\pm x_0)$$

If the two point Wightman function W satisfies positivity, it is the Fourier transform of a positive (tempered) measure $d\mu(k)$ and therefore for $a_0 > 0$, $b_0 > 0$

$$\lim_{a_0 \to +\infty} |W(ia_0 - b_0, x)| \leq \lim_{a_0 \to \infty} \int e^{-k_0 a_0} |d\mu(k)| = 0$$

One can then deform the path of integration in the definition of $S_+(k)$ from the positive x_0 axis to the positive imaginary axis

$$S_+(k_0, k) = \int_0^\infty db_0 \ W(-b_0, x) \ e^{-k_0 b_0} \ e^{ikx} \ d^3k$$

This shows that S_+ is analytic for Re $k_0 > 0$.
A similar argument shows that also $S_-(k)$

$$S_-(k) = -\int_0^\infty db_0 \ W(b_0, x) \ e^{-k_0 b_0}$$

is analytic for Re $k_0 > 0$, so that $S(k)$ is also analytic for Re $k_0 > 0$.
On the other hand from the symmetry of the Schwinger functions in the non-difference variables and the invariance under rotations one gets

$$S(x_0, \vec{x}) = S(-x_0, -\vec{x}) = S(-x_0, \vec{x})$$

i.e.

(A.1) $$S(k_0, \vec{k}) = S(-k_0, \vec{k})$$

and therefore $S(k)$ is analytic for Re $k_0 \neq 0$. The fall off for large Re k_0 follows from the above expressions for $S_\pm(k)$ and eq. (A.1).

LEMMA 2. Under the above assumptions one derives the following relation between the Schwinger and the Wightman two point function

$$(A.2) \qquad iW(q) = \lim \{S(-q^2 - i\varepsilon, \vec{q}) - S(-q^2 + i\varepsilon, \vec{q})\}$$

where q is the Minkowski momentum and $q_0 > 0$.

PROOF. By deforming the path of integration for the Fourier transform of $S(k)$ one gets

$$S(x) = \lim_{\varepsilon \to 0} \left(\int_{c_+} + \int_{c_-} \right) e^{-ikx} S(k) \, d^4k$$

where

$$\int_{c_\pm} e^{-ikx} S(k) \, d^4k = \pm \int d^3k \int_0^\infty dk_0 \, e^{-ix_0(-i|k_0| \pm \varepsilon)} e^{i\vec{k} \cdot \vec{x}} S(k)$$

$$= \pm i \int d^3q \int_0^\infty dq_0 \, e^{-q_0 x_0} e^{i\vec{q} \cdot \vec{x}} S(-iq_0 \mp \varepsilon, \vec{q}) \theta(q_0)$$

A comparison with the expression of $S(x)$ in terms of $W(q)$ gives

$$iW(q) = \lim \{S(-iq_0 + \varepsilon, \vec{q}) - S(-iq_0 - \varepsilon, \vec{q})\}$$

Since by euclidean invariance $S(q_0', \vec{q}')$ is actually a function of $q'^2 = q_0'^2 + \vec{q}'^2$ one gets eq. (A.2).

LEMMA 3. The two point function of $F_{\mu\nu}$ corresponding to the transverse photon propagator (4.1) has a low k singularity which is incompatible with positivity.

PROOF. The two point function of $F_{\mu\nu}$ can be written in terms of a single Lorentz invariant function $W(q)$ whose low q behaviour is provided by transverse propagator (4.1), through eq. (4.2):

$$(A.3) \qquad iW(q^2) = \lim_{\varepsilon \to 0} \left\{ \frac{1}{-q^2 + i\varepsilon} \left(1 - \frac{\alpha(\lambda^2)}{3\pi} \ln \frac{-q^2 - i\varepsilon}{\lambda^2}\right)^{-1} \right.$$

$$\left. - \frac{1}{-q^2 - i\varepsilon} \left(1 - \frac{\alpha(\lambda^2)}{3\pi} \ln \frac{-q^2 + i\varepsilon}{\lambda^2}\right)^{-1} \right\}$$

Putting $s \equiv q^2$, $\alpha \equiv \alpha(\lambda^2/3\pi)$, we get that the low s behaviour of the r.h.s. of eq. (4.3) is given by

$$\lim_{\epsilon \to 0} \frac{-\alpha s \theta_\epsilon(-s)}{(s^2 + \epsilon^2)\alpha^2 (ln|s|)^2} \equiv \lim_{\epsilon \to 0} F_\epsilon(s)$$

where $\theta_\epsilon(x)$ is a smooth approximation of the Heaviside step function. Clearly, the limit $\epsilon \to 0$ has to be properly taken in a distributional sense and therefore some subtraction (or regularization) may in principle be involved.

Clearly, the limit is well defined on test functions $f(s)$ vanishing at $s = 0$, i.e. $f \in S_0$. However, one can prove that there is no extension of such distribution from S_0 to S compatible with positivity. In fact, if such extension F existed, it would be a measure and therefore

(4.4)
$$\left| \int F(s) \, f(s) ds \right| \leq c \, \sup |f|$$

with c a constant, (for example the measure of the interval [-1.1]). One can instead find sequences of functions $f_n \in S_0$, with $\sup |f_n| = 1$, such that

$$\lim_{n \to \infty} \int F(s) \, f_n(s) \, ds = \lim_{n \to \infty} \int F_{\epsilon=0}(s) f_n(s) ds \to \infty,$$

since

$$\int_{-1}^{-\epsilon} \frac{s \, d \, s}{(s^2 + \epsilon^2)(ln|\epsilon|)^2} \sim -ln \, \epsilon^2,$$

and the bound (4.4) is violated.

REFERENCES

[1]   For a review and general discussion of the infrared problem see e.g. G. MORCHIO and F. STROCCHI, Infrared problem, Higgs phenomenon and long range interactions, lectures given by the second author at the Erice School on *"Fundamental Problems of Gauge Field Theory"*, G. VELO and A.S. WIGHTMAN dirs., Plenum Press 1986.

[2]   See e.g. J.M. JAUCH and F. ROHRLICH, *"The Theory of Photons and Electrons"* 2nd exp. ed. 2nd corr. print, Springer Verlag (1980).

[3]   J. FRÖHLICH, Ann. Inst. H. Poincarè, XIX, 1 (1973) and unpublished manuscript.

[4]   D. BUCHHOLZ, Comm. Math. Phys. 42, 269 (1975); ibid. 52, 147 (1977).

[5]   J. FRÖHLICH - G. MORCHIO - F. STROCCHI, Ann. Phys. (N.Y.) 119, 241 (1979) and references therein to previous work by J. Fröhlich.

[6]   J. FRÖHLICH - G. MORCHIO - F. STROCCHI, Phys. Lett. 89B, 61 (1979).

[7]   G. MORCHIO - F. STROCCHI, Nucl. Phys. B211, 471 (1983); B232, 547 (1984).

[8]   G. MORCHIO - F. STROCCHI, Ann. Phys. (N.Y.) 168, 27 (1986).

[9]   G. MORCHIO - F. STROCCHI, Ann. Phys. (N.Y.) 172, 267 (1986).

[10] D. BUCHHOLZ, Comm. Math. Phys. **85**, 49 (1982).

[11] D. BUCHHOLZ, Phys. Lett. **174B**, 331 (1986).

[12] T.W. KIBBLE, J. Math. Phys. **9**, 315 (1986); Phys. Rev. **173**, 1527 (1968); ibid. **174**, 1982 (1968); ibid. **175**, 1624 (1968); Some Applications of Coherent States, in *Cargèse Lectures* 1967, vol. II, M. Levy ed., Gordon and Breach 1968.

[13] P.P. KULISH - L.D. FADDE'EV, Teor. Matem. Fiz. **4**, 153 (1970) [Theor. Math. Phys. **4**, 745 (1971)].

[14] H.J. BORCHERS - D. BUCHHOLZ, Comm. Math. Phys. **97**, 169 (1985).

[15] See e.g. V.G. VAKS, Soviet Physics JETP **13**, 556 (1961).

[16] N.N. BOGOLUBOV - D.V. SHIRKOV, *Introduction to the Theory of Quantized Fields*, 3rd ed. Chap. 9, Interscience New York 1982.

[17] L.D. LANDAU, in *Niels Bohr and the development of physics*, Pergamon Press, London 1955.

[18] V. GRIBOV, Nucl. Phys. **B206**, 103 (1982).

[19] G. ROEPSTORFF, Comm. Math. Phys. **19**, 301 (1970).

[20] R. FERRARI - L.E. PICASSO, Nucl. Phys. **B31**, 316 (1971).

Luci ed ombre nella Scienza del Medioevo: due testi inediti sul fenomeno del magnetismo

LORIS STURLESE

Impegnato in una ricerca sulla storia del magnetismo nel medio evo intrapresa per suggerimento del dedicatario di questa miscellanea, mi sono imbattuto, fra tante curiosità, in due brevi testi manoscritti che forse meritano di essere strappati all'oblio che li avvolge da secoli. Il primo, senza dubbio il più appetitoso, si trova nel manoscritto Royal 12 e XXV della British Library di Londra, una grossa miscellanea scientifico-filosofica che raccoglie insieme a due opuscoli di Tommaso d'Aquino e a scritti di Aristotele una miriade di scritterelli matematici, astronomici e magici [1]. Qui appunto, fra un trattato sui numeri e una tavola per concordare il calendario arabo con quello cristiano, al verso del foglio 148, scritte da una mano inglese databile alla fine del Duecento, compaiono le 16 righe che di seguito fedelmente trascrivo, nelle quali allo "sperimentatore" viene insegnato l'uso della bussola magnetica per trovare il mezzodì e per calcolare la latitudine del luogo nel quale viene effettuata l'osservazione.

Confricando punctum acus ad partem[a] *magnetis meridionalem vel septentrionalem, que transfixa erit*[b] *per medium festuce, et posita in aqua in vase quodam, donec probetur iacere directe inter polos mundi in linea meridionali civitatis in qua erit, experimentator figat in vase apud extremitatem acus meridianam stilum, vel cultellum extra vas, sed melius est intra, aut teneat*[c] *filum in manu, in cuius inferiori parte sit aliquid grave et descendat in aqua perpendiculariter iuxta extremitatem. Si igitur umbra fili intersecat acum ab oriente versus occidens, tunc est ante meridiem. Si e converso, est post. Si directe feratur umbra secundum longitudinem acus, tunc est meridies.*

In hora igitur meridiei accipiatur altitudo solis per aliquod instrumentum, que si sit in equinoctiali, tunc habebitur altitudo Arietis nota. Et notum est, quod

---

(a)	ad partem *in marg.*
(b)	erit *sup. lin.*
(c)	tenear *cod.*

ab orizonte ad ceniht[(d)] *sunt 90 gradus, scilicet quarta circuli. Igitur si altitudo capitis Arietis subtrahatur a 90, residuum ab equinoctiali usque ad ceniht erit notum, et hec est latitudo regionis et altitudo et poli super orizontem. Si vero altitudo sumpta sit minor altitudine capitis Arietis, tunc accipiatur declinatio solis per tabulas*[(e)], *qua addita ad altitudinem solis sumptam habetur altitudo capitis Arietis, que subtrahatur a 90 sicut prius, et habetur latitudo regionis. Si vero altitudo solis fuerit maior altitudine capitis Arietis, subtrahatur declinatio solis et residuum erit altitudo capitis Arietis, que si subtrahatur a 90 patebit intentum.*

Il testo, tradotto, suona: "Soffregando con l'estremità nord o sud del magnete la punta di ago che traverserà nel mezzo un galleggiante, e posto l'ago in un vaso pieno d'acqua, quando giace lungo il meridiano e perpendicolarmente alla linea meridionale del luogo in cui si trova, lo sperimentatore infigga nel vaso uno stilo presso l'estremità sud dell'ago, oppure pianti in questa posizione un coltello fuori dal vaso - ma è meglio lo stilo entro il vaso -, ovvero tenga un filo in mano alla cui estremità inferiore sia legato un piccolo peso e che entri nell'acqua perpendicolarmente presso l'estremità sud dell'ago. Se dunque l'ombra del filo interseca l'ago da est a ovest, è prima di mezzodì. Se al contrario, è dopo mezzodì. Se l'ombra coincide con la lunghezza dell'ago, allora è mezzodì.

"Si misuri dunque a mezzodì l'altezza del sole con un qualche appropriato istrumento, e se il sole è all'equinozio si avrà l'altezza dell'Ariete. Poiché è noto che l'angolo dall'orizzonte allo zenit vale 90 gradi, cioè un quarto del circolo totale, se si sottrarrà da 90 l'altezza del nodo dell'Ariete si otterrà la differenza dal punto equinoziale allo zenit, e questa è la latitudine della regione e l'altezza del polo sull'orizzonte. Se invece l'altezza misurata è minore dell'altezza del nodo dell'Ariete, si desuma la declinazione del sole mediante le tavole, e aggiungendo questa all'altezza del sole già misurata si avrà l'altezza del nodo dell'Ariete. Sottraendo questa da 90 nel modo sopra indicato, si avrà la latitudine della regione. Se invece la detta altezza sarà maggiore dell'altezza del nodo dell'Ariete, si sottragga l'altezza del sole e la differenza sarà l'altezza del nodo dell'Ariete. Sottraendo questa da 90 si avrà quanto si cercava."

Il testo sopra riportato è, per quanto mi consta, inedito, e non reca per certo novità grandi alla storia della scienza, ma contiene nondimeno informazioni di un certo interesse.

Nelle prime righe viene descritto il sistema di calamitazione dell'ago cosiddetto per contatto o strisciamento, attestato da fonti latine ed arabe già alla fine del secolo XII [2]. Se la bussola di cui si parla fosse quella imperniata o quella solo galleggiante, è più difficile giudicare. A prima vista pare trattarsi della semplice e rudimentale bussola galleggiante degli antichi, di cui riferisce in latino ad esempio, e in termini assai simili al nostro testo, Tommaso di Cantimpré nel suo *De naturis rerum*, XIV 7 4 ("naute ... accipiunt acum et

[(d)]   *lege* zenith.
[(e)]   ta. *cod.*

acumine eius ad adamantem lapidem fricato infigunt per transversum in festuca parva inmittuntque vasi pleno aqua. Tunc circumducunt vasi adamantem lapidem, moxque secundum motum lapidis sequitur in circuitu cacumen acus. Rotatum ergo perinde citius per circuitum lapidem subito retrahunt, moxque cacumen acus amisso ductore aciem dirigit contra stellam maris subsistitque statim nec per punctum movetur") [3] e già prima, verso il 1190, aveva scritto in antico francese Guyot de Provins nella sua *Bible*, ("Un art font (li marinier), qui mentir ne peut, / par la vertu de la mannete. / Une pierre laide et brunète / Où li fers volontiers se joint / ont: si esgardent le droit point, / puis qu'une aiguille l'ait touchié, / et en un festu l'ont fichié. / en l'ève la mettent sans plus, / et li festu la tient dessus; / puis se torne la pointe toute / contre l'estoile, si sans doute / que jà por rien ne faussera, / et mariniers nul doutera") [4]. Se riflettiamo tuttavia sul fatto che nella bacinella, e vicino alla punta dell'ago, doveva essere immerso un filo a piombo per potervi far cadere sopra l'ombra del sole, sarà più verosimile supporre che tutto il sistema funzionasse soltanto a patto che l'ago fosse imperniato, secondo che insegnava in dettaglio Pietro di Maricourt nel capitolo 2 della seconda parte del *De magnete* [5], oppure nella guisa più elementare ma non inefficace descritta ad esempio da Francesco da Buti nel *Commento* a Dante, *Par.* XII, vv. 28-30: "Anno li naviganti uno bussolo che nel mezzo è impernato da una rotella di carta leggieri, la quale gira in su detto perno, e la detta rotella àe molte punte et ad una di quelle, che v'è dipinto una stella, è fitta una punta d'ago; la quale punta li naviganti, quando vogliono vedere dove sia la tramontana, imbriacano colla calamita toccandola molto con quella, e poi girano intorno al bussolo la detta calamita, e l'ago seguita la calamita, e quando ànno fatto pigliare lo moto di girare intorno, rimoveno e cessano la calamita, e stanno a vedere quando si posa lo moto della detta rotella..." [6].

Si noti che sia Tommaso di Cantimpré che Guyot condividono l'allora diffusa opinione che l'ago della bussola punti verso la stella polare ("stella maris", "l'estoile"), mentre Ruggero Bacone e Pietro di Maricourt sosterranno che l'ago si orienta verso i poli nord e sud dell'universo per una sorta di generale influenza celeste sul ferro magnetizzato [7]. E con il riferimento ai poli del mondo è definito nel nostro testo il meridiano, ch'è appunto dato dalla linea ideale che passa per i poli indicati dalla bussola (l'esistenza della declinazione magnetica rimase ignota sino al tempo di Cristoforo Colombo). Trovato il mezzodì nel momento della coincidenza dell'ombra del sole col meridiano, il resto del procedimento insegna all'*experimentator* come calcolare il valore della latitudine del luogo dell'osservazione.

Ancora ad esperimenti, questa volta all'*experimentum* dei marinai, fa riferimento il secondo testo che qui pubblico, tratto dal codice vat. lat. 772 della Biblioteca Apostolica Vaticana in Roma. Il codice è una classica miscellanea universitaria, e contiene fra l'altro il *De ente et essentia* di Tommaso d'Aquino, estratti dal commento alla *Fisica* aristotelica di Egidio Romano, questioni di Erveo Natalis, Giacomo di Viterbo e Giovanni di Napoli [8]. Al foglio 108, recto e verso, una mano databile agli inizi del Trecento ha notato una "questione sul movimento di attrazione magnetica, se questo cioè sia naturale o violento".

Seguendo lo schema tradizionale della questione scolastica universitaria, l'anonimo autore adduce prima un certo numero di argomenti in favore della non naturalità del movimento, poi argomenti in vantaggio della tesi contraria. A favore della non naturalità del movimento milita ad esempio il fatto che il movimento naturale deve avere un principio intrinseco, e il magnete è invece principio estrinseco; inoltre, il movimento naturale deve provenire da una forma naturale, e questa nel ferro può essere soltanto la forma della gravità; inoltre, l'attrazione magnetica non rientra nella classificazione tradizionale dei movimenti naturali e così via. Argomenti in favore della naturalità dell'attrazione magnetica: in primo luogo, l'analogia con il movimento naturale per il fatto che in entrambi i movimenti si constata una accelerazione verso la fine; inoltre, naturale è ciò che avviene sempre, è l'attrazione magnetica si verifica sempre: lo prova l'esperimento dei marinai. La risposta muove dal presupposto che il fenomeno dell'attrazione magnetica sia provocata da una forma, emessa dal magnete, che ha connaturata una capacità di muovere il ferro. L'idea era corrente, e la sostiene, fra tanti altri, anche Tommaso d'Aquino. Bisogna distinguere dunque fra una forma naturale indotta dal magnete e una forma naturale permanente (la gravità). In questo senso si può dire che l'attrazione magnetica è naturale e non violenta. Per i dettagli della "dimostrazione" relativa preferisco rinviare direttamente al testo sotto pubblicato.

Se il primo testo, dedicato alla determinazione della latitudine, con la sua misera ed efficace tecnologia illumina un momento certo modesto della scienza medievale, ma che ci fa riflettere sui grandi risultati che sulla medesima via furono raggiunti da un *experimentator* come Pietro di Maricourt, il secondo testo fa entrare il lettore nel cono d'ombra della scienza delle università medievali, che sterilizzò le nuove esperienze che i marinai venivano facendo nella gabbia sempre eguale delle forme, delle essenze, delle cause efficienti e dei luoghi naturali di aristotelica memoria. A passare dall'un testo all'altro si prova un senso di vertigine - lo stesso che coglie quando si confronta uno dei sofisticati portolani pisani del Duecento, identico alla carte odierne, con l'assurda, grossolana "carta delle mondo" che compare ancora in fondo al *De causis proprietatum elementorum* di Alberto il Grande. Eppure, anche la *quaestio de motu ferri ad adamantem* fa parte della scienza medievale, e perciò va letta, e non dimenticata.

Questio est de motu ferri ad adamantem, utrum sit naturalis vel violentus.

Quod non sit naturalis, hoc videtur sic: omnis motus naturalis alicuius est a principio existente in ipso mobili; iste motus ferri, ut communiter dicitur, est ab adamante ut a causa efficiente; ergo non est naturalis.

Ad idem. Omnis motus naturalis est ab aliqua forma naturali ita, quod ab una forma naturali vincitur; set motus naturalis a forma naturali ipsius ferri est motus deorsum eo, quod grave est; ergo motus ferri ad adamantem ab alia causa est quam a forma naturali ipsius mobilis. Item omnis motus naturalis aut est rectus aut circularis aut progressivus vel voluntarius. Motus progressivus solius animalis est, et motus rectus aut est a medio aut ad medium tantum. Motus circularis naturalis solius celi est. Iste motus ferri nec est voluntarius, quia ferrum non est animatum, nec est rectus, quia nec est tantum a medio nec tantum ad medium, sed est indifferenter ascendendo, discendendo et in latus, nec circularis, ut patet. Non ergo est motus naturalis. Item omnis motus naturalis aut est corporis simplicis aut mixti. Sed motus ferri non est motus corporis simplicis. Arguo ergo sic: omnis motus naturalis corporis mixti est a natura corporis simplicis vincentis in eo; corpus simplex vincens in ferro terra est; terre autem motus naturalis est descensus; ergo omnis motus naturalis ipsius erit descensus.

Quod non sit violentus probo. In hoc differt motus naturalis a violento, quod naturalis est in principio tardior et in fine velocior, motus violentus e converso. Age ergo: omnis motus violentus tardior est in fine quam in principio; motus ferri ad adamantem est in fine velocior et in principio tardior; ergo etc. Item in omni motu violento simul sunt movens et motus, ut probat Philosophus VII Phisicorum; in isto motu non simul sunt ferrum et adamas; ergo iste motus non est violentus ita, quod ferrum sit motum et adamas movens.

Quod sit motus violentus videtur sic. Motus violenti quatuor sunt species, vectio, tractio, vertigo, pulsio. Age ergo: omnis tractio est motus violentus; iste motus est tractio; ergo etc. Item quod non sit violentus, sed naturalis, probo. Nihil, quod semper est, violentum est; ista ferri attractio semper est; ergo etc. Quod autem semper sit, patet per experimentum nautarum. Ad hoc dicendum, quod adamas emittit ex se specien, cui innata est hec vis, quod possit movere ferrum. Hec autem species recipitur a ferro et incorporatur post incorporationem. Ergo huiusmodi speciei convenit loqui de ferro dupliciter, scilicet aut de ferro per se et inquantum ferrum est, aut de ferro hac condicione, scilicet cui incorporatur talis species. Loquendo de ferro per se iste motus violentus est, ferrum enim per se motum naturalem non habet preter descensum. Loquendo de ferro, cui unitur hec species emissa ab adamante ita, quod hec duo, ferrum et species, unita cadunt in unum subiectum mobile, sic dico, quod iste motus naturalis est et a principio ipsius subiecti intrinseco. In hoc ergo subiecto due sunt cause intrinsece motive, quarum una est forma resultans ex complexione ferri, in qua complexione dominatur terra, et alia est forma veniens ab adamante ferro incorporata. Per naturam prioris forme movetur ferrum solum descendendo, per naturam posterioris movetur ad adamantem. Nec est inconveniens, quod huiusmodi unius subiecti sint hii duo motus similes, cum sint in ipso due cause intrinsece motive. Et est, quando vincit in ferro virtus forme resultantis ex sui complexione, et tunc movetur deorsum vel quiescit deorsum non obstante presentia adamantis. Et est, quando vincit in ferro virtus forme, que veniens ab adamante ferro incorporatur et pre est virtuti prioris forme, et tunc movetur versus adamantem, scilicet ferrum. Et quia variatur virtus uniuscuiusque corporalis forme tam a magnitudine quam a parvitate corporis, in quo est, quam a diversitate situs in comparatione ad illud corpus a quo exit forma in propinquitate et remotione, propter hoc non pot est omnis ferrum cuiuscumque magnitudinis attrahere nec etiam a quantocumque spatio.

diversificatur etiam ista attractio in velocitate et parvitate ex eisdem differentiis.

Si autem queratur, quare hec forma incorporata ferro veniens ab adamante potius possit trahere ferrum quam forma ferro incorporata veniens aliunde, ut ab igne vel huiusmodi, dicendum, quod hoc est, quia talis virtus ei confertur per naturam. Et forte sic in loco naturali viget potentia attractiva corporis appetentis istum locum naturaliter sicut in centro universi potentia attractiva gravis. Et hoc sensit forte Aristoteles, cum dixit mirabilem esse loci potentiam. Secundum istum modum viget in adamante forma ferri attractiva. Et hoc modo assimilatur motus ferri ad adamantem motui alicuius corporis naturalis ad suum locum naturalem. Et ista attractio tam loci quam adamantis est per hoc, quod ad est in ipso mobili appetens per naturam huius accessum, sicut est in lapide, quod per naturam appetit descensum.

BIBLIOGRAFIA

[1]   Se ne veda la descrizione in *Codices Manuscripti Operum Thomae de Aquino*, II, ed. H.V. Shooner, Roma 1973, pp. 244-246.

[2]   Alexander Neckham, *De naturis rerum* II 98, ed. Th. Wright, London 1863 (rist. anast. Nendeln 1967), p. 183; T. Bertelli, *Sulla epistola di Pietro Peregrino di Maricourt e sopra alcuni trovati e teorie magnetiche del secolo XII. Memoria seconda*, in "Bullettino di bibliografia e di storia delle scienze matematiche e fisiche" 1 (1868), p. 103 sgg.

[3]   Thomas Cantimpratensis, *Liber de natura rerum*, ed. H. Boese, Berlin-New York, p. 357.

[4]   *Les oeuvres de Guyot*, ed. J. Orr, Manchester 1915.

[5]   P. Radelet de Grave - D. Speiser, *Le 'De magnete' de Pierre de Maricourt. Traduction et commentaire*, in "Revue d'histoire des sciences et de leurs applications" 28 (1975), pp. 225-226.

[6]   *Commento sopra la Divina comedia di Dante Allighieri*, ed. C. Giannini, III, Pisa 1862, p. 363.

[7]   Rogerus Bacon, *Opus minus*, in *Opera quaedam hactenus inedita*, ed. J. S. Brewer, I, London 1859 (rist. anast. Nendeln 1965), p. 383-384 3 Petrus Peregrinus, *De magnete* I 10, ed. Radelet de Grave - Speiser p. 219.

[8]   Descrizione in *Codices Vaticani Latini*, II, pars prior, *Codices 679-1134*, rec. A. Pelzer, In Bibl. Vaticana 1931, pp. 78-82.

Gravitational Waves, Supernovae and Quantum Gravity

YUKIO TOMOZAWA (*)

Abstract

I spent a prime time of my youth in Pisa under the scholarship of Luigi Radicati. During this period, I was deeply impressed by his taste for the beauty of the symmetries in nature. I am particularly proud that his discovery of $SU(6)$ symmetry in collaboration with Feza Gürsey, which paved a road to the quark model with color symmetry, was nourished throughout the period of my sojourn in Pisa (1961 - 64). If I am allowed to reflect on a personal matter, it may not be an accident that I came across a theorem in axiomatic field theory, the simple connectedness of the analyticity region of the Wightman functions (the extended tubes) as a result of Lorentz symmetry. When Luigi visited Michigan in a later year, his interest had shifted towards astrophysical problems, such as neutron stars and gravitational waves. My interests also drifted in the same direction lately. In this article which is dedicated to the celebration of Luigi's 70th Birthday, we suggest that quantum effects on gravitational collapse are observable by the detection of gravitational waves. In fact, it will be shown that the gravitational wave signals recorded by the Rome and Maryland detectors at the time of the SN87A explosion, in coincidence with the Mont Blanc neutrino data, may be understood only by the property of quantum mechanical black hole. We wish Luigi have many more happy returns and contribute in unveiling the beauty of symmetry.

1. - Introduction

When SN87A exploded, none of the sensitive gravitational wave (g.w.) detectors was in operation. Four and half hours before the observation of the neutrino signals at the Kamiokande [1] and AMB detectors [2], the room

(*)   Research Institute for Fundamental Physics, Kyoto University, Kyoto 606, Japan and Randall Lab. of Physics, University of Michigan, Ann. Arbor, Michigan 48109 USA

temperature g.w. detectors at Rome and Maryland recorded coincident signals [3] which were also in coincidence with five low energy neutrinos observed by the Mont Blanc group [4]. These data are not well understood because of i) poor statistics, ii) a large energy release in the g.w. signals ($\sim 2400 M_\odot$) and iii) the absence of neutrino signals in the Kamiokande and IMB detectors at the corresponding period.

There are three possibilities for the interpretation of the data,

A) All data in Rome, Maryland and Mont Blanc are false.

B) All or some of the data are significant, but have nothing to do with SN87A.

C) All or some of the data are significant and relevant to SN87A.

. Let us comment on these three cases in turn. Case A is possible for statistical reasons. However, there is a lesson from the history of science which tells us that the lack of theoretical understanding can lead to a misjudgment of observed data, as in the case of the experimental evidence for parity violation in the weak interactions [5]. Case B is unlikely since the probability of having such unusual events in such a tiny time span ($4\frac{1}{2}$ hours) is minute indeed. In order for case C to be realized, a drastic change in the understanding of gravitational collapse and the mechanism of supernova explosions is required. This is related to the difficulty of explaining supernova explosions theoretically. I will suggest that case C is possible if quantum theory is incorporated with gravity.

Quantum effects on gravity are discussed in section 2. Section 3 deals with gravitational collapse and introduces a new concept of black hole oscillation. In section 3, we discuss case C as a possible evidence for a quantum mechanical black hole. Futher discussion on astrophysical observations in AGN and high energy cosmic rays follows in section 5.

2. - Quantum Effects on Gravity

Quantum gravity is important in discussing the early universe and the wave function of the universe, otherwise considered to be irrelevant to any other (astrophysical) observations. If that is the case, little is expected for the development of quantum gravity as a science. In the subsequent sections, it will be argued that the contrary is true and quantum gravity has relevance to astrophysical observations.

In view of the absence of a consistent quantum gravity theory at the present time, however, we will consider quantum effects on gravity as a modest step towards such a goal and show that such effects can be tested by observations. In other words, we aim at an intermediate step towards the ultimate quantum gravity as in the case of the Bohr model for the ultimate quantum theory.

Quantum effects on gravity may be discussed by the equation

(1)
$$R_{\mu\nu} - \frac{1}{2}g_{\mu\nu}R = -8\pi G\langle T_{\mu\nu}\rangle$$

where the right hand side of the Einstein equation is replaced by the expectation value of the quantum energy momentum tensor $T_{\mu\nu}$. Eq(1) is considered to be a semi-classical approximation to quantum mechanical gravity theory and may be called the quantum Einstein equation.

The solution of Eq(1) for the Robertson-Walker metric was obtained numerically [6] and indicates that the universe can collapse and bounce back at a finite radius to a big bang explosion. On the other hand, the solution for a spherically symmetric and static metric [7] indicates that the gravitational potential is repulsive at short distances due to quantum effects. These solutions for both metrics are consistent with each other. In particular, a special case for the latter has the analytic solution

(2)
$$g_{oo} = g_{11}^{-1} = 1 + \frac{r^2}{\xi} - \sqrt{\frac{r^4}{\xi^2} + \frac{4GMr}{\xi}}$$

where ξ is the parameter which represents quantum effects and M is the mass. The metric (2) implies that the Schwarzschild metric is modified by quantum theory. It has no metric singularity, although the curvature singularity at the origin persists. A remarkable feature of the metric (2) is that the corresponding gravitational potential is repulsive near the origin. The metric g_{oo} has two zeros,

(3a)
$$r_+ = GM + \sqrt{(GM)^2 - \frac{\xi}{2}} \simeq 2GM - \frac{\xi}{4GM}$$

(3b)
$$r_- = GM - \sqrt{(GM)^2 - \frac{\xi}{2}} \simeq \frac{\xi}{4GM}.$$

Then, there remains the question how does quantum gravity change the above characteristics of the gravitational potentials. In the following, I will assume that the repulsive nature of the gravitational potential at short distances will remain true after the inclusion of quantum gravity. This is motivated from an analogy in atomic theory or in neutron star physics: Quantum effects play the role of repulsive agencies due to the uncertainty principle to prevent atoms or neutron stars from collapsing.

3. - Gravitational Collapse and Quantum Mechanical Black Hole

The modification of gravity by quantum effects is significant only at short distances. Obviously, it cannot be observed in gravitational systems where motions are dictated by the long range behavior of gravity. The only processes which are influenced by the short distance behavior of gravity are gravitational collapse. The gravitational collapse of a massive object leads to the formation of a black hole. The question, then, is how gravitational collapse or a black hole is influenced by quantum effects. A well known effect is Hawking radiation. But it is a small effect for a massive black hole. Besides, Hawking radiation results from the consideration of quantum effects on a static black hole which is a result of classical general relativity. It does not address to the question of dynamical nature such as gravitational collapse.

In order to find quantum effects on gravitational collapse, let us consider the motion of a test particle around a massive point object (a static black hole). In the classical limit, the Schwarzschild metric which represents the space time structure around a point particle dictates the motion of the test particle. The test particle initially at rest ar $r = R$ starts a radial motion towards the center. It takes an infinite amount of coordinate time to reach the Schwarzschild radius, but in proper time it reaches the center in

$$(4) \qquad\qquad \tau/2 \simeq \pi\sqrt{\frac{R^3}{8GM}}.$$

If the quantum metric (2) is introduced, the test particle makes a bounce back motion and then an oscillation in proper time τ. The question is where this oscillation is observed. A conventional description is that the oscillation takes place in different universes which are best described by the Penrose diagram, Fig. 1. This may be suggested in order to avoid a causality problem, since the coordinate time of the exit from the Schwarzschild radius $r = r_+$ is recorded as $t = -\infty$. Upon neglecting the energy loss by gravitational radiation, the periodic motion requires the existence of an infinite number of universes which contain identical black holes. How can one comprehend an identical black hole in each universe, despite the initial assumption of the existence of a black hole and a test particle in a single universe? The only way to comprehend this situation is to postulate that [A] all universes implied by the oscillation of a test particle in the Penrose diagram are *identical*.

Postulate [A], then, implies that the test particle should exist in each of the other universes, too. The inevitable consequence is that the observer in each universe recognizes an infinite number of oscillations.

The consideration of the infinite set of test particles in the oscillation eliminates the causality problem. Since it takes an infinite amount of coordinate time for the propagation of a signal from the Schwarzschild radius, an observer at $r = R$ receives the signal from the test particle exiting at Schwarzschild radius at a finite time. It is easy to show that the arrival time of the signal is approximately related to the proper time

$$(5) \qquad\qquad t_{\text{exit}} + t_{\text{signal}} \simeq \tau_{\text{exit}} + R, \quad \text{for} \quad r_+ << R.$$

Therefore, the signals sent from the test particle exiting at $r = r_+$ are observed by the observer at $r = R$ in casual sequence.

The same conclusion can be obtained if postulate [A] is replaced by [A']
the universe implied by the Penrose diagram for the oscillation of a test particle
is unique.

In other words, the oscillation of a test particle around a quantum
mechanical black hole can be observed by an observer in our universe. For
the gravitational collapse of a massive object which results in a black hole, a
point particle attached to the surface of the contracting object behaves like a
test particle around a black hole, although it is difficult to draw the Penrose
diagram in this case.

An alternative description is that the horizon is created when the collapsing
surfaces crosses the Schwarzschild radius and disappears when the surface exits
from the Schwarzschild radius in the bounce back motion. None of the internal
points of the collapsing object crosses the horizon. Of course, a signal from
the surface exiting from the Schwarzschild radius behaves like that of a test
particle and reaches an observer at $r = R$ at the time of eq(5). Therefore, we
conclude that gravitational collapse of a massive object leads to an oscillating
black hole which is observable in our universe. We may call it black hole
oscillation (BHO) or a black hole-white hole oscillation. The latter indicates
that the exit from the Schwarzschild radius is referred to a white hole (The
motion inside the Schwarzschild radius is called a wormhole).

4. - Black Hole Oscillation and Gravitational Wave

In the preceding sections, we have shown that a black hole formed
by gravitational collapse should oscillate in the universe. The oscillation (in
proper time) takes place in and out of the Schwarzschild radius. It is extremely
important to find evidence for such an oscillation in astrophysical observations.
Obviously, a multiple gravitational wave may be radiated from an oscillating
BH with spherical asymmetry. A sensitive g.w. detector in the future would find
such signals and provide us with observational evidence for quantum effects on
gravity or quantum gravity.

A BHO will give an enhancement of the gravitational wave signal in a
Weber detector if the proper frequency of the detector matches the period of
the gravitational oscillation (resonance). This can be explained by the formula
for the oscillation amplitude of the detector [8],

(6)
$$\xi(z,t) = \frac{1}{2\pi} \int T(z,\omega) H(\omega) e^{i\omega t} d\omega$$

where

(7)
$$T(z,\omega) = \frac{1}{2} \left(z - \frac{\sin \alpha z}{\alpha \cos \frac{\alpha L}{2}} \right),$$

α is given in terms of the characteristic parameters of the detector [8] and

$$(8) \qquad h(t) = \frac{1}{2\pi} \int H(\omega) e^{i\omega t} d\omega$$

is the gravitational wave amplitude at the detector. Neglecting damping,

$$(9) \qquad \xi_n(z,t) = \xi(z,t) + \xi(z,t+T_g) + \xi(z,t+2T_g) + \cdots \xi(z,t+(n-1)T_g)$$

represents the total amplitude due to gravitational waves which are sent with period

$$(10) \qquad T_g = \frac{2\pi}{\omega_g}.$$

Here we assume that the relaxation time is greater than nT_g. Then

$$(11) \qquad \xi_n(z,t) = \frac{1}{2\pi} \int T(z,\omega) H(\omega) R(\omega) e^{i\omega t} d\omega$$

where

$$R(\omega) = 1 + e^{i\omega T_g} + e^{2i\omega T_g} + \cdots + e^{2i(n-1)\omega T_g}$$

$$(12) \qquad = \frac{1 - e^{in\omega T_g}}{1 - e^{i\omega T_g}} = e^{i(n-1)\omega T_g/2} \frac{\sin \frac{n\omega T_g}{2}}{\sin \frac{\omega T_g}{2}}$$

$$= e^{i(n-1)\pi\omega/\omega_g} \frac{\sin(n\pi\omega/\omega_g)}{\sin(\pi\omega/\omega_g)}.$$

We thus obtain

$$(13) \qquad |R(\omega)|^2 = \left(\frac{\sin(n\pi\omega/\omega_g)}{\sin(\pi\omega/\omega_g)} \right)^2$$

which has the form given in Fig. 2.

The peaks at $\omega = 0,\ w_g, 2\omega_g \cdots$ have height n^2 and half width ω_g/n. Then the effective gravitational wave amplitude is given by $H(\omega) R(\omega)$ and its shape is shown in Fig. 3 for the case of a continuous shape for $H(\omega)$.

For the value $n \simeq 30$, one obtains

$$(14) \qquad n^2 \simeq 10^3$$

This brings down the energy release of a g.w. signal for one oscillation by 10^{-3}. Applied to the case of the coincident signals of the Rome and Maryland detectors, one obtains $2400 M_\odot / 10^3 \propto 2.4 M_\odot$ for one oscillation. We note that the frequencies of the Maryland and Rome detectors are 1660 Hz and 858 Hz respectively. Moreover, if $|H(\omega)|^2$ is peaked around $\omega_g, 2\omega_g, \cdots$ as a result of k

oscillations [9] (let us assume that $k \simeq n$ for somplicity), the energy release for one BHO becomes $\sim 0.08 M_\odot$. The directionality of a g.w. can further reduce this value by $1/5$, resulting in the energy release to be $\sim 0.02 M_\odot$. This is a rather resonable range of energy release for the gravitational collapse of a star with, say, $3 M_\odot$.

Assuming that the energy problem in the g.w. signals for the SN87A explosion is solved as discussed above, how can one understand the coincidence signals of the Rome-Maryland-Mont Blanc detectors? Before constructing a scenario for SN87A, we add a couple of characteristic of BHO.

a) BHO yields not only a multiple g.w. signal, but also emits high energy particles and radiation, in particular, in an expansion phase. This is because the system is equivalent to a black body with a temperature which is decreasing due to expansion.

b) The addition of infalling matter onto a neutron star yields a black hole, an oscillating BH. The emission of high energy particles and radiation results in a damping of the oscillating amplitude (or radius). If enough mass is emitted, an oscillating BH can go back to a neutron star. This is a novel feature of a quantum mechanical black hole.

Now we are ready to present a possible scenario for the SN87A explosion.

1. When the core of the progenitor of SN87A collapsed, the explosion did not occur immediately. The addition of infalling matter lead to the formation of a black hole. At least this is the result of various numerical calculations for the SN87A (non)explosion [10].

2. Due to quantum theory, this black hole oscillates and emits a multiple gravitational wave as well as particles and radiation. The former was detected by the Rome and Maryland detectors due to resonance effects on the Weber bars. There were several g.w. bursts [3] observed in two hours.

3. The cross section for neutrino-nucleon reaction is an increasing function of energy. A large density due to infalling matter makes the escape of high energy neutrino difficult. This results in a severe cut off of high energy neutrinos. Since the Mont Blanc detector has a somewhat lower threshold than those of the Kamiokande and IMB detectors, the observation of five low energy neutrinos ($6 \sim 8$ MeV) only by the Mont Blanc group implies a severe cut off in the neutrino energy spectrum. The 1.4 sec delay of the neutrino signals after the g.w. signals may be understood in terms of neutrino mass ($\lesssim 7$ MeV) or neutrino diffusion [4].

4. The release of enrgy by the emission of high energy particles, radiation and neutrinos from the oscillatng BH to the high density environment yields the explosion of the envelope $4\frac{1}{2}$ hour after the collapse. This releases the neutrino bursts which were recorded by the Kamiokande and IMB detectors. The small size of the Mont Blanc detector explains the absence of signals at the time of the Kamiokande and IMB detection.

5. An oscillating BH sometimes becomes a neutron star after the explosion takes place. This neutron star was detected 2 years later.

5. - Other Astrophysical Observations

An oscillating BH can leave a trace in other astrophysical phenomena.

I. High Energy Cosmic Rays.
The energy spectrum of high energy particles emitted in the expansion phase of a BHO has been computed [11, 7]. It gives the spectrum

$$(15) \qquad\qquad \sim E^{-3}$$

by a dimensional argument. This spectrum resembles that of high energy cosmic rays above the knee energy. The absence of a suitable model for the origin of high energy cosmic rays makes a BHO an attractive candidate.

II. Active Galactic Nuclei.
Due to the large energy release from distant objects and relatively small variability time for x-ray luminosities, the energy source for AGN is considered to be a massive ($\sim 10^8 M_\odot$) BH. However, the meaning of the variability remains a mystery. We have proposed that the variability is naturally understood in terms of BHO [12].

Acknowledgment

It is a great pleasure to thank many people who contributed in forming the theoretical framework of this article through critical discussions and helpful suggestions. To name a few, they are Humitaka Sato, Roger Penrose, Kenichi Maeda, Joe Weber, Jim Wilson and Paul Anderson. Thanks are also due to David N. Williams for reading the manuscript. This work is supported in part by the US Department of Energy and in part by the Research Institute for Fundamental Physics, Kyoto University. The author is deeply indebted to Kazuhiko Nishijima for the warm hospitality of RIFP.

REFERENCES

[1] K. HIRATA et. al., Phys. Rev. Lett. **58**, 1490 (1987).

[2] R.M. BIONTA et. al., Phys. Rev. Lett. **58**, 1494 (1987)

[3] E. AMALDI et. al., Europhys, Lett. **3** 1325 (1987); E. AMALDI et. al., in Proc. the fourth George Mason Astrophysics Workshop (ed. M. Kafatos and A.G. Michalitsianos, Cambridge University Press, 1987) p. 453.

[4] M. AGLIETTA et. al., Europhys. Lett. **3**, 1315 (1987).

[5] R.T. COX et. al., Proc. N.A.S. **14**, 544 (1928); C.T. CHASE, Phys. Rev. **34**, 1069 (1929); **36** 984, 1060 (1930).

[6] P. ANDERSON, Phys. Rev. **D28**, 271 (1983); **D29**, 615 (1984).

[7] Y. TOMOZAWA, Proc. INS International Symposium on Composite Models of Quarks and Leptons, (ed H. Terazawa and M. Yasue 1985) p. 386; Lectures in the Second Workshop on Fundamental Physics, Univ. of Puerto Rico, Humacao (ed E. Esteben, 1986) p. 144; Proc. of the 26th International Astrophys. Colloquium, Liege (1986) p. 137.

[8] E. AMALDI - G. PIZZELLA, in Relativity, Quanta and Cosmology (ed M. Pantaleo and F. DeFinis, 1979) p. 9.

[9] T. NAKAMURA et. al., Progr. Theoret. Phys. Suppl. No.90, (1987).

[10] J. WILSON, in the Aspen Conference on Supernovae, (1988); W. HILLEBRANDT, Proc. of the Yamada Conference on Big Bang, Active Galactic Nuclei and Supernova (ed K. Sato, 1988).

[11] Y. TOMOZAWA, in Quantum Field Theory (ed. Mancini, 1985) p. 241.

[12] Y. TOMOZAWA, Proc. of the Conference on Active Galactic Nuclei (ed. R. Miller and P. Wiita) Spring Verlag 1988, p. 236; Proc. of the Yamada Conference on Big Bang, Active Galactic Nuclei and Supernova (ed. K. Sato, 1988) p. 577; the 5th Marcel Grossman Meeting on Gravity, Perth, Australia 1988.

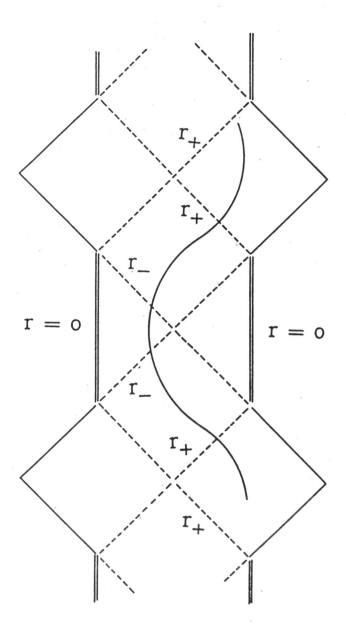

Fig. 1 - Oscillation of a test particle in the quantum metric described by a
Penrose diagram. (Similar to that of the Reissner-Nordstom metric).

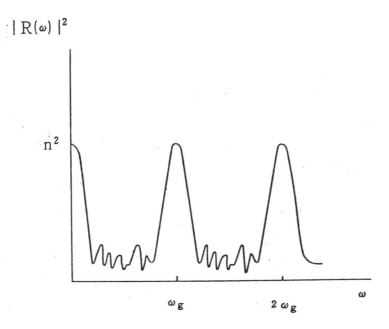

Fig. 2 - *Resonance signal in a Weber bar detector by a multiple g.w. due to quantum effects.*

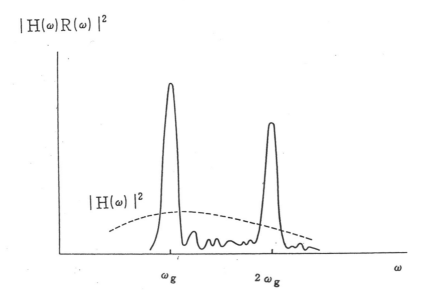

Fig. 3 - *The effective g.w. amplitude squared in a Weber bar detector.*

The Penetration of an Elastic Wedge

PIERO VILLAGGIO (*)

Abstract

The elementary formula of applied mechanics, relating the forces acting on a rigid prism of triangular cross section wedged into a seat whose opening is of the same width, is revised by assuming the material constituting the wedge to be elastic. The solution is consequently able to specify the stress distribution inside the wedge; an evaluation which is not possible when the material is regarded as rigid.

1. - Introduction

One of the most classical problems in the theory of friction is that of calculating the force P that must be exerted on a wedge of opening 2α (Fig. 1), inserted between two jaws whose opening is of the same angle 2α in such a way as to permit its penetration or oppose its extraction when, between the surfaces in contact, a frictional force T arises related to the normal compressive force N by a relation of the type

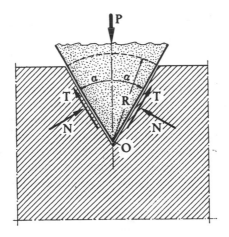

Fig. 1

(*)   Pisa

(1.1) $$T = \mu N,$$

where μ is the friction factor between the two bodies.

Generally both the wedge and its seat are regarded as rigid bodies so that, neglecting local interactions, the representative quantities of the problem are the resultants N and T which the two bodies trasmit to each other across the faces of the wedge. Therefore, if equilibrium at the vertical translation among the various forces is imposed, the equation

(1.2) $$P = 2N \sin \alpha + 2T \cos \alpha$$

must necessarily hold. If the friction angle $\rho_0 = $ artg μ is introduced, it then becomes

(1.3) $$P = 2N \frac{\sin(\alpha + \rho_0)}{\cos \rho_0}.$$

This formula also applies when the force P, instead of facilitating penetration, help to maintain the wedge fixed, but in this case the frictional force changes its orientation and therefore ρ_0 must be replaced by $-\rho_0$ and the formula (1.3) turns into

(1.4) $$P = 2N \frac{\sin(\alpha - \rho_0)}{\cos \rho_0},$$

which proves that, for $\alpha \leq \rho_0$, the so called "self-stoppage" occurs, that is the wedge can be extracted only by applying a force directed upwards (Szabó [1, § 22]).

The equilibrium between resultants is however unable to characterize the detailed distribution of stresses along the walls of the wedge nor can it reveal the law by which the axial force varies along the axis of the wedge as the vertex is approached. In order to obtain this result it is necessary to restate the physical model by assuming, for instance, that the wedge is made of an elastic material, while the body into which it penetrates is still regarded as perfectly rigid. Under these assumptions it is relatively easy to find an elastic solution which satisfies the point for point conditions of contact with friction along the region of adherence as well as those of global equilibrium between exterior load and stress resultants.

The conclusion is that the classical formulae of technical mechanics are nearly completely confirmed even within the framework of elasticity theory, consequently justifying their popularity. Conversely, the behaviour of stresses near the vertex is highly influenced by the friction factor and the angle of the opening.

2. - The Problem of the Elastic Wedge

The most realistic way to describe the fastening of a wedge is by assuming that the penetrating body is a dihedron, whose cross section is drawn in figure 1. By indicating the axis perpendicular to the plane of the figure as z and imagining the dihedron to be sufficiently long in its transverse direction z, it is natural to assume that the displacements in each plane perpendicular to the edge are independent of z. A strain of this type is called "plane" since all strain and stress components are independent of z.

Due to symmetry, it is convenient to refer the generic section of the wedge to a system of plane polar coordinates with origin at the vertex in such a way that it occupies the plane region defined by the inequalities $0 \leq r \leq R, -\alpha \leq \varphi \leq \alpha$.

The material constituting the wedge is assumed to be linearly elastic, homogeneous and isotropic, that is characterized by Young modulus E and Poisson ratio σ; the material of the sorrounding body is instead perfectly rigid. In the polar system with origin at the vertex let u be the radial component of displacement and v the one perpendicular to the radius, and let

$$(2.1) \qquad \varepsilon_r = \frac{\partial u}{\partial r}, \ \ \varepsilon_\varphi = \frac{u}{r} + \frac{1}{r}\frac{\partial v}{\partial \varphi}, \ \ \varepsilon_{r\varphi} = \frac{1}{2}\left(\frac{1}{r}\frac{\partial u}{\partial \varphi} + \frac{\partial v}{\partial r} - \frac{v}{r}\right)$$

be the corresponding strain components.

But, since the state of strain is planar, it is inopportune to take the displacements as unknowns. It is instead better to introduce a function $F = F(r, \varphi)$, biharmonic in the considered domain, such that the stress components $\sigma_r, \sigma_\varphi, \tau_{r\varphi}$ are derivable from it through the relations

$$(2.2) \qquad \sigma_r = \frac{1}{r}\frac{\partial F}{\partial r} + \frac{1}{r^2}\frac{\partial^2 F}{\partial \varphi^2}, \ \ \sigma_\varphi = \frac{\partial^2 F}{\partial r^2}, \ \ \tau_{r\varphi} = -\frac{\partial}{\partial r}\left(\frac{1}{r}\frac{\partial F}{\partial \varphi}\right).$$

The function $F(r, \varphi)$ is also known as Airy's "stress function" (Love [2, Art. 56]).

A significant fact to be considered is that, once the function $F(r, \varphi)$ is constructed, it is possible to obtain the displacement components u, v by integration of (2.1), since strains are related to stresses by the constitutive equations

$$(2.3) \qquad \varepsilon_r = \frac{1}{\overline{E}}(\sigma_r - \overline{\sigma}\sigma_\varphi), \ \ \varepsilon_\varphi = \frac{1}{\overline{E}}(\sigma_\varphi - \overline{\sigma}\sigma_r), \ \ \varepsilon_{r\varphi} = \frac{1+\overline{\sigma}}{\overline{E}}\tau_{r\varphi},$$

where, being in the state of plane strain, the new elastic constants $\overline{E}$ and $\overline{\sigma}$ are defined in the terms of the physical constants E and σ by (cf. Worch [3, pag. 6])

$$(2.4) \qquad\qquad \overline{E} = \frac{E}{1-\sigma^2}, \ \ \overline{\sigma} = \frac{\sigma}{1-\sigma}.$$

The simplest way to find the function $F(r, \varphi)$ is that of superimposing particular biharmonic solutions in order to satisfy the boundary conditions. The latter must be set as follows: along the two sides $\varphi = \pm\alpha$ the material adheres to the walls and, therefore, the v-component of displacement must vanish; again, by the effect to friction on $\varphi = \alpha$, the normal stress σ_r and the tangential stress $\tau_{r\varphi}$ must be related by a condition of the form

$$(2.5) \qquad\qquad \tau_{r\varphi} = -\mu\sigma_\varphi,$$

where μ is the factor od friction and the negative sign cosiders the fact that, if the force P is directed downwards, the stresses σ_φ are compressive, that is negative, whereas the stresses $\tau_{r\varphi}$ are positive since directed along the increasing values of the radius; the same condition as (2.5) must hold on $\varphi = -\alpha$, but without the minus sign before μ. On the part of the boundary $r = R$ the surface tractions are not specified; it is only known that these are statically equivalent to the force P.

In order to satisfy the boundary conditions it is convenient to choose $F(r, \varphi)$ of the form

$$(2.6) \qquad\qquad F(r, \varphi) = Ar^k \cos k\varphi + Br^k \cos(k-2)\varphi,$$

where A and B are indeterminate constants. A function $F(r, \varphi)$ of this type has been introduced by Williams [4] to characterize the singularities of elastic solutions in the neighbourhood of the vertex of an angle. From the function $F(r, \varphi)$ the stresses are immediately obtained by differentiation:

$$\sigma_r = -Ak(k-1)r^{k-2} \cos k\varphi - B(k-1)(k-4)r^{k-2} \cos(k-2)\varphi,$$
$$(2.7) \qquad \sigma_\varphi = Ak(k-1)r^{k-2} \cos k\varphi + Bk(k-1)r^{k-2} \cos(k-2)\varphi,$$
$$\tau_{r\varphi} = Ak(k-1)r^{k-2} \sin k\varphi + B(k-1)(k-2)r^{k-2} \sin(k-2)\varphi,$$

whereas it is possible to show, though less directly (cf. Worch [3, pag. 23]), that the displacement components are

$$(2.8) \qquad \begin{aligned} \overline{E}u &= -A(1+\overline{\sigma})kr^{k-1} \cos k\varphi - B[\overline{\sigma}k + (k-4)]r^{k-1} \cos(k-2)\varphi, \\ \overline{E}v &= A(1+\overline{\sigma})kr^{k-1} \sin k\varphi + B[(1+\overline{\sigma})(k-2) + 4]r^{k-1} \sin(k-2)\varphi; \end{aligned}$$

these displacements are in effect defined to within a rigid motion, which can nonetheless be disregarded.

The boundary conditions on the sides $\varphi = \pm\alpha$, along which v is zero and the normal and tangential stresses are related by the law of friction, impose A and B to satisfy the two equations

$$A(1+\overline{\sigma})k \sin k\alpha + B[(1+\overline{\sigma})(k-2) + 4] \sin(k-2)\alpha = 0,$$
$$(2.9) \quad Ak(k-1) \sin k\alpha + B(k-1)(k-2) \sin(k-2)\alpha = -\mu[Ak(k-1) \cos k\alpha$$
$$+ Bk(k-1) \cos(k-2)\alpha].$$

These two equations admit non trivial solutions A and B when the determinant of the coefficients vanishes, that is when

$$-4 \sin k\alpha \sin(k-2)\alpha + \mu(1+\bar{\sigma})k \sin k\alpha \cos(k-2)\alpha$$

(2.10)
$$- \mu[(1+\bar{\sigma})(k-2)+4] \cos k\alpha \sin(k-2)\alpha = 0,$$

and this is a trascendental equation which permits the determination of the exponent k, or better, a sequence of exponents. These roots can also be complex and generate solutions oscillatory along the radius, a property already known for problems of this type (cf. Parton and Perlin [5, VIII]). It is also useful to put the equation into the form

$$4[\cos(2k-2)\alpha - \cos 2\alpha] - 2\mu(1-\bar{\sigma}) \sin(2k-2)\alpha$$

(2.11)
$$+ 2\mu[(1+\bar{\sigma})k + (1-\bar{\sigma}] \sin 2\alpha = 0.$$

But, before solving (2.11), it is important to observe that not all solutions are significant, but only those of the type $k = k_1 + ik_2$, of which the real part k_1 is strictly greater than one in order that the stresses, though singular at the vertex, generate finite strain energy. In the absence of friction, that is for $\mu = 0$, it is immediately verifiable that the first root of (2.11) obeying the required condition of regularity is $k_1 = 2, k_2 - 0$. For values of μ other than zero calculation is much more difficult. However, for sufficiently small values of μ, an approximate expression of k_1 may be obtained by setting $k_1 = 2 + \varepsilon$ and regarding μ and ε as infinitesimal quantities. Neglecting higher order terms, (2.11) becomes

$$4[-2\varepsilon \sin 2\alpha] - 2\mu[-2(1+\bar{\sigma})] \sin 2\alpha = 0,$$

and therefore $\varepsilon = \frac{1}{2}\mu(1+\bar{\sigma})$, $k_1 = 2 + \frac{1}{2}\mu(1+\bar{\sigma})$, while the imaginary part k_2 is still zero.

Having found k_1, the equations (2.9) permit the calculation of A and B, which are, of course, defined to within an arbitrary constant C:

(2.12)
$$\begin{aligned} A &= -C[(1+\bar{\sigma})(k_1-2)+4] \sin(k_1-2)\alpha, \\ B &= C(1+\bar{\sigma})k_1 \sin k_1\alpha. \end{aligned} \Big\}$$

The constant C is determined by imposing the last, still unexploited, boundary condition, that is that the vertical resultant of surface tractions exerted on the arc $r = R, -\alpha \leq \varphi \leq \alpha$, be equal to P, or better to $-P$, since the force is directed further downwards. In terms of the stresses $\sigma_r, \tau_{r\varphi}$, this condition can be written

$$R \int_{-\alpha}^{\alpha} (\sigma_r \cos\varphi - \tau_{r\varphi} \sin\varphi)d\varphi = -P,$$

and hence, expressing σ_r and $\tau_{r\varphi}$ thorough (2.7) and evaluating the integrals, it follows that

$$2R^{k_1-1}[-Ak_12\sin(k_1-1)\alpha - B(k_1-1)\sin(k_1-3)\alpha + B\sin(k_1-\alpha)] = -P.$$

Now, assigning A and B their values as functions of C, the equation above becomes

$$Ck_1R^{k_1-1}\{[(1+\bar{\sigma})(k_1-1)+4](\cos\alpha - \cos(2k_1-3)\alpha)$$
$$-(1+\bar{\sigma})(\cos 3\alpha - \cos(2k_1-3)\alpha)\} = -P,$$

whence, by using known trigonometric identities, the constant C is derived:

$$C = -\frac{P}{2k_1R^{k-1}}\{[(1+\bar{\sigma})(k_1-1)+4]\sin(k_1-1)\alpha\sin(k_1-2)\alpha$$

(2.13)
$$- (1+\bar{\sigma})\sin k_1\alpha\sin(k_1-3)\alpha\}^{-1}.$$

Once C is found, the complete state of stress within the wedge-shaped region is determined. Since k_1, in the form valid for small values of μ, is equal to $2+\frac{1}{2}\mu(1+\bar{\sigma})$, the stresses vanish at the vertex if μ is strictly positive, that is when the force P tends to draw the wedge in, they are instead infinite if μ is strictly negative, that is when the force P tends to prevent the expulsion of the wedge. This kind of-behaviour is rather surprising because it shows that the stress diffusion around the vertex radically changes when the virtual motion of the wedge is reversed.

3. - The Forces on the Walls

From the stresses it is easy to evaluate the forces N and T, exerted across the faces, by integration of the components σ_φ and $\tau_{r\varphi}$, their integrals being finite since, for small values of μ, the factor k_1 is strictly greater than one. Having put

$$\left.\begin{array}{l} N = -\int_0^R \sigma_\varphi\,dr = -Ak_1R^{k_1-1}\cos k_1\alpha - Bk_1R^{k_1-1}\cos(k_1-2)\alpha, \\[2mm] T = \int_0^R \tau_{r\varphi}\,dr = Ak_1R^{k_1-1}\sin k_1\alpha + Bk_1R^{k_1-1}\sin(k_1-2)\alpha, \end{array}\right\}$$
(3.1)

and again expressing A and B in terms of C, these forces can be written as

$$N = Ck_1R^{k_1-1}\{[(1+\bar{\sigma})(k_1-2)+4]\sin(k_1-2)\alpha\cos k_1\alpha - (1+\bar{\sigma})\sin k_1\alpha$$

(3.2)
$$\cos(k_1-2)\alpha\},$$

$$T = -Ck_1R^{k_1-1}\{[(1+\bar{\sigma})(k_1-2)+4]\sin(k_1-2)\alpha\sin k_1\alpha - (1+\bar{\sigma})\sin k_1\alpha$$
$$\sin(k_1-2)\alpha\}.$$

According to the signs, N is positive if acting in the direction of compression and T is positive if directed along the increasing values of the radius. The sum of their components along the x-axis is thus $(2N\sin\alpha + 2T\cos\alpha)$, and a simple calculations shows that this quantity may be put in the form

$$-2Ck_1R^{k_1-1}\{[(1+\bar{\sigma})(k_1-2)+4]\sin(k_1-2)\alpha\sin(k_1-1)\alpha$$
$$-(1+\bar{\sigma})\sin k_1\alpha\sin(k_1-3)\alpha\},$$

which, when the value of C is obtained by equation (2.13), is exactly equal to P, confirming the property that the forces on the walls balance the external load.

Alternatively, the force P can be expressed as a function of the single component N as

$$P = 2N\left(\sin\alpha + \frac{T}{N}\cos\alpha\right)$$

or also, using (3.2),

$$(3.3)\qquad P = 2N\frac{-[(1+\bar{\sigma})(k_1-2)+4]\sin(k_1-2)\alpha\sin(k_1-1)\alpha+(1+\bar{\sigma})\sin k_1\alpha\sin(k_1-3)\alpha}{[(1+\bar{\sigma})(k_1-2)+4]\sin(k_1-2)\alpha\cos k_1\alpha-(1+\bar{\sigma})\sin k_1\alpha\cos(k_1-2)\alpha}.$$

It is thus quite clear that the solution to the problem of an elastic wedge differs structurally from that of the corresponding rigid model, although the numerical values of the coefficients relating P to N are not substantially different. For instance, for $\alpha = \frac{\pi}{4}$, $\bar{\sigma} = 0$, $\mu = \frac{1}{10}$, k_1 is equal to $2,05$ and formula (3.3) furnishes the relation $P = 1,68N$; formula (1.3) yields instead $P = 2,18N$.

REFERENCES

[1] I. SZABÓ, *Einführung in die Technische Mechanik*, Berlin-Göttingen-Heidelberg: Springer (1963).

[2] A.E.H. LOVE, *A Treatise on the Mathematical Theory of Elasticity*, Cambridge: The University Press (1927).

[3] G. WORCH, "Elastiche Scheiben" Beton-Kalender 1967, Vol. II. Berlin-München: W. Ernst und Sohn.

[4] M.E. WILLIAMS "Stress singularities resulting from various boundary conditions in angular corners of plates in extension" Journ. Appl. Mech. Vol. 19, pp. 526-528 (1952).

[5] V.Z. PARTON - P.I. PERLIN, *Mathematical Methods of the Theory of Elasticty*, Moscow: Mir (1984).

Properties of Quantum 2×2 Matrices[†]

S. VOKOS (*) - B. ZUMINO (*) - J. WESS (**)

1. - Introduction

The theory of quantum groups has attracted great interest recently. In mathematics quantum groups are related to Hopf algebras, non commutative geometry and the theory of knots and links. In physics they are relevant for the theory of integrable systems, certain problems in statistical physics and the study of conformal field theories in two dimensions.

One approach to the study of quantum groups, followed especially by Faddeev and collaborators, defines them in terms of their basic representation by matrices. Thus the quantum version of $SL(2, C)$, denoted by $SL_q(2, C)$, is defined by giving quantization relations for the elements of the 2×2 $SL(2, C)$ matrix, as described briefly in Section 2. Similarly, for other Lie groups, one starts from the basic representation and quantizes the matrix elements of the classical matrix. Higher representations of the same quantum group can be obtained by multiplying and reducing quantum representations.

It can happen that the basic representation of a quantum group possesses special interesting properties. In this paper we describe the special properties we have found for the 2×2 representation of $SL_q(2, C)$. Although the results can be stated very simply, the proofs are usually somewhat lengthy and involved. We shall only sketch the basic ideas of the proofs here and will describe the details in a longer paper.

The literature on quantum groups is very extensive. At the end we list only a few papers where numerous other mathematical and physical references can be found.

We are deeply indebted to Ludwig Faddeev and to Vaughan Jones for introducing us to the theory of quantum groups.

[†] This work was supported in part by DOE contract DE-AC03-76SF00098 and in part by NSF grant PHY-85/15857.
(*) University of California, Berekeley, CA, USA.
    Present address, LAPP, BP.110, 74941 Annecy-le-Vieux Cedex, France.
(**) Karlsruhe University, Karlsruhe, Germany.

It is a pleasure to dedicate this paper to Luigi Radicati on the occasion of his 70th birthday.

2. - Quantum $SL(2,C)$

We review the usual definition of quantum $SL(2,C)$, i.e. $SL_q(2,C)$, in terms of its two dimensional representation. We consider first the general linear group in two dimensions. A matrix

(2.1)
$$A = \begin{pmatrix} a & b \\ c & d \end{pmatrix}$$

is said to belong to the quantum linear group $GL_q(2,C)$ if its matrix elements, instead of being complex numbers, are non commuting quantities (which can be realized as operators in a Hilbert space) satisfying the commutation relations

(2.2)
$$\left. \begin{aligned} ab &= qba \\ ac &= qca, \quad bc = cb \\ bd &= qdb, \quad ad - da = \left(q - \frac{1}{q} \right) bc \\ cd &= qdc \end{aligned} \right\}.$$

Here q is a complex number, the quantum parameter. Matrices like A have the following remarkable property which can be taken as the definition of a quantum group. Let

(2.3)
$$A' = \begin{pmatrix} a' & b' \\ c' & d' \end{pmatrix}$$

be a matrix of the same type, i.e. let its elements satisfy commutation relations similar to (2.2)

(2.4)
$$\left. \begin{aligned} a'b' &= qb'a' \\ a'c' &= qc'a' \\ \text{etc.} \end{aligned} \right\}.$$

Let also a', b', c', d' commute with a, b, c, d. Then the matrix $A'' = AA'$

(2.5)
$$A'' = \begin{pmatrix} a'' & b'' \\ c'' & d'' \end{pmatrix} = \begin{pmatrix} aa' + bc' & ab' + bd' \\ ca' + dc' & cb' + dd' \end{pmatrix}$$

is also of the same type, i.e.

(2.6)
$$\left.\begin{array}{c} a''b'' = qb''a'' \\ a''c'' = qc''a'' \\ \text{etc.} \end{array}\right\}.$$

With reference to matrices like A, A' and A'' one talks of a quantum group, the corresponding classical group being obtained in the limit $q \to 1$, when the matrix elements commute. A quantum group is not a group; a better epression would be quantized group.

The quantum determinant of the matrix A is defined as

(2.7)
$$D_q = \det_q A = ad - qbc = da - \frac{1}{q} bc.$$

It reduces to the usual determinant for $q = 1$. Using (2.2) it is easy to verify that D_q is central, i.e. it commutes with $a, b, c,$ and d. Using the quantum determinant one obtains the (both right and left) inverse matrix

(2.8)
$$A^{-1} = \frac{1}{D_q} \begin{pmatrix} d & -\frac{1}{q} b \\ -qc & a \end{pmatrix}.$$

Notice that A^{-1} is a quantum matrix which corresponds to the quantum parameter q^{-1}. Indeed, from (2.2),

(2.9)
$$d\left(-\frac{1}{q} b\right) = \frac{1}{q} \left(-\frac{1}{q} b\right) d \quad \text{etc.}$$

Similarly the matrix

(2.10)
$$A^2 = \begin{pmatrix} a^2 + bc & ab + bd \\ ca + dc & cb + d^2 \end{pmatrix}$$

corresponds to the quantum parameter q^2, a fact which can also be easily verified using the commutation relations (2.2). In general one can show that the matrix A^n is a quantum matrix corresponding to the quantum parameter q^n, as we discuss in Sections 3 and 4. This fact was also noticed and proven by Corrigan and Tunstall [7].

One can impose the condition

(2.11)
$$D_q = 1$$

which restricts $L_q(2, C)$ to $SL_q(2, C)$. In addition one can impose reality conditions on the matrix A. One choice is that it be unitary

(2.12)
$$\bar{a} = d \quad \bar{b} = -qc \quad \bar{c} = -\frac{1}{q} b.$$

These relations restrict $SL_q(2, C)$ to $SU_q(2)$. They require for consistency that q be real. Another choice is that A be real

$$(2.13) \qquad \bar{a} = a \quad \bar{b} = b \quad \bar{c} = c \quad \bar{d} = d.$$

This gives $SL_q(2, R)$. For consistency with (2.2) it must now be $|q| = 1$.

The commutation relations (2.2) can be interpreted as quantum symplectic conditions on A. Define the quantum epsilon matrix

$$(2.14) \qquad \varepsilon_q = \begin{pmatrix} 0 & \frac{1}{\sqrt{q}} \\ -\sqrt{q} & 0 \end{pmatrix}$$

which satisfies

$$(2.15) \qquad \varepsilon_q^2 = -1.$$

One has

$$(2.16) \qquad \varepsilon_q A^T \varepsilon_q^{-1} = \begin{pmatrix} d & -\frac{b}{q} \\ -qc & a \end{pmatrix} = D_q A^{-1},$$

where A^T is the transposed of the matrix A. (2.16) can be written as

$$(2.17) \qquad A^T \varepsilon_q A = A \varepsilon_q A^T = D_q \varepsilon_q.$$

For $D_q = 1$ this the quantum analogue of the usual conditions for a matrix to be symplectic. The two conditions (2.17) are equivalent to (2.2) plus (2.7).

3. - Properties of 2×2 quantum matrices

As we mentioned in Section 2, the $n - th$ power A^n of the matrix A is a quantum matrix corresponding to the quantum parameter q^n. In this section we sketch a proof of this fact in the case when $n \gtrless 0$ is an integer, $n \in \mathbb{Z}$. As we shall see in the next section the result is valid for continuous values of n. We are dealing here with special properties of 2×2 quantum matrices. It would be interesting to see if and how they generalize to higher dimensional representations of $GL_q(2, C)$ or to other quantum groups.

Let us call a_n, b_n, c_n, d_n the matrix elements of the $n - th$ power of the matrix A in (1)

$$(3.1) \qquad A^n = \begin{pmatrix} a_n & b_n \\ c_n & d_n \end{pmatrix}$$

and let

$$(3.2) \qquad D_n = \det_{q^n} A^n = a_n d_n - q^n b_n c_n.$$

The following relations are valid

$$a_n b_m - q^m b_n a_m = -q^m D_m b_{n-m}$$

$$a_n c_m - q^n c_n a_m = -q^{n-m} D_m c_{n-m}$$

$$a_n d_m - q^{n+m} d_n a_m = D_m a_{n-m} - q^{n+m} D_m d_{n-m}$$

(3.3) $\qquad b_n c_m - q^{n-m} c_n b_m = 0 \qquad\qquad\qquad\qquad \Big\} .$

$$b_n d_m - q^n d_n b_m = q^m D_m b_{n-m}$$

$$c_n d_m - q^m d_n c_m = D_m c_{n-m}$$

$$D_m = D^m$$

Notice that (3.3) are not *commutation* relations except for $n = m$, in which case they imply that A^n is a quantum matrix of quantum parameter q^n (see below). We have proven (3.3) by double induction. First for $m = 1$ by induction in n, then for fixed n, by induction in m. This induction proof shows that (3.3) are valid for $n, m \in \mathbb{Z}$. In the course of the induction proof we also show that

(3.4) $\qquad\qquad a_n d_m - q^m b_n c_m = D_m a_{n-m}.$

Using (3.3), (3.4) can be rewritten as

(3.5) $\qquad\qquad a_n d_m - q^n c_n b_m = D_m a_{n-m},$

or as

(3.6) $\qquad\qquad d_n a_m - q^{-n} b_n c_m = d_n a_m - q^{-m} c_n b_m = D_m d_{n-m}.$

We shall not reproduce here the induction proof which is rather lengthy and tedious, although relatively straightforward. As mentioned in the introduction, we intend to give it in a longer paper together with the detailed proof of the statements of the next section.

Set $n = m$ in (3.3). Since $a_0 = d_0 = 1$, $b_0 = c_0 = 0$, we find

$$a_n b_n - q^n b_n a_n = 0$$

$$a_n c_n - q^n c_n a_n = 0$$

$$a_n d_n - q^{2n} d_n a_n = (1 - q^{2n}) D_n$$

(3.7) $\qquad b_n c_n - c_n b_n = 0 \qquad\qquad\qquad\qquad \Big\} .$

$$b_n d_n - q^n d_n b_n = 0$$

$$c_n d_n - q^n d_n c_n = 0$$

On the other hand, setting $n = m$ in (3.4), (3.5) and (3.6) we have

(3.8) $\qquad\qquad a_n d_n - q^n b_n c_n = D_n$

and

$$(3.9) \qquad d_n a_n - q^{-n} b_n c_n = D_n.$$

The third equation in (3.7) is a consequence of (3.8) and (3.9). Subtracting (3.9) from (3.8) we obtain

$$(3.10) \qquad a_n d_n - d_n a_n = (q^n - q^{-n}) b_n c_n.$$

We have now all relations stating that A^n is a quantum matrix corresponding to q^n.

4. - Exponential description

The fact that A^n corresponds to the quantum parameter q^n suggest the ansatz

$$(4.1) \qquad A = e^{hM}, \quad q = e^h$$

where the matrix elements of the 2×2 matrix M should satisfy commutation relations independent of h. The commutation relations (2.2) for the elements of A should be a consequence of those for the elements of M. If this is the case, the matrix $A_1 = e^{h_1 M}$ would have quantum parameter $q_1 = e^{h_1}$ and the matrix $A A_1 = e^{(h+h_1)M}$ would have quantum parameter $q q_1 = e^{h+h_1}$. In particular this would imply that A^n has quantum parameter $q^n = e^{nh}$, not only for integer n but also for continuous values of n, as long as e^{nhM} has a meaning. All this is actually true,. at least formally, and furthermore the properties of M are extremely simple. It turns out that for our quantum matrices the usual relation between determinant and trace is valid for the quantum determinant

$$(4.2) \qquad D_q = \det_q A = \exp \, \mathrm{tr} \, (hM),$$

where tr denotes the *ordinary* trace. Therefore we can limit ourselves at first to the case $D_q = 1$ when M is traceless

$$(4.3) \qquad M = \begin{pmatrix} \lambda & \mu \\ \nu & -\lambda \end{pmatrix}$$

and introduce later a non trivial central trace to account for a determinant different from one. The correct commutation relations are simply

$$(4.4) \qquad \left. \begin{array}{c} \lambda \mu - \mu \lambda = \mu, \quad \lambda \nu - \nu \lambda = \nu \\ \mu \nu - \nu \mu = 0 \end{array} \right\}.$$

For the matrix elements of A, given by (4.1), the commutation relations (4.4), together with (4.3), imply (2.2). More precisely they imply that

(4.5)
$$\varepsilon_q A^T \varepsilon_q^{-1} = A^{-1}.$$

Comparing with (2.16) we see that also $D_q = 1$. We sketch now the proof of these statements.

The condition (4.5), or

(4.6)
$$A^T = \varepsilon_q^{-1} A^{-1} \varepsilon_q,$$

can be written more explicitly, using (4.1). Since

(4.7)
$$\varepsilon_q^{-1} M \varepsilon_q = \begin{pmatrix} -\lambda & -\frac{\nu}{q} \\ -q\mu & \lambda \end{pmatrix},$$

(4.6) becomes

(4.8)
$$\left(\exp\left[h \begin{pmatrix} \lambda & \mu \\ \nu & -\lambda \end{pmatrix} \right] \right)^T = \exp\left[h \begin{pmatrix} \lambda & e^{-h}\nu \\ e^{h}\mu & -\lambda \end{pmatrix} \right].$$

We have verified that this equation is correct to all orders in h by expanding the exponentials. In spite of the great simplicity of the commutation relations (4.4) the proof is not trivial. The crucial point for the validity of (4.8) is of course that, for quantum matrices like M, it is not true that $(M^n)^T = (M^T)^n$. Instead, left and right hand side are related exactly in such a way as to account for the extra factors e^h and e^{-h} occuring in the right hand side of (4.8).

If the matrix M is not traceless

(4.9)
$$M = \begin{pmatrix} \alpha & \mu \\ \nu & \beta \end{pmatrix}$$

the commutation relations (4.4) must be replaced by the slightly more general relations

(4.10)
$$\left. \begin{array}{c} \alpha\mu - \mu\alpha = \mu, \quad \alpha\nu - \nu\alpha = \nu \\ \mu\nu - \nu\mu = 0 \\ \beta\mu - \mu\beta = -\mu, \quad \beta\nu - \nu\beta = -\nu \\ \alpha\beta - \beta\alpha = 0 \end{array} \right\}.$$

It is clear from these relations that

(4.11)
$$\operatorname{tr} M = \alpha + \beta$$

is central, i.e. it commutes with α, β, μ and ν. From (4.2) we see that

(4.12)
$$D_q = \det_q A = \exp\left[h(\alpha + \beta) \right].$$

It is now also obvious that

$$(4.13) \qquad \det_{q^n} A^n = (\det_q A)^n.$$

5. - Concluding remarks

We restrict ourselves to the case $D_q = 1$. Since A, given by (4.1), (4.3) and (4.4), is an $SL_q(2, C)$ matrix, it behaves as described in Section 2 under multiplication. This means that, if

$$(5.1) \qquad A' = e^{hM'}, \quad M' = \begin{pmatrix} \lambda' & \mu' \\ \nu' & -\lambda' \end{pmatrix}$$

is a matrix of the same type as A

$$(5.2) \qquad \left. \begin{array}{c} \lambda'\mu' - \mu'\lambda' = \mu', \; \lambda'\nu' - \nu'\lambda' = \nu' \\ \mu'\nu' - \nu'\mu' = 0 \end{array} \right\}$$

and furthermore λ, μ, ν commute with λ', μ', ν', then

$$(5.3) \qquad A'' = AA' = e^{hM''}, \quad M'' = \begin{pmatrix} \lambda'' & \mu'' \\ \nu'' & -\lambda'' \end{pmatrix}$$

is also an $SL_q(2, C)$ matrix, i.e.

$$(5.4) \qquad \left. \begin{array}{c} \lambda''\mu'' - \mu''\lambda'' = \mu''', \; \lambda''\nu'' - \nu''\lambda'' = \nu'' \\ \mu''\nu'' - \nu''\mu'' = 0 \end{array} \right\}.$$

One can find explicit expressions for λ'', μ''' and ν'' by means of the Baker-Campbell-Hausdorff formula, which gives

$$(5.5) \qquad hM'' = hM + hM' + \frac{1}{2}h^2[M, M'] + \frac{1}{12}h^3[M, [M, M']]$$

$$-\frac{1}{12}h^3[M', [M, M']] + \dots$$

To the order indicated one finds

$$\lambda'' = \lambda + \lambda' + \frac{h}{2}(\mu\nu' - \nu\mu') +$$

$$+ \frac{h^2}{6}\left[2\lambda\mu'\nu' + 2\mu\nu\lambda' - \mu\nu'\lambda' - \nu\lambda\mu' - \nu\mu'\lambda' - \mu\lambda\nu'\right] + \dots$$

$$\mu'' = \mu + \mu' + h\left(\lambda\mu' - \mu\lambda'\right) +$$

(5.6)
$$+ \frac{h^2}{6}\left[\mu\mu'\nu' + \mu\nu\mu' - \nu\mu'^2 - \mu^2\nu' + 2\mu\lambda'^2 + 2\lambda^2\mu'\right.$$

$$\left. - \lambda\mu'\lambda' - \mu\lambda\lambda' - \lambda\lambda'\mu' - \lambda\mu\lambda'\right] + \dots$$

$$\nu'' = \nu + \nu' + h\left(\nu\lambda' - \lambda\nu'\right) +$$

$$+ \frac{h^2}{6}\left[\nu\mu'\nu' + \mu\nu\nu' - \mu\nu'^2 - \nu^2\mu' + 2\nu\lambda'^2 + 2\lambda^2\nu'\right.$$

$$\left. - \lambda\nu'\lambda' - \nu\lambda\lambda' - \lambda\lambda'\nu' - \lambda\nu\lambda'\right] + \dots$$

Notice that, when the matrix elements of two matrices don't commute, the trace of the commutator does not vanish in general. However M'' is traceless, because the traces of terms of a given order in h cancel. For instance in (5.5) the two terms in h^3 separately have non zero trace, but the sum of the traces is zero.

In (5.6) the ordering of non commuting quantities is important. Formulas (5.6) give a realization of the quantum group in terms of exponential quantum group parameters. However, in spite of the great simplicity of the commutation relations (4.4) the group composition laws (5.6) are relatively complicated because of the ordering. We emphasize that, to obtain (5.6), we have used explicitly the 2×2 representation. The reason is that we are dealing here with linear combinations with non commuting coefficients of the generators of a Lie algebra: in general the commutator of two such quantities is not an object of the same kind unless additional algebraic relations (such as nilpotency, square equal to the unit matrix, etc.) are imposed on the Lie algebra generators.

Finally, we consider the limit $q \to 1$, $h \to 0$. In this limit the quantities a, b, c, d of Section 2 commute. With the usual relation between Poisson brackets (for which we use round brackets) and commutators we find

(5.7)
$$(a, b) = \lim_{h \to 0} \frac{[a, b]}{h} = ab$$

and similarly

(5.8)
$$\left.\begin{array}{ll} (a, c) = ac, & (b, c) = 0 \\ (b, d) = bd, & (a, d) = 2bc \\ (c, d) = cd & \end{array}\right\}.$$

If a', b', c', d' have similar Poisson brackets

(5.9)
$$\left.\begin{array}{l}(a', b') = a'b' \\ (a', c') = a'c' \\ \text{etc.}\end{array}\right\},$$

then a'', b'', c'', d'' given by (2.5) also do, i.e.

(5.10)
$$\left.\begin{array}{l}(a'', b'') = a''b'' \\ (a'', c'') = a''c'' \\ \text{etc.}\end{array}\right\}.$$

In this case one speaks of a Poisson group.

For the exponential description we must first rescale the variables and introduce new quantities $\hat{\lambda}, \hat{\mu}, \hat{\nu}$

(5.11)
$$\hat{\lambda} = h\lambda, \quad \hat{\mu} = h\mu, \quad \hat{\nu} = h\nu,$$

so that

(5.12)
$$A = \exp \begin{pmatrix} \hat{\lambda} & \hat{\mu} \\ \hat{\nu} & -\hat{\lambda} \end{pmatrix}.$$

In the Poisson limit $\hat{\lambda}, \hat{\mu}$ and $\hat{\nu}$ commute and their Poisson brackets are

(5.13)
$$\left.\begin{array}{l}(\hat{\lambda}, \hat{\mu}) = \hat{\mu}, \quad (\hat{\lambda}, \hat{\nu}) = \hat{\nu} \\ (\hat{\mu}, \hat{\nu}) = 0\end{array}\right\}.$$

In the limit the composition laws (5.6) become the usual composition laws for the (commuting) exponential parameters of the Lie group and depend now only on the Lie algebra, not on additional properties of the particular representation. In terms of the rescaled variables the composition laws do not contain h. They have, of course, the usual non linear structure typical of exponential parameters but preserve exactly the very simple Poisson relations (5.13).

REFERENCES

[1] L.D. FADDEEV, *Integrable models in* $(1 + 1)$-*dimensional quantum field theory* (Les Houches lectures, 1982), Elsevier Science Publishers, Amsterdam, 1984.

[2] L.D. FADDEEV - N. YU RESHETIKHIN - L.A. TAKHTAJAN, *Quantization of Lie groups and Lie algebras*, LOMI preprint E-14-87, to appear in M. Sato's 60th birthday volume.

[3] V.G. DRINFELD, *Quantum groups*, Proc. Internat. Congr. Math., Berkeley, 1986, vol. 1, 798-820.

[4] YU I. MANIN, *Quantum groups and non commutative geometry*, report of the Centre de Recherches Mathématiques, Montréal University, 1988.

[5] L. ALVAREZ-GAUMÉ - C. GOMEZ - G. SIERRA, *Duality and quamtum groups*, preprint CERN-TH. 5369/89, UGVA-DPT-3/605/89, as well as earlier CERN and UGVA preprints.

[6] G. MOORE - N. SEIBERG, *Classical and quantum conformal field theory*, preprint IASSNS-HEP-88/39, as well as earlier IASSNS-HEP preprints.

[7] E. CORRIGAN - C. TUNSTALL, private communication.

The Black Hole Revisited

JOHN ARCHIBALD WHEELER (*)

Luigi Radicati has thrown new light on the symmetries that pervade physics. As a small tribute to a great scientist and great human being on his 70th birthday, it may therefore be appropriate to revisit an object of great symmetry, the black hole.

Nothing is more remarkable about black hole physics than the beauty of the subject, attested by now in many articles and books. No matter how diverse happen to be the structures, the particles and the radiations that pour in together to make it, and no matter how conspicuous the waves, spouts, jets and other "hairs" that accompany the process of formation, the black hole quickly ends up with no hair. Completely collapsed, it requires for its external description three parameters and - so all present evidence argues - three alone: mass, charge and angular momentum.

Regarding each known black hole in the Milky Way and other galaxies, this is not the place to report the ever fuller details on where it is and on what companion - or companions, in the case of the three-and-a-half solar-mass black hole at the center of our galaxy - circulates in orbit around it. Nor do we need to recall here that for the powerhouse of the quasar no mechanism is known that is more plausible, and none that explains more and predicts more, than accretion on a rotating black hole.

Behind the ideal symmetry, however, of the ideal horizon of the ideally isolated black hole hide issues of principle of ever greater depth and interest, issues which younger colleagues have, one by one, been bringing to light and clarifying.

The Penrose mechanism by which a particle acquires energy from a rotating black hole [1]: what can you say about its efficiency? I put this question to Demetrios Christodoulou, a young Princeton graduate student. By 1972, when he was still 21, he had the answer.

In black hole physics, Christodoulou showed, the Penrose process allows of both reversible and irreversible transformations [2]. For the case of greatest interest and simplicity, the black hole has a zero or negligible electric charge. In this case a reversible transformation of black hole angular momentum into

(*)   Physics Departments, Princeton University and University of Texas at Austin

particle energy signals itself in a simple way. The mass, M, of the black hole and its angular momentum, J, however much they change individually, still comport with unchanged irreducible mass M_{ir} - irreducible mass as it is defined in Christodoulou's beautiful equation,

$$M^2 = M_{ir}^2 + J^2/4M_{ir}^2.$$

By contrast, in an irreversible interchange of angular momentum and energy between rotating black hole and approaching particle, the irreducible mass always increases.

The irreducible mass lends itself to immediate visualization. It measures the surface area, A, of the horizon of the black hole,

$$A = 16\pi M_{ir}^2.$$

In an irreversible transformation the black hole's area increases. In a reversible one it doesn't.

One day another graduate student, Jacob Bekenstein, was visiting in my office. I spoke to him of the bad conscience I had always felt when I put a hot cup of tea in thermal contact with a cold cup of tea. The energy which the one cup loses, the other one gains, to be sure; but no such conservation law applies to entropy. It increases. That increase, moreover, is irreversible. "My crime, Jacob," I went on, "echoes down the corridors of history to the end of time. But now let a black hole swim by, and let me drop into it both teacups. Haven't I concealed my crime forever?"

Bekenstein, soul of integrity, did not say yes, did not say no, but went off looking deeply troubled. Two or three days later he came back with his conclusion: You can't hide entropy. The black hole, too, has entropy; and not only entropy, but also temperature. The entropy of the black hole is given, up to a factor of order unity, by the surface area of the horizon, this area being expressed in units of the Planck length.

Brandon Carter tells us how, when he and Stephen Hawking read Bekenstein's conclusion that a black hole has entropy [3], it seemed so preposterous that they resolved to prove it wrong. The more they worked, the more they became convinced that he was right, after all. The outcome was a discovery: Hawking's mechanism [4] for the creation of particles by a black hole.

Subsequently, Kip Thorne and Wociech H. Zurek counted up all the ways in which it is possible to make a black hole of a given mass, charge and angular momentum. This number, translated to the language of entropy, is identical, they show [5] to the Bekenstein-Hawking result.

"Look *into* the black hole," however, by adopting a different slicing (Figure 1) of the same spacetime and come to a different conclusion about the entropy [6]! How come? Because in that frame of reference we *know* more. Entropy in this sense proves to be a quantity dependent on choice of global reference system. To revisit the black hole is to discover a new question about the entropy of a black hole: Entropy as seen **by whom**.

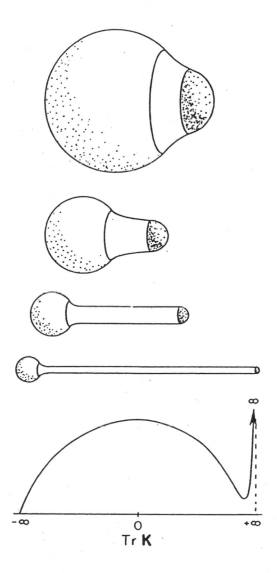

Fig. 1 - The 3-geometry of four constant-crunch-time slices through the spacetime of the Qadir-J.A.W. thick-and-thin suture model universe, from near the phase of maximum expansion (first, or top, frame) to near crunch (fourth frame); detailed calculations as summarized in free-hand sketches. An axis of 2-sphere symmetry, not shown, it to be imagined as cutting horizontally through each frame. Bottom: Proper distance, as measured in the 3-geometry, from pole to pole of the model cosmology, in its dependence on crunch time.

REFERENCES

[1] R. PENROSE, *Gravitation collapse: the role of general relativity*, Riv. Nuovo Cimento, I (1969) 252-276.

[2] D. CHRISTODOULOU, *Reversible and irreversible transformations in black-hole physics*, Phys. Rev. Lett. **25** (1970) 1596-1597.

[3] J.D. BEKENSTEIN, *Black holes and entropy*, Phys. Rev. D, **7** (1973) 2333-2346.

[4] S.W. HAWKING, *Particle creation by black holes*, Comm. Math. Phys., **43** (1975) 199-220.

[5] W.H. ZUREK - K.S. THORNE, *Statistical mechanical origin of the entropy of rotating, charged black hole*, Phys. Rev. Lett. **54**, (1985) 2171-2175.

[6] A. QADIR - J.A. WHEELER, *Late stages of crunch*, pp. 345-348 in *Spacetime Symmetries: International Symposium on Spacetime Symmetries in Commemoration of the 50th Anniversary of Eugene Paul Wigner's Fundamental Paper on the Inhomogenous Lorentz Group*, Y.S. KIM and W.W. ZACHARY, eds., North Holland, Amsterdam (1989).

I Chimici e la Fisica

GIAN CARLO WICK

Questo articoletto, che vorrei dedicare al caro amico Luigi Radicati, in omaggio alle sagge opinioni da lui udite in numerose occasioni, si basa sui miei ricordi ormai lontani, ma ridestati dalla recente frenesia di interesse nella "fusione fredda" e dalle riserve e dai dubbi che per ora accompagnano questa scoperta. Riserve ben comprensibili, in vista anche del modo un po' caotico in cui i risultati sono stati annunciati e delle informazioni incomplete sinora pubblicate. L'aspetto più interessante, ma anche un po' sconcertante, della scoperta sta nella sua natura del tutto imprevista, tant'è vero che anche ora che si sa, o si crede di sapere, che l'effetto realmente esiste, ancora si stenta a capire teoricamente come esso avvenga. In un certo senso ogni vera scoperta è il risultato di una scommessa: si sono spesi tempo e sforzi e spesso anche molto denaro, senza alcuna certezza di arrivare a un risultato importante. Il fatto che la scoperta di un fenomeno fisico nuovo sia stata annunciata anzitutto da chimici mi riporta alla memoria la scoperta della scissione e altre cose che riguardano il ruolo dei chimici nella fisica nucleare degli anni 30.

Anni di cui conservo un ricordo ancor fresco, o almeno così mi pare. Dopo un'attesa di circa un anno, ero entrato nell'autunno del '32 a far parte del gruppo romano di fisici come assistente teorico di Fermi. È assai discutibile se io possa vantarmi di essere stato uno dei "ragazzi di via Panisperna"; in quanto teorico, non ebbi che una parte occasionale e direi trascurabile nella ricerca sulle reazioni indotte da neutroni, a cui il gruppo deve la sua fama. Ma da alcuni aspetti teorici di quei problemi trassi occasione per alcuni articoli abbastanza interessanti. In qualunque modo non sarebbe stato possibile per me convivere fianco a fianco con quel gruppo, senza venire coinvolto nel loro entusiasmo e nella loro convinzione di partecipare a una grande avventura.

Tra i molti ricordi, che si collegano a questa esperienza, mi se ne riaffacciano ora alcuni che riguardano le relazioni tra fisici e chimici e un certo complesso di superiorità abituale nei primi rispetto ai secondi, complesso al quale io stesso partecipavo. Quei ricordi mi consigliano una certa prudenza nel giudicare alcune ingenuità, che nelle questioni di fisica nucleare spesso si insinuano nelle opinioni dei chimici; una prudenza che si riassume in sostanza in una regola; non dimenticare mai che nella ricerca non sempre il saperla troppo lunga è un vantaggio.

Un primo ricordo si riferisce a certe osservazioni di Oscar D'Agostino, un giovane chimico che Fermi aveva persuaso ad associarsi al suo gruppo negli studi sulla radioattività indotta da bombardamento con neutroni. L'aiuto di un chimico era prezioso, per non dire indispensabile, nel lavoro di identificazione dei radioelementi così prodotti. In un primo tempo questi venivano individuati in base al periodo di dimezzamento caratteristico, che ogni radioelemento possiede. Ma questo non significava conoscere la sua natura chimica, né la reazione nucleare che l'aveva prodotto: poteva trattarsi di un processo (n, γ) - cattura di un neutrone con emissione di raggi gamma - oppure (n, p) - cattura accompagnata da emissione di un protone - e così via. Insomma per ogni attività di dato periodo osservata era necessario poi stabilire se si trattava di un isotopo instabile dell'elemento borbardato o invece di un isotopo di un altro elemento vicino nel sistema periodico. Questo si doveva fare studiando le proprietà chimiche del radioelemento, in particolare il suo comportamento nella precipitazione di un qualche suo composto da una soluzione. In vari casi il procedimento più semplice per studiare questo tipo di separazione consisteva nell'irradiare coi neutroni l'elemento voluto sotto la forma di un suo composto già in soluzione acquosa e in seguito precipitare l'attività prodotta, dalla soluzione stessa.

D'Agostino, dopo aver preso parte attiva a molte di queste misure, ne trasse l'*impressione* che l'attivazione di certi elementi fosse nettamente più intensa quando essa avveniva in soluzione; richiesto quale interpretazione volesse dare a questo fatto, rispondeva di credere a un "effetto del legame chimico sulla reazione nucleare", destando così le nostre canzonature, poiché l'idea di un effetto non trascurabile del legame chimico sulla reazione era evidentemente risibile. All'impressione che ci fosse un effetto era facile rispondere che un confronto quantitativo tra le attività indotte nei due casi (in soluzione o rispettivamente in condizioni ordinarie - per es. irradiando un cilindretto di diametro tale da poter venire infilato su un contatore) non aveva molto senso, non essendo possibile calcolare l'effetto della differente "geometria" delle due esperienze. Ciononostante, mesi più tardi, colla scoperta delle proprietà dei neutroni lenti si capì che l'effetto osservato da D'Agostino esisteva realmente; non se l'era sognato, anche se l'effetto era dovuto alla presenza dell'acqua e non al "legame chimico". Come il bambino nella nota novella di Andersen, l'occhio senza pregiudizi del chimico aveva notato ciò che noi non volevamo vedere.

Non vi fu gran danno; più tardi altri effetti curiosi vennero notati e infine la scoperta avvenne ugualmente a Roma. Ma ricordo un altro caso, in cui l'aver ignorato le osservazioni di un chimico fece perdere forse ai romani un'occasione importante. (Mi si dirà forse giustamente che esagero: bisogna pure lasciare qualcosa da scoprire anche agli altri!) Mi riferisco in particolare a Ida Noddack, la quale non molto dopo l'osservazione da parte del gruppo di Roma della produzione di vari radioelementi nel bombardamento dell'uranio pubblicò una nota in "Angewandte Chemie", sostenendo che le proprietà chimiche di questi radioelementi si potevano interpretare assai meglio come quelle di frammenti di massa atomica pari a circa la metà di quella del nucleo di uranio. Nessuno le prestò molta attenzione, né a Roma né, mi pare, altrove forse anche perché la Noddack non aveva prestato a sua volta alcuna attenzione, se ben ricordo, al serio problema della barriera di potenziale Coulombiana, che nel caso delle reazioni da lei postulate ha un'importanza essenziale. Si può anche aggiungere che nel caso dell'uranio il problema dell'identificazione chimica dei radioelementi era assai complicato. In realtà solo le esaurienti ricerche successive di Hahn, Meitner e Strassman portarono a conclusioni sicure. Ricordo ancora l'ammirazione provata nel leggere, nel dicembre 1939 l'affermazione di Hahn: "mi si dice che il risultato è poco comprensibile, secondo i fisici, ma io come chimico devo concludere che questo radioelemento è un isotopo del bario". Quanto sarebbe rassicurante leggere oggi un'affermazione altrettanto autorevole riguardo alla fusione fredda! Non occorre qui ricordare che di sfuggita come il risultato della scoperta venisse poi confermato dall'esperienza di Frish e Meitner, constringendo infine i fisici a riesaminare la dinamica della reazione sotto nuovi punti di vista, una riesamina iniziata dagli stessi autori e in seguito approfondita dallo stesso Bohr in collaborazione con Wheeler.

In tutta questa vicenda vi è senza dubbio una lezione per i fisici, ma disgraziatamente non me la sento di dire con sicurezza quale sia questa lezione, salvo la conclusione piuttosto banale che "la cautela non è mai troppa". Una regola più esplicita di condotta, che valga in tutti i casi, non mi pare che ci sia. Volendo fare dello spirito, la regola vera, buona in tutti i casi è che avere un po' di fortuna non guasta!

Saranno così fortunati i vari Pons, Fleischmann, Jones, ecc. che con tanto coraggio si sono buttati a cercare un fenomeno sulla cui esistenza ben pochi avrebbero scommesso?

In un certo senso c'è da augurarselo, perché senza dubbio i loro risultati hanno agito come una folata di aria fresca in un campo appesantito da anni di ricerca senza risultati veramente eccitanti. Ma mentre scrivo queste righe, non posso fare a meno di concedermi altre riflessioni un po' nostalgiche, e meno rassicuranti. Quanto è diversa l'atmosfera della scienza oggi da quella in cui si svolgevano le ricerche sui neutroni! Senza dubbio vi era anche allora un senso di urgenza nel pubblicare i risultati; a questo scopo Fermi aveva scoperto che "La Ricerca Scientifica", pubblicata a Roma dal C.N.R., era il veicolo ideale per la velocità di accettazione dei manoscritti e l'immediata distribuzione. Ma si trattava purtuttavia di una pubblicazione strattamente scientifica. I mass-media erano completamente all'oscuro di quanto avveniva. Niente interviste, conferenze stampa, ecc. e con quale vantaggio per la tranquillità dei fisici e la serietà della scienza non occorre insistere. Qualcuno potrà dire che coll'uso invalso in questi anni il grande pubblico è assai meglio informato sui progressi della scienza. E si potrà anche dire che dopotutto il pubblico ha anche il diritto di essere così informato. Ma tutto questo è vero? In realtà io vedo, in quello che appare sui giornali o in televisione una ridda di notizie che si contraddicono. Più che di informazione, si può parlare di disinformazione; anche chi ne sa qualcosa, cosa che forse non si applica più a me, data la mia età (ma io sento anche cosa dicon gli altri) ebbene sembra sia assai difficile farsi un'idea chiara.

Con tutto questo, la ricerca rimane pur sempre una gran bella cosa. Vuol dire che, oggi più che mai, ci vuole serietà e pazienza. Se Dio vuole, le nebbie si diraderanno e qualcosa di più sapremo.

List of Participants

B. Alles

G. Alzetta

E. Amaldi

B. Andreotti

D. Arcoya

F.T. Arecchi

E. Arrigoni

M. Baldo-Ceolin

R. Barbieri

P. Barocchi

M. Barsanti

F. Bassani

M.A.B. Bég

G. Bellettini

E. Beltrametti

C. Bemporad

G. Bernardini

L. Bertanza

G. Bertin

N. Beverini

A. Bigi

P. Biscari

M. Bismut

A. Bondi

F. Bonetto

P.L. Braccini

R. Buczko

F. Busnelli

G. Caglioti

M.V. Calahorrano

M. Calvetti

L. Caneschi

S. Caracciolo

G. Careri

R. Castaldi

M. Cavenago

C. Ceolin

P. Christillin

M. Ciafaloni

M. Cini

R. Colle

F. Conti

G. Curci

R.H. Dalitz

N. Dallaporta

P. De Falco

E. De Giorgi

M. Dell'Orso

E. D'Emilio

A. Di Giacomo

F. Dyson

L.D. Faddeev

G. Ferraro

F. Ferrini

E. Fiorini

V. Flaminio

L. Foà

P. Franzini

G. Gentili

G. Giacomin

C. Giannessi

A. Giazotto

G. Giovannini

F. Giuliani

A. Gozzini

E. Guadagnini

B. Guerrini

F. Gürsey

E. Iacopini

M. Inguscio

T.D. Lee

R. Leonardi

F. Ligabue

L. Lovitch

P. Maestrini

G. Maino

N. Majlis

M. Mangano

I. Mannelli

A. Marino

A. Martin

F. Mauri

P. Menotti

A. Michalec

L. Michel

M. Mintchev

G. Montalenti

G. Morpurgo

S. Mortola

G. Moruzzi

M.K.V. Murthy

K. Nishijima

G. Occhialini

G. Paffuti

P. Paolicchi

F. Papoff

G. Parisi

G. Pastori Parravicini

F. Pegoraro

R. Peierls

A. Pelissetto

G.M. Piacentino

L.E. Picasso

G. Pierazzini

E. Polacco

M.C. Prati

A. Profeti

G. Puppi

A. Quattropani

R. Rattazzi

E. Remiddi

R.A. Ricci

G. Ricciardi

M. Rosa-Clot

S. Rosati

G.C. Rota

G. Salvini

R. Sancisi

A. Scribano

S. Selzer

F. Sette

S. Spagnolo

D. Speiser

M. Stiavelli

G. Strinati

S. Stringari

F. Strocchi

F. Strumia

A. Tani

Y. Tomozawa

G. Torelli

V. Tortorelli

R. Tripiccione

C. Vannini

R. Vergara Caffarelli

P. Villaggio

J.A. Wheeler

G.C. Wick

P. Zanella

Finito di stampare per conto della "CompoMat",
dalla Nuova Grafica 86 nell'aprile 2005